Advanced Electromagnetics and Scattering Theory

AF768460

Kasra Barkeshli

Advanced Electromagnetics and Scattering Theory

Edited by

Sina Khorasani

Springer

Author
Kasra Barkeshli
School of Electrical Engineering
Sharif University of Technology
Tehran
Iran

Editor
Sina Khorasani
School of Electrical Engineering
Sharif University of Technology
Tehran
Iran

The Author “Kasra Barkeshli” is Deceased.

ISBN 978-3-319-36472-8 ISBN 978-3-319-11547-4 (eBook)
DOI 10.1007/978-3-319-11547-4

Springer Cham Heidelberg New York Dordrecht London

© Springer International Publishing Switzerland 2015
Softcover reprint of the hardcover 1st edition 2015
This work is subject to copyright. All rights are reserved by the Publisher, whether the whole or part of the material is concerned, specifically the rights of translation, reprinting, reuse of illustrations, recitation, broadcasting, reproduction on microfilms or in any other physical way, and transmission or information storage and retrieval, electronic adaptation, computer software, or by similar or dissimilar methodology now known or hereafter developed. Exempted from this legal reservation are brief excerpts in connection with reviews or scholarly analysis or material supplied specifically for the purpose of being entered and executed on a computer system, for exclusive use by the purchaser of the work. Duplication of this publication or parts thereof is permitted only under the provisions of the Copyright Law of the Publisher's location, in its current version, and permission for use must always be obtained from Springer. Permissions for use may be obtained through RightsLink at the Copyright Clearance Center. Violations are liable to prosecution under the respective Copyright Law.
The use of general descriptive names, registered names, trademarks, service marks, etc. in this publication does not imply, even in the absence of a specific statement, that such names are exempt from the relevant protective laws and regulations and therefore free for general use.
While the advice and information in this book are believed to be true and accurate at the date of publication, neither the authors nor the editors nor the publisher can accept any legal responsibility for any errors or omissions that may be made. The publisher makes no warranty, express or implied, with respect to the material contained herein.

Printed on acid-free paper

Springer is part of Springer Science+Business Media (www.springer.com)

Foreword

Kasra Barkeshli (1961–2005)

It is our utmost pleasure to bring the attention of our dear readers to one of the most comprehensive monographs ever written on the subject of electromagnetic radiation and scattering. The *Advanced Electromagnetic and Scattering Theory* by the late Prof. Kasra Barkeshli compiles two decades of intensive studies, research, and teaching in the United States and Iran.

The late Prof. Kasra Barkeshli was born on August 11, 1961 in Tehran, Iran. He was initially admitted to the Tehran Polytechnique University at the age of 16 prior to joining the University of Kansas where he received his Bachelor's degree. He later entered University of Michigan at Ann Arbor, where he received his Master's and Doctorate degrees in Electrical Engineering under supervision of the renowned scientist, Prof. John L. Volakis (now at Ohio State University), along with a graduate degree in Applied Mathematics. Soon after his graduation in 1991, Dr. Barkeshli returned to Iran and started his career with the Department of Electrical Engineering at Sharif University of Technology as an Assistant Professor. During his relatively brief career at Sharif University of Technology, he served as the Department Chair (1996–2000) and Vice-President for Student Affairs (2001–2003), founded the Numerical Electromagnetics Laboratory, and designed and offered new official graduate courses on Advanced Electromagnetics, Scattering, and Numerical Electromagnetics for the first time across the nation. He was promoted to the rank of Full Professorship in 2005, when he was on a sabbatical visit to Polytechnique University of New York.

In 2004, the late Prof. Kasra Barkeshli was diagnosed with a terminal illness, which prevented him from finishing many of his scientific plans he had in mind. At the moment of his untimely death on June 26, 2005, he had been elevated to the rank of the IEEE Senior Membership, had supervised 15 graduate projects, had published around 55 scientific journal and conference papers, and had almost completed drafting of the present book, *Advanced Electromagnetics and Scattering Theory*.

Unfortunately, the most updated existing electronic LaTeX source of this book (in 267 pages) did not match the most updated print (in 308 pages). Many figures and/or subsections of the book had been left for future elaboration, and the text needed major editing or insertion of new material. We had no information about existence of such an unfinished manuscript until 2010, when we received a rough collected draft of the book from his wife, Mrs. Paimaneh Hastaie, and the LaTeX files from his brother, Dr. Sina Barkeshli. We then initiated a subsequent activity toward its completion. This task would not have been possible without the dedicated efforts of my colleagues at Sharif University of Technology, Mr. Seyed Armin Razavi, Mr. Shahriar Aghaei Meibodi, and Dr. Farhad Azadi Namin, who undertook an extensive task to bring Prof. Kasra Barkeshli's initial plan to end. I feel proud to have had the privilege of supervising this taskforce in the meantime.

In parallel to his academic activities, the late Prof. Kasra Barkeshli also had a strong passion for his family and private life. As once told by his wife, he would have certainly dedicated this book to his children, Zoha and Mohammad Mehdi, without whom this project would never have formed.

On behalf of my colleagues, I hereby would like to extend our humblest regards and most sincere thanks to the family of the late Prof. Kasra Barkeshli, and in particular to his mother, Mrs. Sorour Katouzian for her indefinite support, as well as the School of Electrical Engineering at Sharif University of Technology for hosting this effort. We moreover wish our readers a most pleasant journey along, with the wonderful world of Electromagnetics.

Tehran, Iran, December 2013 Sina Khorasani

Preface

These are the lecture notes used in two courses I have offered in the past ten years at Sharif University of Technology, namely *Advanced Electromagnetics and Scattering Theory*. These courses are usually taken by the first- and second-year graduate students in the communications group. The prerequisite for the sequence is vector calculus and electromagnetic fields and waves. Some familiarity with Green's functions and integral equations is desirable but not necessary.

This manuscript is meant to provide a brief but concise introduction to classical topics in the field. It is divided into two parts including annexes. Part I covers principles of *Electromagnetic Theory*. The discussion starts with a review of the Maxwell's equations in differential and integral forms and basic boundary conditions. The solution of inhomogeneous wave equation and various field representations, including Lorentz's potential functions and the Green's function method, are discussed next. The solution of Helmholtz equation and wave harmonics follow. Plane wave propagation in dielectric and lossy media and various wave velocities are presented next. This part concludes with a general discussion of planar and circular waveguides.

Part II presents basic concepts of electromagnetic *Scattering Theory*. After a brief discussion of radar equation and scattering cross-section, we review the canonical problems in scattering. These include the cylinder, the wedge, and the sphere. The edge condition for the electromagnetic fields in the vicinity of geometric discontinuities are discussed. We also present the low frequency Rayleigh and Born approximations. The integral equation method for the formulation of scattering problems is presented next, followed by an introduction to scattering from periodic structures.

In preparing these notes, I have benefited from the contribution of two respectable groups of individuals. First, I have been fortunate enough to learn electromagnetic theory under a number of distinguished scholars at The University of Michigan. I am indebted to Profs. Chen-To Tai, Dipak Sengupta, and Thomas B. A. Senior. I am also thankful to my research advisor at Radiation Laboratory, Prof. John L. Volakis. I have been influenced directly by their style of approach, although all the shortcomings and mistakes which may be encountered in this work are those of mine.

Next, I am indebted to my students at Sharif University of Technology whose enthusiasm has contributed significantly to the refinement of my original notes.

Tehran, Iran, December 2002 Kasra Barkeshli

Contents

Part I Electromagnetic Theory

Biographies of Contributors

Shahriar Aghaei Meibodi was born in Tehran, Iran. He received his high school diploma from National Organization for Development of Exceptional Talents, and was distinguished as a Silver Medal Winner of National Physics Olympiad in 2009. Since then, he has been a member of National Elite Foundation. He also was ranked sixth in Nationwide Universities Entrance Exam in Iran in 2010. He has majored in Electrical Engineering and Physics at Sharif University of Technology. His research interests include Electromagnetics, Nano-structures, and Optical Devices. He is now with the University of Maryland as a graduate student of Nanophotonics.

Seyed Armin Razavi is born in Tehran, Iran in 1990. In 2009 he was ranked 28th in the National University Exam and was awarded a 4-year fellowship from National Elites Foundation. He has received his B.Sc. degree in Electrical Engineering from Sharif University of Technology in 2014. He is interested in Communications and Electromagnetics. He is now with the University of California at Los Angeles as a graduate student of Electrical Engineering.

Farhad Azadi Namin received the B.S. and M.S. degrees in Electrical Engineering from the University of Texas at Dallas, and his Ph.D. from The Pennsylvania State University in 2012. From December 2012 until August 2013 he was a post-doctoral research fellow with the Computational Electromagnetics and Antenna Research Lab (CEARL), Department of Electrical Engineering, at The Pennsylvania State University. He has coauthored a chapter on ultra-wideband antenna arrays, and has authored six journal publications and 14 conference papers. His research interests include metamaterials, aperiodic plasmonic nanoparticle arrays, photonic crystals, dielectric filters, ultra-wideband aperiodic antenna arrays, evolutionary optimization algorithms, and wireless ad hoc networks. Mr. Namin is the recipient of the 2010 and 2011 Society of Penn State Electrical Engineers (SPSEE) Graduate Fellowship and the 2012 Monkowski Graduate Fellowship in Electrical Engineering. He is also the winner of the 2012 Dr. Nirmal K. Bose Dissertation Excellence Award. His Ph.D. research is funded through an Exploratory and Foundational Program Fellowship from the Applied Research Laboratory (ARL), The Pennsylvania State University. He is currently an Assistant Professor of Electrical Engineering at Amirkabir University of Technology, Tehran, Iran.

Sina Khorasani was born in Tehran, Iran, on November 25, 1975. He received the B.Sc. degree in Electrical Engineering from the Abadan Institute of Technology, Tehran, in 1995, and the M.Sc. and Ph.D. degrees in Electrical Engineering from the Sharif University of Technology, Tehran, in 1996 and 2001, respectively. He is currently a Tenured Associate Professor of electrical engineering with the School of Electrical Engineering, Sharif University of Technology. He has been with the School of Electrical and Computer Engineering, Georgia Institute of Technology, Atlanta, GA, USA, as a Post-Doctoral, from 2002 to 2004, and a Research Fellow form 2010 to 2011. His current research interests include quantum optics, photonics, and quantum electronics. He is a Senior Member of IEEE.

Part I
Electromagnetic Theory

Chapter 1
Maxwell's Equations

This chapter is devoted to the review of Maxwell's equations in differential and integral forms as well as the electromagnetic boundary conditions. We will also discuss the electromagnetic potentials and wave equations. The solution to inhomogeneous wave equation using field representations by the Green's function will be presented.

1.1 Differential Form

The electromagnetic field is governed in the macroscopic domain by the classical field equations of Maxwell. Maxwell's equations constitute a first order linear coupled differential equations to describe the general behavior of electromagnetic fields in free space and material media. Maxwell introduced these equations based on the works of his predecessors like Faraday and Ampere. The equations are as follows

$$\nabla \times \mathcal{E} = -\frac{\partial \mathcal{B}}{\partial t} \quad \text{Faraday's Law} \tag{1.1}$$

$$\nabla \times \mathcal{H} = \mathcal{J} + \frac{\partial \mathcal{D}}{\partial t} \quad \text{Maxwell-Ampere's Law} \tag{1.2}$$

These are supplemented by the continuity relation

$$\nabla \cdot \mathcal{J} = -\frac{\partial \rho}{\partial t} \tag{1.3}$$

which is also known as the basic law of conservation of Charge. The field quantities are

$\mathcal{E}$ *electric field intensity* (V/m)
$\mathcal{H}$ *magnetic field intensity* (A/m)
$\mathcal{D}$ *electric flux density* (C/m^2)

© Springer International Publishing Switzerland 2015

K. Barkeshli, *Advanced Electromagnetics and Scattering Theory*,
DOI 10.1007/978-3-319-11547-4_1

$\mathcal{B}$ *magnetic flux density* (T)
$\mathcal{J}$ *volumetric current density* (A/m^2)
ρ *electric charge density* (C/m^3)

Another independent physical law is the Lorentz[1] equation of force

$$\mathbf{F} = q(\mathcal{E} + \mathbf{U} \times \mathcal{B}) \tag{1.4}$$

which describes the force $\mathbf{F}$ experienced by a moving point charge q of velocity $\mathbf{U}$ in an electromagnetic field $(\mathcal{E}, \mathcal{B})$.

We will concentrate on the first three equations here. From (1.2), we have

$$\nabla \cdot (\nabla \times \mathcal{H}) = \nabla \cdot \mathcal{J} + \frac{\partial}{\partial t}(\nabla \cdot \mathcal{D}) \equiv 0 \tag{1.5}$$

and from the continuity relation (1.3)

$$\frac{\partial}{\partial t}(\nabla \cdot \mathcal{D} - \rho) = 0 \tag{1.6}$$

which implies upon integration

$$\nabla \cdot \mathcal{D} - \rho = C(x, y, z) \tag{1.7}$$

Also, from (1.1)

$$\nabla \cdot (\nabla \times \mathcal{E}) = -\frac{\partial}{\partial t}(\nabla \cdot \mathcal{B}) \equiv 0 \tag{1.8}$$

yielding

$$\nabla \cdot \mathcal{B} = C'(x, y, z) \tag{1.9}$$

where C and C' are constants of integration (with respect to time); they may be functions of space coordinates. If C is not zero, it must be a static charge. In that case, it could be absorbed in ρ. Also, C' is zero since free magnetic charges, if they exist, are of no engineering significance. Thus, setting $C = C' \equiv 0$, we have

$$\nabla \cdot \mathcal{D} = \rho \tag{1.10}$$
$$\nabla \cdot \mathcal{B} = 0 \tag{1.11}$$

which are sometimes referred to as Gauss' laws of electric and magnetic type, although it is more appropriate to refer to the latter as the conservation of magnetic flux.

---

[1] Hendrik A. Lorentz (1853–1928), Dutch physicist.

Table 1.1 Maxwell's differential equations

Name	Boundary relation	Equation number
Faraday's law	$\nabla \times \mathcal{E} = -\dfrac{\partial \mathcal{B}}{\partial t}$	(1.1)
Ampere's law	$\nabla \times \mathcal{H} = \mathcal{J} + \dfrac{\partial \mathcal{D}}{\partial t}$	(1.2)
Conservation of charge	$\nabla \cdot \mathcal{J} = -\dfrac{\partial \rho}{\partial t}$	(1.3)
Gauss' law	$\nabla \cdot \mathcal{D} = \rho$	(1.10)
Magnetic flux continuity	$\nabla \cdot \mathcal{B} = 0$	(1.11)

The general expression of the fundamental laws of electromagnetics can thus be taken as (1.1)–(1.3), or equivalently, as (1.1), (1.2), (1.10) and (1.11). Both sets of equations are compatible.

The differential forms of Maxwell's equations are tabulated in Table 1.1.

1.2 Constitutive Relations

The fundamental equations of electromagnetics include five vector and one scalar functions of space and time, for a total of sixteen unknowns, but allowing for the fact that the first two equations are vector ones, we have only seven equations. Obviously, the system of equations is indeterminate. To make the system determinate, nine more equations are needed, and these are from a consideration of the medium in which the fields exist.

The constitutive relations are vector functional relations between $\mathcal{D}$, $\mathcal{B}$, $\mathcal{J}$ and $\mathcal{E}$, $\mathcal{H}$. They are of the general form

$$\begin{aligned}
\mathcal{D} &= \mathbf{F}_1\left(\mathcal{E}, \frac{\partial \mathcal{E}}{\partial t}, \frac{\partial^2 \mathcal{E}}{\partial t^2}, \ldots, \mathcal{H}, \frac{\partial \mathcal{H}}{\partial t}, \frac{\partial^2 \mathcal{H}}{\partial t^2}, \ldots\right) \\
\mathcal{B} &= \mathbf{F}_2\left(\mathcal{E}, \frac{\partial \mathcal{E}}{\partial t}, \frac{\partial^2 \mathcal{E}}{\partial t^2}, \ldots, \mathcal{H}, \frac{\partial \mathcal{H}}{\partial t}, \frac{\partial^2 \mathcal{H}}{\partial t^2}, \ldots\right) \\
\mathcal{J} &= \mathbf{F}_3\left(\mathcal{E}, \frac{\partial \mathcal{E}}{\partial t}, \frac{\partial^2 \mathcal{E}}{\partial t^2}, \ldots, \mathcal{H}, \frac{\partial \mathcal{H}}{\partial t}, \frac{\partial^2 \mathcal{H}}{\partial t^2}, \ldots\right)
\end{aligned} \tag{1.12}$$

where $\mathbf{F}_i$ may be function of time. They provide the additional equations necessary to make Maxwell's equations deterministic. The precise form of constitutive relations depends on the medium.

For a stationary medium, the properties and, hence, the functionals $\mathbf{F}_i$ are independent of time. The Maxwell's equations presented in the previous section already assumes stationary media. Thus, in principle, moving media are ruled out.

In a linear medium, each functional $\mathbf{F}_i$ depends on $\mathcal{E}$ or $\mathcal{H}$, but not both, and is independent of the time derivatives of the fields. In other words, $\mathbf{F}_i$ actually depends linearly on $\mathcal{E}$ or $\mathcal{H}$. This is a drastic restriction, and rules out dependence on powers of $\mathcal{E}$ or $\mathcal{H}$ above the first. Thus, nonlinear problems, even those where the nonlinearity arises because of high power concentrations, for example, in optics or ionospheric modification experiments in microwaves, are eliminated, though it may still be possible to treat some of these problems using perturbation techniques.

When a linear dielectric is exposed to an electric field, the electron orbits of the various atoms and molecules are perturbed, leading to the creation of a secondary field. The molecules describing this situation are microscopic even though a macroscopic theory is used to compute the fields. It is customary to write

$$\mathcal{D} = \epsilon_0 \mathcal{E} + \mathcal{P} \tag{1.13}$$

where $\mathcal{E}$ is the actual field at a point in a dielectric, and $\mathcal{P}$ is the electric polarization. It is a measure of the volumetric dipole moment density caused by the applied electric field. Also, ϵ_0 is the electric permittivity of the vacuum. The vector $\mathcal{P}$ may or may not be aligned with $\mathcal{E}$ depending whether the medium is isotropic or anisotropic. In general, we may write

$$\mathcal{P} = \bar{\chi_e} \epsilon_0 \mathcal{E} \tag{1.14}$$

where $\bar{\chi_e}$ is the electric susceptibility tensor. Hence,

$$\mathcal{D} = \epsilon_0 (I + \bar{\chi_e}) \mathcal{E} \tag{1.15}$$

where I is the unitary tensor. Defining the electric permittivity of the medium as

$$\bar{\epsilon} = \epsilon_0 (I + \bar{\chi_e}) \tag{1.16}$$

we have

$$\mathcal{D} = \bar{\epsilon} \mathcal{E} \tag{1.17}$$

Similarly, in a magnetic material

$$\mathcal{B} = \mu_0 \mathcal{H} + \mathcal{M} \tag{1.18}$$

where $\mathcal{M}$ is the magnetic polarization, and μ_0 is the magnetic permeability of the vacuum. In general, $\mathcal{M}$ is expressed as

$$\mathcal{M} = \bar{\chi_m} \mu_0 \mathcal{H} \tag{1.19}$$

where $\bar{\chi_m}$ is the magnetic susceptibility tensor. Then

$$\mathcal{B} = \mu_0(I + \bar{\chi_m})\mathcal{H} \tag{1.20}$$

Defining the magnetic permeability tensor of the medium as

$$\bar{\mu} = \mu_0(I + \bar{\chi_m}) \tag{1.21}$$

we write

$$\mathcal{B} = \bar{\mu}\mathcal{H} \tag{1.22}$$

As a result of linearity, we may also write

$$\mathcal{J} = \bar{\sigma}\mathcal{E} \tag{1.23}$$

where $\bar{\sigma}$ is the conductivity of the medium. Its unit is Siemens/meter.

The explicit form of $\bar{\epsilon}$ is

$$\begin{aligned}\bar{\epsilon} = {} & \epsilon_{xx}\widehat{x}\widehat{x} + \epsilon_{xy}\widehat{x}\widehat{y} + \epsilon_{xz}\widehat{x}\widehat{z} \\ & + \epsilon_{yx}\widehat{y}\widehat{x} + \epsilon_{yy}\widehat{y}\widehat{y} + \epsilon_{yz}\widehat{y}\widehat{z} \\ & + \epsilon_{zx}\widehat{z}\widehat{x} + \epsilon_{zy}\widehat{z}\widehat{y} + \epsilon_{zz}\widehat{z}\widehat{z}\end{aligned} \tag{1.24}$$

with a similar definition for $\bar{\mu}$ and $\bar{\sigma}$. With the above definition, Eq. (1.17) can be interpreted as follows

$$\mathcal{D}_x = \epsilon_{xx}\mathcal{E}_x + \epsilon_{xy}\mathcal{E}_y + \epsilon_{xz}\mathcal{E}_z \tag{1.25}$$

with similar expressions for the y and z components. Thus, each component of $\mathcal{D}$ is related to all three components of $\mathcal{E}$.

Examples of anisotropic media include problems in crystal optics, including birefringence where ϵ is a tensor and in the ionosphere at microwave frequencies where the earth's magnetic field creates an anisotropy.

For an *isotropic medium*, the constitutive parameters are independent of direction and ϵ, μ and σ are scalar functions of position (instead of tensors). For an isotropic medium

$$\mathcal{D} = \epsilon\mathcal{E} \quad \mathcal{B} = \mu\mathcal{H} \quad \mathcal{J} = \sigma\mathcal{E} \tag{1.26}$$

where ϵ, μ and σ may depend on position. In such media $\mathcal{D}$ and $\mathcal{E}$, and $\mathcal{B}$ and $\mathcal{H}$ are in general parallel to each other, respectively.

In an isotropic medium, Maxwell's equations take the form

$$\nabla \times \mathcal{E} = -\mu \frac{\partial \mathcal{H}}{\partial t} \tag{1.27}$$

$$\nabla \times \mathcal{H} = (\sigma + \epsilon \frac{\partial}{\partial t}) \mathcal{E} \tag{1.28}$$

These are two coupled partial differential equations in two (vector) unknowns, and they are deterministic.

For a *homogeneous medium*, the constitutive parameters are independent of position, and ϵ, μ and σ are constant scalars.

The simplest possible medium is a linear, isotropic and homogeneous one, and for this reason we sometimes refer to it as a *simple medium*. A special case is free space (or vacuum) for which the constants have clearly defined values:

$$\epsilon_0 = 8.8544 \times 10^{-12} (\mathrm{F/m}) \quad \mu_0 = 4\pi \times 10^{-7} (\mathrm{H/m}) \quad \sigma = 0 (\mho/\mathrm{m}) \tag{1.29}$$

Constitutive parameters are in general functions of frequency. In such cases the medium is referred to as dispersive. For a non-dispersive medium, ϵ, μ and σ are independent of frequency.

1.3 Integral Form

The integral forms of the Maxwell's equations are obtained by applying Stokes and Divergence theorems to the curl and divergence equations, respectively. If S is an open surface bounded by a regular closed contour C as shown in Fig. 1.1, we have for any differentiable vector function **F**

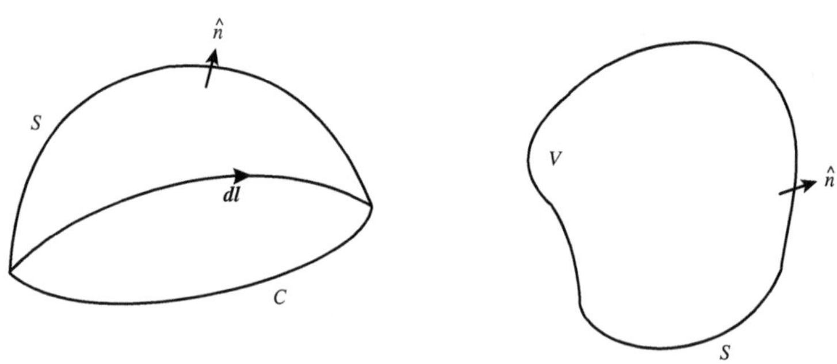

Fig. 1.1 Open and closed surfaces

$$\int_S (\nabla \times \mathbf{F}) \cdot \mathbf{ds} = \oint_C \mathbf{F} \cdot d\ell, \quad \text{Stokes theorem} \tag{1.30}$$

Also, for a closed surface S surrounding a volume V, we have

$$\int_V \nabla \cdot \mathbf{F} dv = \oint_S \mathbf{F} \cdot \mathbf{ds}, \quad \text{Divergence theorem} \tag{1.31}$$

Applying Stokes theorem to (1.1) and (1.2), we obtain

$$\oint_C \mathcal{E} \cdot d\ell = -\int_S \frac{\partial \mathcal{B}}{\partial t} \cdot \mathbf{ds} \tag{1.32}$$

and

$$\oint_C \mathcal{H} \cdot d\ell = \int_S \left(\mathcal{J} + \frac{\partial \mathcal{D}}{\partial t} \right) \cdot \mathbf{ds} \tag{1.33}$$

Applying the divergence theorem to (1.10) and (1.11), we get

$$\oint_S \mathcal{D} \cdot \mathbf{ds} = \int_V \rho dv \tag{1.34}$$

and

$$\oint_S \mathcal{B} \cdot \mathbf{ds} = 0 \tag{1.35}$$

Equation (1.32) is known as the experimental Faraday's law. It relates the time rate of change of magnetic flux linking a closed contour to the circulation of the electric field around the contour. Recognizing the left hand side as the induced electromotive force, we can deduce the Lenz's law

$$\mathcal{V}_{\text{emf}} = -\frac{d\Phi}{dt} \tag{1.36}$$

where Φ is the magnetic flux linking the closed loop C.

Equation (1.33), on the other hand, is the extended Ampere's law and includes the contribution of displacement current as well as the conduction current to the magnetic field. Also, (1.34) states that free electric charges serve as the source of electric field, while (1.35) denies the existence of independent magnetic charges. The latter is also known as the conservation of magnetic flux.

The integral form of the continuity relation is given by

$$\oint_S \boldsymbol{\mathcal{J}} \cdot \mathbf{ds} = -\int_V \frac{\partial \rho}{\partial t} dv \tag{1.37}$$

which relates the net electric current flux out of a closed surface to the time rate of change in electric charge enclosed within.

An alternative form of Faraday and Ampere laws can be obtained by applying the curl theorem

$$\int_V \nabla \times \mathbf{F} dv = \oint_S (\widehat{n} \times \mathbf{F}) ds \tag{1.38}$$

to (1.1) and (1.2) giving

$$\oint_S (\widehat{n} \times \boldsymbol{\mathcal{E}}) ds = -\int_V \frac{\partial \boldsymbol{\mathcal{B}}}{\partial t} dv \tag{1.39}$$

and

Table 1.2 Maxwell's equations in integral form

Name	Integral equation	Equation number
Faraday's law	$\oint_C \boldsymbol{\mathcal{E}} \cdot d\ell = -\int_S \frac{\partial \boldsymbol{\mathcal{B}}}{\partial t} \cdot \mathbf{ds}$	(1.32)
	$\oint_S (\widehat{n} \times \boldsymbol{\mathcal{E}}) ds = -\int_V \frac{\partial \boldsymbol{\mathcal{B}}}{\partial t} dv$	(1.39)
Ampere's law	$\oint_C \boldsymbol{\mathcal{H}} \cdot d\ell = \int_S \left(\boldsymbol{\mathcal{J}} + \frac{\partial \boldsymbol{\mathcal{D}}}{\partial t} \right) \cdot \mathbf{ds}$	(1.33)
	$\oint_S (\widehat{n} \times \boldsymbol{\mathcal{H}}) ds = \int_V (\mathbf{J} + \frac{\partial \boldsymbol{\mathcal{D}}}{\partial t}) dv$	(1.40)
Conservation of charge	$\oint_S \boldsymbol{\mathcal{J}} \cdot \mathbf{ds} = -\int_V \frac{\partial \rho}{\partial t} dv$	(1.37)
Gauss' law	$\oint_S \boldsymbol{\mathcal{D}} \cdot \mathbf{ds} = \int_V \rho dv$	(1.34)
Magnetic flux continuity	$\oint_S \boldsymbol{\mathcal{B}} \cdot \mathbf{ds} = 0$	(1.35)

$$\oint_S (\widehat{n} \times \mathcal{H}) ds = \int_V (\mathbf{J} + \frac{\partial \mathcal{D}}{\partial t}) dv \tag{1.40}$$

These forms are useful in deriving the boundary relations. The Maxwell's equations in integral form are tabulated in Table 1.2.

1.4 Boundary Relations

Maxwell's equations are in differential form and valid at points of continuity in the material properties of the medium. In any practical electromagnetic problem, however, we are concerned with the field in some region that is bounded—perhaps by an impenetrable surface such as a perfect conductor, or simply by the surface of another medium.

A discontinuous change in the properties of the medium may produce discontinuity in some components of the field. In general, the electromagnetic fields, $\mathcal{E}$, $\mathcal{D}$, $\mathcal{B}$, $\mathcal{H}$ and $\mathcal{J}$ may be discontinuous at the interface of two different media or on a surface carrying a surface charge or current. Thus, in order to uniquely specify the fields at the points of discontinuity, boundary conditions need to be imposed.

Boundary conditions are always independent of the differential equations, and cannot be derived from them. We, therefore, use the integral form of Maxwell's equations under certain assumptions to derive boundary conditions.

These assumptions are made on physical grounds and are stated as follows:

I Maxwell's equations in integral form are valid in all space including the discontinuities, and

II Field quantities are bounded functions of position.

The approach presented here is due to Schelkunoff. An alternative approach due to Lorentz is based on the postulation of the *required* boundary conditions in order to *extend* the integral form of the Maxwell's equations to regions containing interfaces.

It should be noted that the second assumption above is not strictly necessary, but it is good enough for many problems involving dielectric interfaces. However, if edges are present, such as a half plane or a dielectric cone, this requirement is relaxed, because in such cases the field quantities *can* in fact go to infinity; but their rate of growth is limited by physical energy requirements.

1.4.1 Derivation

Consider, as shown in Fig. 1.2, the interface S between two general media with the unit normal $\widehat{n}$ directed from region 2 to region 1.

In order to derive the boundary relations, we consider the pillbox volume at the interface and employ the integral form of the Maxwell's equations. Applying the alternative form of Faraday's law, (1.39), for the volume V, we obtain

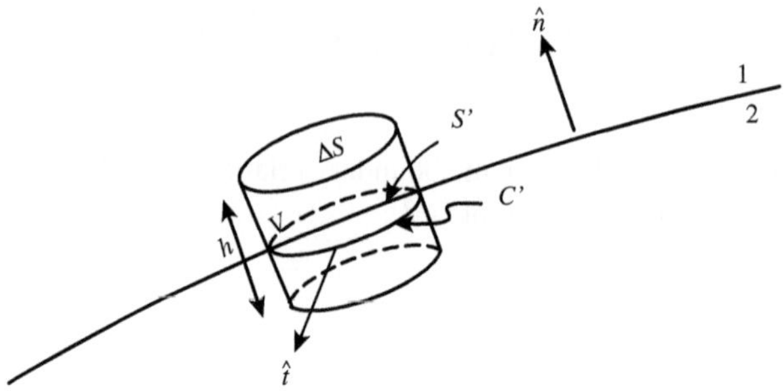

Fig. 1.2 The pillbox used for deriving the boundary relations

$$\widehat{n} \times (\mathcal{E}_1 - \mathcal{E}_2)\Delta S + \int_{\text{side}} (\widehat{t} \times \mathcal{E})ds = -\Delta S \int_{-h/2}^{h/2} \frac{\partial \mathcal{B}}{\partial t} d\xi \tag{1.41}$$

where $\widehat{t}$ denotes the unit normal to the cylindrical side surface and ξ is the coordinate normal to the surface. Taking the limit as the thickness h goes to zero, the following boundary relation is derived as we approach the interface

$$\widehat{n} \times (\mathcal{E}_1 - \mathcal{E}_2) = 0 \tag{1.42}$$

where use has been made of the assumption that the electric field is finite at the interface, such that the lateral electric flux through the shrinking cylindrical surface vanishes in the limit.

Applying now the extended Ampere's law (1.40) to the pillbox volume, we obtain

$$\widehat{n} \times (\mathcal{H}_1 - \mathcal{H}_2)\Delta S + \int_{\text{side}} (\widehat{t} \times \mathcal{H})ds = \Delta S \int_{-h/2}^{h/2} (\mathcal{J} + \frac{\partial \mathcal{D}}{\partial t}) d\xi \tag{1.43}$$

where ξ coordinate is picked parallel to the direction of the normal to the interface boundary, $\widehat{n}$. We will now take the limit as the thickness h goes to zero. Let the surface current density be defined as

$$\mathcal{K} \equiv \lim_{h \to 0} \int_{-h/2}^{h/2} \mathcal{J} d\xi \tag{1.44}$$

Again, noting that the fields are finite at the interface, Eq. (1.43) reduces to

$$\widehat{n} \times (\mathcal{H}_1 - \mathcal{H}_2) = \mathcal{K} \tag{1.45}$$

We may also obtain boundary relations for the electric and magnetic flux densities by employing the dependent Eqs. (1.34) and (1.35). Thus, applying (1.34) to the pillbox volume, we have

$$\widehat{n} \cdot (\mathcal{D}_1 - \mathcal{D}_2)\Delta S + \int_{\text{side}} (\widehat{t} \cdot \mathcal{D}) ds = \Delta S \int_{-h/2}^{h/2} \rho d\xi \tag{1.46}$$

Defining the surface charge density as

$$\rho_s \equiv \lim_{h \to 0} \int_{-h/2}^{h/2} \rho d\xi \tag{1.47}$$

we have in the limit as the thickness of the cylinder goes to zero

$$\widehat{n} \cdot (\mathcal{D}_1 - \mathcal{D}_2) = \rho_s \tag{1.48}$$

Similarly, from (1.35), we arrive at

$$\widehat{n} \cdot (\mathcal{B}_1 - \mathcal{B}_2) = 0 \tag{1.49}$$

Finally, application of the continuity relation (1.37) yields

$$\widehat{n} \cdot (\mathcal{J}_1 - \mathcal{J}_2)\Delta S + \int_{\text{side}} (\widehat{t} \cdot \mathcal{J}) ds = -\Delta S \int_{-h/2}^{h/2} \frac{\partial \rho}{\partial t} d\xi \tag{1.50}$$

In the limit as the height of the cylinder collapses to zero, the second term on the left hand side of (1.50) can be written as

$$\oint_{C'} (\lim_{h \to 0} \int_{-h/2}^{h/2} \mathcal{J} d\xi) \cdot \widehat{t} d\ell = \oint_{C'} \mathcal{K} \cdot \widehat{t} d\ell \tag{1.51}$$

where C' is the contour enclosing the projection of the cylindrical surface on the interface plane and $\widehat{t}$ is now taken as the unit vector normal to the contour C' in the interface plane. In the above, use has been made of the definition (1.44). But from the two-dimensional form of the divergence theorem, we have

$$\oint_{C'} \mathcal{K} \cdot \widehat{t} d\ell = \int_{S'} \nabla_s \cdot \mathcal{K} ds \tag{1.52}$$

where ∇_s is the planar del operator. Therefore, we may write

$$\widehat{n} \cdot (\mathcal{J}_1 - \mathcal{J}_2) + \nabla_s \cdot \mathcal{K} = -\frac{\partial \rho_s}{\partial t} \tag{1.53}$$

The boundary relations (1.42) and (1.49) imply that the tangential component of the electric field intensity and the normal component of the magnetic flux density are always continuous at the boundary interface. The boundary relations (1.45) and (1.48), on the other hand, imply that the tangential component of the magnetic field intensity as well as the normal component of the electric flux density may be discontinuous if surface current and charge densities exist at the interface, respectively.

It should be noted that all field quantities above are time varying; the integrals involving the time derivative $\partial/\partial t$ go to zero due to flux going to zero. The surface charge density ρ_s in general exists when one medium is perfectly conducting, although it can also exist at lossy interfaces. The surface current density **K**, however, can exist only if one medium is of infinite conductivity; under all other circumstances, $\mathcal{K}$ is identically zero.

It is also noted that since the divergence Eqs. (1.10) and (1.11) are dependent on the two curl Eqs. (1.1) and (1.2), the above boundary relations are not independent from each other. Thus, the simultaneous specification of the tangential component of $\mathcal{E}$ and the normal component of $\mathcal{B}$ at a boundary surface in a time-varying situation would be redundant and may yield contradictory results.

The above five relations are the most general form of boundary relations in electrodynamics. They form the basis for the theory of electromagnetic reflection and refraction of waves at media interfaces. In fact, they should be referred to as *transition relations*, since they describe the field behavior in the vicinity of interfaces between two different media.

The boundary relations are tabulated in Table 1.3.

Table 1.3 The electromagnetic transition relations

Name	Boundary relation	Equation number
Faraday's law	$\widehat{n} \times (\mathcal{E}_1 - \mathcal{E}_2) = 0$	(1.42)
Ampere's law	$\widehat{n} \times (\mathcal{H}_1 - \mathcal{H}_2) = \mathcal{K}$	(1.45)
Gauss' law	$\widehat{n} \times (\mathcal{D}_1 - \mathcal{D}_2) = \rho_s$	(1.48)
Magnetic flux continuity	$\widehat{n} \times (\mathcal{B}_1 - \mathcal{B}_2) = 0$	(1.49)
Conservation of charge	$\widehat{n} \cdot (\mathcal{J}_1 - \mathcal{J}_2) + \nabla_s \cdot \mathcal{K} = -\frac{\partial \rho_s}{\partial t}$	(1.53)

1.4.2 Special Cases

The above boundary relations take simple forms in some special cases. The following cases are of particular interest.

1.4.2.1 Conducting Interface

Suppose that S is the interface between a perfect conductor and a dielectric medium. If (say) $\sigma_2 = \infty$, we have

$$\mathcal{J}_2 = \lim_{\sigma_2 \to \infty} \sigma_2 \mathcal{E}_2 \tag{1.54}$$

and in order to have a finite current density, we must have $\mathcal{E}_2 = 0$ inside the conductor. A perfect conductor, therefore, cannot support a time-varying electrodynamic field inside. In other words

$$\widehat{n} \times \mathcal{E} = 0 \tag{1.55}$$

$$\widehat{n} \times \mathcal{H} = \mathcal{K} \tag{1.56}$$

$$\widehat{n} \cdot \mathcal{D} = \rho_s \tag{1.57}$$

$$\widehat{n} \cdot \mathcal{B} = 0 \tag{1.58}$$

$$\nabla_s \cdot \mathcal{K} = -\frac{\partial \rho_s}{\partial t} \tag{1.59}$$

where the symbols refer to fields external to the conducting medium. Note that in this case, $\mathcal{J}_1$ and $\mathcal{J}_2$ are both zero inside the dielectric and perfectly conducting regions.

From (1.55) and (1.58), it is seen that in the vicinity of a perfect conductor, the electric field is normal to the boundary while the magnetic field is tangential to it. Also, (1.56) and (1.57) now *define* the surface current and charge densities, respectively. Thus, the electric field points away from the conductor surface when the surface charges are positive, and points to the surface otherwise.

Since the fields inside the conducting region are zero, we need only one condition. Either (1.55) or (1.58) suffices and it is customary to choose the former.

In this case only, the transition conditions are truly boundary conditions in the mathematical sense of specifying a field quantity on the boundary. In other cases, however, the conditions are actually transition conditions relating two fields, which is why we need two conditions in general.

In the static case, the equations for $\mathcal{E}$ and $\mathcal{H}$ are uncoupled. Inside the conducting medium, $\mathcal{E} = 0$ due to Coulomb forces acting at equilibrium. However, $\mathcal{H}$ could be nonzero inside a conductor if a steady electric current flows. Thus

$$\widehat{n} \times \boldsymbol{\mathcal{E}}_1 = 0 \tag{1.60}$$
$$\widehat{n} \times (\boldsymbol{\mathcal{H}}_1 - \boldsymbol{\mathcal{H}}_2) = \boldsymbol{\mathcal{K}} \tag{1.61}$$
$$\widehat{n} \cdot \boldsymbol{\mathcal{D}}_1 = \rho_s \tag{1.62}$$
$$\widehat{n} \cdot (\boldsymbol{\mathcal{B}}_1 - \boldsymbol{\mathcal{B}}_2) = 0 \tag{1.63}$$
$$\nabla_s \cdot \boldsymbol{\mathcal{K}} = 0. \tag{1.64}$$

1.4.2.2 Dielectric Interfaces

At the interface between two lossless dielectric media, assume the surface charge and current densities are identically zero. Thus

$$\widehat{n} \times (\boldsymbol{\mathcal{E}}_1 - \boldsymbol{\mathcal{E}}_2) = 0 \tag{1.65}$$
$$\widehat{n} \times (\boldsymbol{\mathcal{H}}_1 - \boldsymbol{\mathcal{H}}_2) = 0 \tag{1.66}$$
$$\widehat{n} \cdot (\boldsymbol{\mathcal{D}}_1 - \boldsymbol{\mathcal{D}}_2) = 0 \tag{1.67}$$
$$\widehat{n} \cdot (\boldsymbol{\mathcal{B}}_1 - \boldsymbol{\mathcal{B}}_2) = 0 \tag{1.68}$$

When S is the interface between two lossy dielectrics, the surface current density $\boldsymbol{\mathcal{K}}$ is identically zero because $\boldsymbol{\mathcal{E}}$ is finite in both media. The surface charge density ρ_s, however, may or may not be zero. Therefore

$$\widehat{n} \times (\boldsymbol{\mathcal{E}}_1 - \boldsymbol{\mathcal{E}}_2) = 0 \tag{1.69}$$
$$\widehat{n} \times (\boldsymbol{\mathcal{H}}_1 - \boldsymbol{\mathcal{H}}_2) = 0 \tag{1.70}$$
$$\widehat{n} \cdot (\boldsymbol{\mathcal{D}}_1 - \boldsymbol{\mathcal{D}}_2) = \rho_s \tag{1.71}$$
$$\widehat{n} \cdot (\boldsymbol{\mathcal{B}}_1 - \boldsymbol{\mathcal{B}}_2) = 0 \tag{1.72}$$
$$\widehat{n} \cdot (\boldsymbol{\mathcal{J}}_1 - \boldsymbol{\mathcal{J}}_2) = -\frac{\partial \rho_s}{\partial t} \tag{1.73}$$

In order to study the condition under which $\rho_s = 0$, we write

$$\begin{aligned} \epsilon_1 \mathcal{E}_{1n} - \epsilon_2 \mathcal{E}_{2n} &= \rho_s \\ \sigma_1 \mathcal{E}_{1n} - \sigma_2 \mathcal{E}_{2n} &= -\frac{\partial \rho_s}{\partial t} \end{aligned} \tag{1.74}$$

Let $\rho_s = 0$, then the above system of equations will have a nontrivial solution if the determinant of the coefficients of $\mathcal{E}_{1n}$ and $\mathcal{E}_{2n}$ on the left hand side is zero. That is

$$\epsilon_1 \sigma_2 - \epsilon_2 \sigma_1 = 0 \tag{1.75}$$

or

$$\frac{\epsilon_1}{\sigma_1} = \frac{\epsilon_2}{\sigma_2} \tag{1.76}$$

which means that the relaxation times in the two media are the same. If the relaxation times are different, a surface charge accumulates at the boundary.

In static case, the interface between two dielectric media cannot support any surface charge or current densities unless one medium is a perfect conductor. Even a lossy dielectric cannot maintain a surface current or charge for a long period of time. Thus, ρ_s, $\mathcal{K}$, and $\mathcal{J}$ are zero at the interface of two dielectric media in static case. All four Eqs. (1.65)–(1.68), therefore, represent usable transition relations. We need two conditions and we choose continuity of tangential components of the electric and magnetic field intensities at the interface.

1.4.3 Other Boundary Conditions

In addition to the boundary relations stated above, other boundary conditions exist in electromagnetics, which are applied in various situations. They include the radiation condition (Sect. 1.11.5), impedance boundary condition (Chap. 4), edge condition (Chap. 6), and resistive sheet boundary condition (Chap. 7).

1.5 The Wave Equation

Consider a simple, lossless medium of permittivity ϵ and permeability μ. Our aim is to integrate Maxwell's equations in order to solve for the field quantities $\mathcal{E}$ and $\mathcal{H}$.

Taking the curl of Faraday's equation, we have

$$\nabla \times \nabla \times \mathcal{E} = -\nabla \times \frac{\partial \mathcal{B}}{\partial t} \tag{1.77}$$

or

$$\nabla(\nabla \cdot \mathcal{E}) - \nabla^2 \mathcal{E} = -\frac{\partial}{\partial t}(\nabla \times \mu \mathcal{H}) \tag{1.78}$$

Thus,

$$\nabla^2 \mathcal{E} - \mu\epsilon \frac{\partial^2 \mathcal{E}}{\partial t^2} = \mu \frac{\partial \mathcal{J}}{\partial t} + \nabla(\frac{\rho}{\epsilon}) \tag{1.79}$$

A similar analysis yields the corresponding equation for the magnetic field $\mathcal{H}$.

$$\nabla^2\mathcal{H} - \mu\epsilon\frac{\partial^2\mathcal{H}}{\partial t^2} = -\nabla \times \mathcal{J} \tag{1.80}$$

Equations (1.79) and (1.80) form a system of coupled differential equations. They are inhomogeneous wave equations and are sometimes written as

$$\Box^2 \begin{pmatrix} \mathcal{E} \\ \mathcal{H} \end{pmatrix} = \begin{pmatrix} \mu\frac{\partial \mathcal{J}}{\partial t} + \nabla(\frac{\rho}{\epsilon}) \\ -\nabla \times \mathcal{J} \end{pmatrix} \tag{1.81}$$

where $\Box^2 = \nabla^2 - \frac{1}{v^2}\frac{\partial}{\partial t^2}$ is the *wave* operator.[2] The parameter $v = 1/\sqrt{\mu\epsilon}$ is the velocity of wave propagation and is equal to the speed of light in the medium.

We have converted Maxwell's first order coupled differential equations involving two field quantities to second order differential equations each involving one field quantity.

From the above equations, the *effective sources* for electric and magnetic fields are

$$\mu\frac{\partial \mathcal{J}}{\partial t} + \nabla(\frac{\rho}{\epsilon}) \quad \text{for } \mathcal{E} \tag{1.82}$$

$$-\nabla \times \mathcal{J} \quad \text{for } \mathcal{H} \tag{1.83}$$

Note that from the conservation of charge, we can ascribe the electric field to the current alone as well.

Example 1.1 Consider a source free region. In this case, the fields satisfy the homogeneous wave equation

$$\nabla^2\mathcal{E} - \mu\epsilon\frac{\partial^2\mathcal{E}}{\partial t^2} = 0 \tag{1.84}$$

$$\nabla^2\mathcal{H} - \mu\epsilon\frac{\partial^2\mathcal{H}}{\partial t^2} = 0 \tag{1.85}$$

Let us seek a solution to the wave Eq. (1.84) of the form

$$\mathcal{E} = \mathbf{E}_0 f(z,t) \tag{1.86}$$

where $\mathbf{E}_0$ is a constant vector. Substituting in the wave equation, we have

$$\frac{\partial^2 f}{\partial z^2} - \frac{1}{v^2}\frac{\partial^2 f}{\partial t^2} = 0 \tag{1.87}$$

[2] Otherwise known as the d'Alembert operator.

The solution of the above equation can be easily shown to have the general form

$$f(z,t) = g(z - vt) + h(z + vt) \tag{1.88}$$

where g and h are arbitrary functions. The solution $g(z - vt)$ moves in the positive z-direction for increasing t, while the solution $h(z + vt)$ moves in the negative z-direction for increasing t. The shapes of the functions g and h are undistorted as they move along. Considering the solution

$$\boldsymbol{\mathcal{E}} = \mathbf{E}_0 g(z - vt) \tag{1.89}$$

we seek the condition under which the above wave can actually be an electric field satisfying Maxwell's equations. Noting the fact that the second order wave equation was derived based on the source free assumption, we impose the condition $\nabla \cdot \boldsymbol{\mathcal{E}} = 0$. Hence we find that

$$\begin{aligned}\nabla \cdot \boldsymbol{\mathcal{E}} &= \nabla \cdot [\mathbf{E}_0 g(z - vt)] \\ &= (\mathbf{E}_0 \cdot \widehat{z}) \frac{\partial g}{\partial z} = 0\end{aligned} \tag{1.90}$$

This implies that $\mathbf{E}_0$ should be perpendicular to the direction of propagation $\widehat{z}$. This allows solutions of the form

$$\boldsymbol{\mathcal{E}} = \widehat{x} E_x(z,t), \quad \boldsymbol{\mathcal{E}} = \widehat{y} E_y(z,t) \tag{1.91}$$

but not

$$\boldsymbol{\mathcal{E}} = \widehat{z} E_z(z,t) \tag{1.92}$$

because the latter violates $\nabla \cdot \boldsymbol{\mathcal{E}} = 0$ unless E_z is independent of z. The magnetic field vector can be found from the Maxwell's equation to be

$$\boldsymbol{\mathcal{H}} = Y(\widehat{z} \times \mathbf{E}_0) g(z - vt) \tag{1.93}$$

where $Y = \sqrt{\epsilon/\mu}$ is the intrinsic admittance of the medium. The wave is propagating in the z-direction, and the electric and magnetic fields are transverse to the direction of propagation. Such a wave is known as **T**ransverse **E**lectro **M**agnetic wave or TEM wave. □

In usual circumstances, it is not an easy task to solve inhomogeneous vector wave Eq. (1.81). Although in the time harmonic case, they may be uncoupled by using the equation of continuity, still the inhomogeneous terms would be quite complicated. Potential concepts are introduced to simplify the solution process.

1.6 Electromagnetic Potentials

Solution of vector wave equations satisfied by the electromagnetic fields is a difficult task. Each equation represents three scalar equations for the three orthogonal components. This is made yet more difficult by the complicated nature of the forcing function in each wave equation. In order to simplify the equations, we introduce the concept of potential functions. These are extensions of the electrostatic and magnetostatic potentials to the electrodynamic situation.

1.6.1 Lorenz's Potentials

Consider a simple, nondispersive and lossless medium. The magnetic flux density is a divergenceless vector field and it can always be written as

$$\mathcal{B} = \nabla \times \mathcal{A} \tag{1.94}$$

where $\mathcal{A}$ is called the magnetic vector potential. From Maxwell's equation

$$\nabla \times \left(\mathcal{E} + \frac{\partial \mathcal{A}}{\partial t} \right) = 0 \tag{1.95}$$

Thus, the quantity in the parenthesis is curl-free and can be expressed as the gradient of a scalar field

$$\mathcal{E} + \frac{\partial \mathcal{A}}{\partial t} = -\nabla \Phi \tag{1.96}$$

The function Φ is referred to as the electric scalar potential. Therefore

$$\mathcal{E} = -\frac{\partial \mathcal{A}}{\partial t} - \nabla \Phi \tag{1.97}$$

We have expressed the electromagnetic field quantities in terms of auxiliary potential functions.

We will now derive the equations in which $\mathbf{A}$ and Φ hold. Substituting (1.94) and (1.97) into (1.2), we get

$$\nabla \times \nabla \times \mathcal{A} = \mu \mathcal{J} - \mu\epsilon \frac{\partial^2 \mathcal{A}}{\partial t^2} - \mu\epsilon \nabla \frac{\partial \Phi}{\partial t} \tag{1.98}$$

and substituting (1.97) into Gauss' law, we obtain

$$\nabla^2 \Phi + \frac{\partial}{\partial t} \nabla \cdot \mathcal{A} = -\frac{\rho}{\epsilon} \tag{1.99}$$

The above equations are the two equations the vector and scalar potentials must satisfy.

Note that in the quasistatic case $\partial/\partial t \simeq 0$ and $\boldsymbol{\mathcal{E}} = -\nabla\Phi$, while (1.99) reduces to Poisson's equation. The physical interpretation of Φ in this case is simply that of the energy required to move a unit point charge in the static electric field

$$\begin{aligned}\Delta V = W_{21}/q &= -\int_{P_1}^{P_2} \boldsymbol{\mathcal{E}} \cdot \mathbf{d}\boldsymbol{\ell} = -\int_{P_1}^{P_2} (-\nabla\Phi) \cdot \mathbf{d}\boldsymbol{\ell} \\ &= \Phi_2 - \Phi_1 \end{aligned} \tag{1.100}$$

This quantity is known as the voltage difference between P_2 and P_1. The scalar potential Φ in this case is independent of the path of integration. In the time-varying case, on the other hand, we may write

$$\begin{aligned} W_{21}/q &= -\int_{P_1}^{P_2} \boldsymbol{\mathcal{E}} \cdot \mathbf{d}\boldsymbol{\ell} = -\int_{P_1}^{P_2} (-\nabla\Phi + \tfrac{\partial \boldsymbol{\mathcal{A}}}{\partial t}) \cdot \mathbf{d}\boldsymbol{\ell} \\ &= \Phi_2 - \Phi_1 - \tfrac{\partial}{\partial t} \int_{p_1}^{P_2} \boldsymbol{\mathcal{A}} \cdot \mathbf{d}\boldsymbol{\ell} \end{aligned} \tag{1.101}$$

Thus, in general, W_{21} depends on the path of integration and the concept of a unique voltage between pairs of points is not valid in the time-varying case.

To interpret the magnetic vector potential $\boldsymbol{\mathcal{A}}$, consider the magnetic flux through an open surface S

$$\int_S \boldsymbol{\mathcal{B}} \cdot \mathbf{ds} = \int_S (\nabla \times \boldsymbol{\mathcal{A}}) \cdot \mathbf{ds} = \oint_C \boldsymbol{\mathcal{A}} \cdot \mathbf{d}\boldsymbol{\ell} \tag{1.102}$$

where C is the closed contour enclosing the surface S. Thus, the circulation of the vector potential $\boldsymbol{\mathcal{A}}$ around a closed loop gives the magnetic flux through the loop.

Using the vector identity $\nabla\times\nabla\times\boldsymbol{\mathcal{A}} = \nabla(\nabla\cdot\boldsymbol{\mathcal{A}}) - \nabla^2\boldsymbol{\mathcal{A}}$ in (1.98) and rearranging terms, we get

$$\nabla^2\boldsymbol{\mathcal{A}} - \mu\epsilon\frac{\partial^2\boldsymbol{\mathcal{A}}}{\partial t^2} = -\mu\boldsymbol{\mathcal{J}} + \nabla(\nabla\cdot\boldsymbol{\mathcal{A}} + \mu\epsilon\frac{\partial\Phi}{\partial t}) \tag{1.103}$$

Also (1.99) can be written as

$$\nabla^2\Phi - \mu\epsilon\frac{\partial^2\Phi}{\partial t^2} = -\rho/\epsilon - \frac{\partial}{\partial t}(\nabla\cdot\boldsymbol{\mathcal{A}} + \mu\epsilon\frac{\partial\Phi}{\partial t}) \tag{1.104}$$

where we have subtracted the term $\mu\epsilon\partial^2\Phi/\partial t^2$ from both sides of (1.99).

1.6.2 Lorenz's Gauge

We have so far specified the curl of the vector potential $\mathcal{A}$ and its divergence has not been specified yet. In general, however, a vector field is specified up to a constant vector by defining both its curl and divergence. We may use this specification in order to simplify the system of Eqs. (1.103) and (1.104).

It is clear that if the parenthesis in the right hand sides of (1.103) and (1.104) are set to zero, the equations become decoupled. That is, if we choose $\nabla \cdot \mathbf{A}$ so that it satisfies the *Lorenz's gauge*[3]

$$\nabla \cdot \mathcal{A} + \mu\epsilon \frac{\partial \Phi}{\partial t} = 0 \tag{1.105}$$

the equations satisfied by $\mathcal{A}$ and Φ become

$$\nabla^2 \mathcal{A} - \mu\epsilon \frac{\partial^2 \mathcal{A}}{\partial t^2} = -\mu \mathcal{J} \tag{1.106}$$

and

$$\nabla^2 \Phi - \mu\epsilon \frac{\partial^2 \Phi}{\partial t^2} = -\rho/\epsilon \tag{1.107}$$

Using the wave operator, the above equations are sometimes written as

$$\Box^2 \begin{pmatrix} \mathcal{A} \\ \Phi \end{pmatrix} = -\begin{pmatrix} \mu \mathcal{J} \\ \rho/\epsilon \end{pmatrix} \tag{1.108}$$

It can be easily verified that $\mathcal{J}$ and ρ above are related through the continuity relation (1.3). In other words, solutions of (1.106) and (1.107) subject to the Lorenz's gauge describe electromagnetic fields arising from conserved charges. When $\mathcal{A}$ and Φ satisfy the Lorenz's condition, they are called *Lorenz's potentials*. The potentials are both causal in this case and they both satisfy the wave equation.

Note that the forcing functions in (1.108) contain no derivatives, implying a reduction of the order of singularities at the source at the expense of requiring a subsequent differentiation of the potentials to find the fields. Once (1.106) and (1.107) are solved for $\mathbf{A}$ and Φ, the fields are given by (1.94) and (1.97). In other words, the derivatives have been transferred from the source to the field point. We will examine the solution of the inhomogeneous wave equation later.

[3] Ludvig Lorenz (1829–1891), Danish physicist.

1.6.3 Gauge Transformation

The potentials introduced above, bear some degree of arbitrariness in their definition. It is noted that the above representation (1.94) and (1.97) of fields is invariant under the following transformations

$$\begin{aligned} \boldsymbol{\mathcal{A}}' &= \boldsymbol{\mathcal{A}} + \nabla\Psi \\ \Phi' &= \Phi - \frac{\partial \Psi}{\partial t} \end{aligned} \tag{1.109}$$

Thus, the two potentials $\boldsymbol{\mathcal{A}}'$ and Φ' yield the same fields as the potentials $\boldsymbol{\mathcal{A}}$ and Φ do under the above transformations. Such transformations are called gauge transformations and fields described by these potentials are said to be gauge invariant. That is, an infinite number of potentials satisfying (1.109) yield the same fields. The function Ψ is called the gauge function.

Now, we examine the condition under which the gauge transformations (1.109) yield potentials $\boldsymbol{\mathcal{A}}'$ and Φ' which satisfy the Lorenz's gauge (1.105)

$$\nabla \cdot \boldsymbol{\mathcal{A}}' + \mu\epsilon \frac{\partial \Phi'}{\partial t} = \nabla \cdot \boldsymbol{\mathcal{A}} + \mu\epsilon \frac{\partial \Phi}{\partial t} + (\nabla^2 \Psi - \mu\epsilon \frac{\partial^2 \Psi}{\partial t^2}) \tag{1.110}$$

Thus, $\boldsymbol{\mathcal{A}}'$ and Φ' satisfy Lorenz's condition if the gauge function Ψ is any solution of the homogeneous wave equation

$$\nabla^2 \Psi - \mu\epsilon \frac{\partial^2 \Psi}{\partial t^2} = 0 \tag{1.111}$$

Under this condition, $\boldsymbol{\mathcal{A}}'$ and Φ' also satisfy (1.106) and (1.107), respectively.

1.6.4 Coulomb's Gauge

A simple choice for the divergence of $\boldsymbol{\mathcal{A}}$ is to set it equal to zero. Thus

$$\nabla \cdot \boldsymbol{\mathcal{A}} = 0 \tag{1.112}$$

This is referred to as the *Coulomb's gauge*. Clearly, the scalar potential Φ satisfies the familiar Poisson's equation

$$\nabla^2 \Phi = -\rho/\epsilon \tag{1.113}$$

However, the equation for the magnetic vector potential $\boldsymbol{\mathcal{A}}$ remains complicated. Examining the solution to (1.113), we find analogous to the static case that

$$\Phi = \frac{1}{4\pi\epsilon}\int_V \frac{\rho(\mathbf{r}', t)}{R}dv' \tag{1.114}$$

This is a noncausal solution. However, $\boldsymbol{\mathcal{E}}$ may not be specified by Φ alone; the vector potential $\boldsymbol{\mathcal{A}}$ is needed as well. Coulomb's gauge is particularly useful in source-free regions as well as in the static case where only the scalar potential function Φ is needed. In a source-free region, Φ satisfies Laplace's equation under Coulomb's gauge

$$\nabla^2\Phi = 0 \tag{1.115}$$

The Lorenz's gauge reduces to the Coulomb's gauge under static conditions.

1.6.5 Hertz Potential

From the observation that $\boldsymbol{\mathcal{J}}$ and ρ are related to each other by the equation of continuity, it is possible to combine the vector and scalar potentials and the Lorenz's condition to define a *single* vector potential to describe electromagnetic fields. Let us first define the polarization charge density ρ_p and polarization current density $\mathbf{J}_p$ as

$$\rho_p = -\nabla\cdot\boldsymbol{\mathcal{P}} \quad \text{and} \quad \mathbf{J}_p = \frac{\partial\boldsymbol{\mathcal{P}}}{\partial t} \tag{1.116}$$

consistent with the equation of continuity. In the above, $\boldsymbol{\mathcal{P}}$ is the electric polarization vector. Therefore, $\boldsymbol{\mathcal{A}}$ and ϕ satisfy

$$\nabla^2\boldsymbol{\mathcal{A}} - \mu\epsilon\frac{\partial^2\boldsymbol{\mathcal{A}}}{\partial t^2} = -\mu\frac{\partial\boldsymbol{\mathcal{P}}}{\partial t} \tag{1.117}$$

and

$$\nabla^2\Phi - \mu\epsilon\frac{\partial^2\Phi}{\partial t^2} = \frac{\nabla\cdot\boldsymbol{\mathcal{P}}}{\epsilon} \tag{1.118}$$

Now, define the vector π such that

$$\boldsymbol{\mathcal{A}} = \mu\epsilon\frac{\partial\pi}{\partial t} \quad \text{and} \quad \Phi = -\nabla\cdot\pi \tag{1.119}$$

Then it is easy to see that the Lorenz's condition is automatically satisfied. Using (1.119), it can be shown that both (1.117) and (1.118) lead to the single vector equation

$$\nabla^2 \pi - \mu\epsilon \frac{\partial^2 \pi}{\partial t^2} = -\frac{\mathcal{P}}{\epsilon} \tag{1.120}$$

The vector function π is called the electric Hertz potential. Once π is determined from (1.120), the electromagnetic fields can be obtained as follows

$$\mathcal{E} = \nabla(\nabla \cdot \pi) - \mu\epsilon \frac{\partial^2 \pi}{\partial t^2} \tag{1.121}$$

$$\mathcal{H} = \epsilon \nabla \times \frac{\partial \pi}{\partial t} \tag{1.122}$$

The Hertz potential displays the desirable symmetry in field representations, particularly if one introduces the concept of fictitious magnetic sources as will be shown later in the discussion of the duality principle. It is therefore referred to as a *super* potential.

The above analysis is applicable to homogeneous media. For inhomogeneous media, it is more convenient to work with the Maxwell's equations directly.

1.7 Energy Flow

Any propagating field transports energy, and we now seek an expression for the power flow. Writing Maxwell's equations in a homogeneous medium as

$$\nabla \times \mathcal{E} = -\mu \frac{\partial \mathcal{H}}{\partial t} \tag{1.123}$$

$$\nabla \times \mathcal{H} = \mathcal{J} + \epsilon \frac{\partial \mathcal{E}}{\partial t} \tag{1.124}$$

we consider the identity

$$\nabla \cdot (\mathcal{E} \times \mathcal{H}) \equiv \mathcal{H} \cdot \nabla \times \mathcal{E} - \mathcal{E} \cdot \nabla \times \mathcal{H} \tag{1.125}$$

Thus

$$\nabla \cdot (\mathcal{E} \times \mathcal{H}) = -\mu \frac{\partial}{\partial t}(\frac{1}{2}\mathcal{H} \cdot \mathcal{H}) - \epsilon \frac{\partial}{\partial t}(\frac{1}{2}\mathcal{E} \cdot \mathcal{E}) - \mathcal{E} \cdot \mathcal{J} \tag{1.126}$$

The vector

$$\mathcal{S} = \mathcal{E} \times \mathcal{H} \tag{1.127}$$

Fig. 1.3 The Poynting theorem

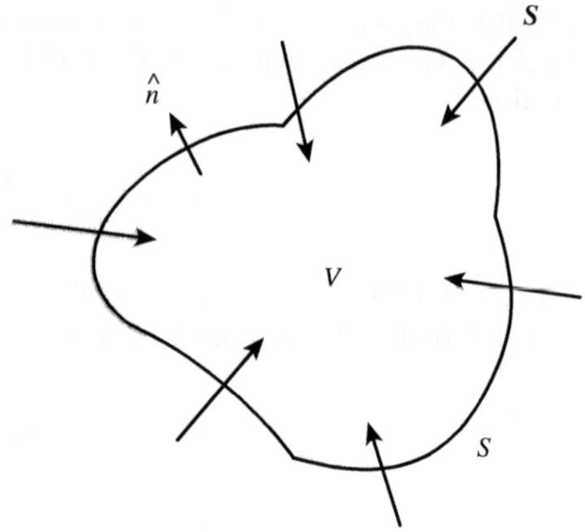

has the dimensions of Watt per meter squared [W/m^2] and is called the *Poynting vector*.[4] In the static case, the terms $\frac{1}{2}\epsilon\mathcal{E}\cdot\mathcal{E}$ and $\frac{1}{2}\mu\mathcal{H}\cdot\mathcal{H}$ denote the electric and magnetic energy densities, respectively. We extend the interpretation to the dynamic situation. Also, The last term on the right hand side is the power loss per unit volume. We thus have

$$\nabla\cdot\mathcal{S}+\frac{\partial}{\partial t}[w_e+w_m]+p_\ell=0 \tag{1.128}$$

which is a statement of the *Poynting theorem.* In order to examine the physical significance of this theorem, we integrate both sides over a finite volume V. Using the divergence theorem, we have

$$-\oint_S \widehat{n}\cdot\mathcal{S}ds=\frac{\partial}{\partial t}\int_V[w_e+w_m]dv+\int_V\mathcal{J}\cdot\mathcal{E}dv \tag{1.129}$$

The left hand side is the total power flux into the closed surface S (Fig. 1.3), while the right hand side is the sum of the time rate of increase in the stored electromagnetic energy and the total power dissipated in the enclosed volume V. In other words, (1.129) is simply a statement of the conservation of energy.

Example 1.2 For the propagating field

$$\mathcal{E}=\mathbf{E}_0 g(z-vt),\quad \mathcal{H}=Y(\widehat{z}\times\mathbf{E}_0)g(z-vt)$$

[4] After John H. Poynting (1852–1914).

where Y is the intrinsic admittance of the medium, we have

$$\begin{aligned}\boldsymbol{\mathcal{S}} &= \boldsymbol{\mathcal{E}} \times \boldsymbol{\mathcal{H}} \\ &= \widehat{z} Y E_0^2 [g(z - vt)]^2\end{aligned}$$

Also

$$w_e + w_m = \frac{1}{2}\epsilon E_0^2 g^2(z - vt) + \frac{1}{2}\epsilon E_0^2 g^2(z - vt) = \epsilon E_0^2 g^2(z - vt)$$

Therefore,

$$\boldsymbol{\mathcal{S}} = \widehat{z} Y E_0^2 [g(z - vt)]^2 = \widehat{z} v (w_e + w_m)$$

and the Poynting theorem is easily verified. The above analysis shows that the velocity times the total energy density stored equals the power density flow in a plane wave. □

It is noted that the Poynting vector $\boldsymbol{\mathcal{S}}$ is perpendicular to both $\boldsymbol{\mathcal{E}}$ and $\boldsymbol{\mathcal{H}}$ and gives the instantaneous surface power density over S. It is a measure of the rate of energy or instantaneous power flow density at a point.

1.7.1 Uniqueness Conditions

As an application of the Poynting theorem, consider an isotropic region V bounded by surface S. Assume that two solution sets exist for the field equations, namely $(\boldsymbol{\mathcal{E}}_1, \boldsymbol{\mathcal{H}}_1)$ and $(\boldsymbol{\mathcal{E}}_2, \boldsymbol{\mathcal{H}}_2)$ which are identical at all points in V and on S at time $t = t_0$. We seek the minimum number of conditions to be imposed on the field components at the boundary S so that the two solution sets remain identical at all times $t > t_0$.

Let us define the difference fields $\boldsymbol{\mathcal{E}} = \boldsymbol{\mathcal{E}}_1 - \boldsymbol{\mathcal{E}}_2$ and $\boldsymbol{\mathcal{H}} = \boldsymbol{\mathcal{H}}_1 - \boldsymbol{\mathcal{H}}_2$. The difference fields satisfy the homogeneous equations

$$\begin{aligned}\nabla \times \boldsymbol{\mathcal{E}} &= -\mu \frac{\partial \boldsymbol{\mathcal{H}}}{\partial t} \\ \nabla \times \boldsymbol{\mathcal{H}} &= \epsilon \frac{\partial \boldsymbol{\mathcal{E}}}{\partial t} + \sigma \boldsymbol{\mathcal{E}}\end{aligned}$$

From Poynting's theorem, we write

$$\int_V \sigma \boldsymbol{\mathcal{E}} \cdot \boldsymbol{\mathcal{E}} dv + \frac{\partial}{\partial t} \int_V [\frac{1}{2}\epsilon \boldsymbol{\mathcal{E}} \cdot \boldsymbol{\mathcal{E}} + \frac{1}{2}\mu \boldsymbol{\mathcal{H}} \cdot \boldsymbol{\mathcal{H}}] dv + \oint_s (\widehat{n} \cdot \boldsymbol{\mathcal{S}}) ds = 0 \tag{1.130}$$

We know that at $t = t_0$, $\mathcal{E} = 0$ and $\mathcal{H} = 0$ in V and on S. To find the necessary and sufficient condition under which $\mathcal{E}$ and $\mathcal{H}$ remain identically zero for $t > 0$ in V, we deduce from the above equation that the surface integral must vanish. That is

$$\oint_S (\widehat{n} \cdot \boldsymbol{\mathcal{S}}) ds = 0 \tag{1.131}$$

But this can be true if and only if

$$\widehat{n} \times \mathcal{E} = 0, \quad t > 0 \quad \text{on } S \tag{1.132}$$

or

$$\widehat{n} \times \mathcal{H} = 0, \quad t > 0 \quad \text{on } S \tag{1.133}$$

or

$$\widehat{n} \times \mathcal{E} = 0, \quad \text{on } S_1 \quad \text{and} \quad \widehat{n} \times \mathcal{H} = 0, \quad \text{on } S_2 \tag{1.134}$$

where S_1 and S_2 are subsurfaces of S and $S = S_1 \bigcup S_2$. Thus, the electromagnetic field is uniquely determined within region V bounded by S at all times $t > t_0$ by the initial values of field throughout V and the values of either tangential $\mathcal{E}$ or $\mathcal{H}$ on S for $t > 0$.

1.8 Time Harmonic Fields

Any physically realizable time variation can be decomposed into a spectrum of time harmonic functions represented by the complex exponential function $e^{j\omega t}$ using a Fourier integral, that is

$$f(t) = \frac{1}{2\pi} \int_{-\infty}^{\infty} g(\omega) e^{j\omega t} d\omega \tag{1.135}$$

and

$$g(\omega) = \int_{-\infty}^{\infty} f(\omega) e^{-j\omega t} dt \tag{1.136}$$

So, without loss of generality we may consider time harmonic fields.

Consider, for the sake of illustration, the electric field vector

$$\boldsymbol{\mathcal{E}}(\mathbf{r}, t) = \Re\mathrm{e}[\mathbf{E}e^{j\omega t}] \tag{1.137}$$

where **E** is the electric field *phasor* which is a complex vector function of the spatial coordinates. For example, in Cartesian system of coordinates, we write

$$\mathbf{E}(\mathbf{r}) = \mathcal{E}_x\widehat{x} + \mathcal{E}_y\widehat{y} + \mathcal{E}_z\widehat{z} \tag{1.138}$$

where each component is a complex function like

$$\mathcal{E}_x\widehat{x} = E'_x + jE''_x \tag{1.139}$$

Therefore, the x component of the actual time-varying field can be expressed in terms of the corresponding phasor as

$$\mathcal{E}_x(\mathbf{r}, t) = E'_x(\mathbf{r})\cos\omega t - E''_x(\mathbf{r})\sin\omega t \tag{1.140}$$

Using phasor techniques, Maxwell's equations can be easily solved for a time-harmonic signal. For example, letting

$$\boldsymbol{\mathcal{E}} = \Re\mathrm{e}[\mathbf{E}e^{j\omega t}], \quad \boldsymbol{\mathcal{H}} = \Re\mathrm{e}[\mathbf{H}e^{j\omega t}] \tag{1.141}$$

and substituting in (1.1), we have

$$\Re\mathrm{e}[\nabla \times \mathbf{E}e^{j\omega t}] = -\Re\mathrm{e}[\frac{\partial}{\partial t}\mu\mathbf{H}e^{j\omega t}] \tag{1.142}$$

Replacing $\frac{\partial}{\partial t}$ by $j\omega$ and removing the $\Re\mathrm{e}$ operator, we obtain

$$\nabla \times \mathbf{E} = -j\omega\mu\mathbf{H} \tag{1.143}$$

where the time dependence $e^{j\omega t}$ is understood and suppressed. Similarly, from (1.2), (1.10), and (1.11) we have

$$\nabla \times \mathbf{H} = j\omega\epsilon\mathbf{E} + \mathbf{J} \tag{1.144}$$

$$\nabla \cdot \mathbf{E} = \rho/\epsilon \tag{1.145}$$

$$\nabla \cdot \mathbf{H} = 0 \tag{1.146}$$

Taking the curl of (1.143), we have

$$\nabla \times \nabla \times \mathbf{E} = -j\omega\mu\nabla \times \mathbf{H} \tag{1.147}$$

and substituting from (1.144), we arrive at

$$\nabla^2 \mathbf{E} + k^2 \mathbf{E} = j\omega\mu \mathbf{J} + \nabla(\frac{\rho}{\epsilon}) \tag{1.148}$$

where $k = \omega\sqrt{\mu\epsilon}$ is the wavenumber. The corresponding equation for the magnetic field $\mathbf{H}$ is

$$\nabla^2 \mathbf{H} + k^2 \mathbf{H} = -\nabla \times \mathbf{J} \tag{1.149}$$

These are the inhomogeneous Helmholtz wave equations for the time harmonic fields.

In terms of vector and scalar potentials, the time harmonic fields can be written as

$$\mathbf{E} = -j\omega \mathbf{A} - \nabla\Phi \tag{1.150}$$

$$\mathbf{H} = \frac{1}{\mu}\nabla \times \mathbf{A} \tag{1.151}$$

The Lorenz's gauge in the harmonic case is given by

$$\nabla \cdot \mathbf{A} + j\omega\mu\epsilon\Phi = 0 \tag{1.152}$$

so that the electric field can be written as

$$\mathbf{E} = \frac{1}{j\omega\mu\epsilon}[\nabla(\nabla \cdot \mathbf{A}) + k^2 \mathbf{A}] \tag{1.153}$$

In this case, the potentials $\mathbf{A}$ and Φ satisfy the Helmholtz equations

$$(\nabla^2 + k^2)\mathbf{A} = -\mu \mathbf{J} \tag{1.154}$$

$$(\nabla^2 + k^2)\Phi = -\rho/\epsilon \tag{1.155}$$

The Hertz potential also satisfies the Helmholtz equation

$$(\nabla^2 + k^2)\pi = -\frac{\mathbf{J}}{j\omega\epsilon} \tag{1.156}$$

and the fields are expressed as

$$\mathbf{E} = \nabla(\nabla \cdot \pi) + k^2 \pi \tag{1.157}$$

$$\mathbf{H} = j\omega\epsilon\nabla \times \pi \tag{1.158}$$

For a lossy medium, the current density vector in (1.144) is given by $\sigma\mathbf{E}$ and we have

$$\nabla \times \mathbf{H} = j\omega(\epsilon - j\frac{\sigma}{\omega})\mathbf{E} \tag{1.159}$$

Defining the effective *complex permittivity* of the medium

$$\epsilon_c = \epsilon - j\frac{\sigma}{\omega} \tag{1.160}$$

we write

$$\nabla \times \mathbf{H} = j\omega\epsilon_c\mathbf{E} \tag{1.161}$$

Notice that the above equation is identical to (1.144) except for the complex permittivity instead of the real permittivity.

In general, a dielectric material exhibits polarization damping losses due to bound electrons in addition to a possible finite conductivity, so even though σ may be zero, ϵ is still complex and of the form $\epsilon' - j\epsilon''$. Thus,

$$\epsilon_c = \epsilon - j\frac{\sigma}{\omega} = \epsilon' - j\epsilon'' - j\frac{\sigma}{\omega} \tag{1.162}$$

It is a usual practice to represent the significance of both types of losses through the *loss tangent* of the medium defined as

$$\tan\delta = \frac{\epsilon''}{\epsilon'} + \frac{\sigma}{\omega\epsilon'} \tag{1.163}$$

The loss tangent is a measure of the goodness of the dielectric. If $\tan\delta \ll 1$, the medium is referred to as a *good dielectric*, while if $\tan\delta \gg 1$, it is called a *good conductor*.

Similarly, magnetic polarization damping losses are characterized by a complex permeability.

1.9 Complex Poynting Theorem

Consider the vector $\mathbf{E} \times \mathbf{H}^\star$. The divergence of this vector is given by

$$\nabla \cdot (\mathbf{E} \times \mathbf{H}^\star) = \mathbf{H}^\star \cdot \nabla \times \mathbf{E} - \mathbf{E} \cdot \nabla \times \mathbf{H}^\star \tag{1.164}$$

Substituting from the Maxwell's equations (1.143) and (1.144) into (1.164), we have

$$\begin{aligned}\nabla\cdot(\mathbf{E}\times\mathbf{H}^\star) &= -j\omega\mathbf{H}^\star\cdot\mathbf{B}+j\omega\mathbf{E}\cdot\mathbf{D}^\star-\mathbf{E}\cdot(\mathbf{J}_a^\star+\sigma\mathbf{E}^\star)\\ &= -j\omega[\mathbf{H}^\star\cdot\mathbf{B}-\mathbf{E}\cdot\mathbf{D}^\star]-\mathbf{E}\cdot(\mathbf{J}_a^\star+\sigma\mathbf{E}^\star)\end{aligned} \tag{1.165}$$

where we have included both the excitation and conduction current terms explicitly. Clearly, the time average power flux density is given by

$$\langle\boldsymbol{\mathcal{S}}\rangle = \langle\boldsymbol{\mathcal{E}}\times\boldsymbol{\mathcal{H}}\rangle = \frac{1}{2}\Re\mathrm{e}[\mathbf{E}\times\mathbf{H}^\star] = \Re\mathrm{e}\mathbf{S} \tag{1.166}$$

where

$$\mathbf{S} = \frac{1}{2}\mathbf{E}\times\mathbf{H}^\star \tag{1.167}$$

is referred to as *the complex Poynting vector*. We may also define the time average input power density as

$$p_i = \frac{1}{2}\mathbf{E}\cdot\mathbf{J}_a^\star \tag{1.168}$$

Equivalently, $-p_i$ would be the power density emitted by the source. Also, the time average dissipated power density in the medium is defined as

$$p_\ell = \frac{1}{2}\sigma\mathbf{E}\cdot\mathbf{E}^\star \tag{1.169}$$

The time average energy densities stored in electric and magnetic fields are defined respectively as

$$w_e^c = \frac{1}{2}\frac{\mathbf{E}\cdot\mathbf{D}^\star}{2} \tag{1.170}$$

and

$$w_m^c = \frac{1}{2}\frac{\mathbf{H}^\star\cdot\mathbf{B}}{2} \tag{1.171}$$

Therefore, the point form of the complex Poynting theorem reads

$$\nabla\cdot\mathbf{S}+j2\omega[w_m^c-w_e^c]+p_i+p_\ell=0 \tag{1.172}$$

Integrating over the volume V, and applying the divergence theorem, we obtain

$$\oint_S \mathbf{S} \cdot \mathbf{ds} = -j\omega 2 \int_V \frac{1}{4}[\mu|\mathbf{H}|^2 - \epsilon|\mathbf{E}|^2]dv - \int_V \frac{1}{2} E \cdot \mathbf{J}_a^{\star} dv - \int_V \frac{1}{2}\sigma|\mathbf{E}|^2 dv \tag{1.173}$$

This is the statement of the Poynting theorem for time harmonic fields. Comparing with the real form of the Poynting theorem (1.129), the above involves the difference of the stored energy terms rather than the sum.

If μ and ϵ are both real, then by equating the real and imaginary parts of the two sides of (1.173), we find that

$$\Re\mathrm{e} \oint_S \mathbf{S} \cdot \widehat{n} ds = -\int_V \frac{1}{2}[\mathbf{E} \cdot \mathbf{J}_a^{\star} + \sigma|\mathbf{E}|^2]dv \tag{1.174}$$

and

$$\Im\mathrm{m} \oint_S \mathbf{S} \cdot \widehat{n} ds = -2\omega \int_V \frac{1}{4}[\mu|\mathbf{H}|^2 - \epsilon|\mathbf{E}|^2]dv \tag{1.175}$$

The first equation corresponds to the dissipated power in V. The right-hand side is the negative of the time-averaged power dissipated in the source plus that dissipated in V. Therefore, $\Re\mathrm{e}\{\mathbf{S}\}$ is a measure of the outward time-averaged power flow per unit area of the enclosing surface. However, **S** is a complex vector and its imaginary part is related to the time-averaged power stored in the medium. This is known as the reactive power which is related to the difference in the stored magnetic and electric energies. Hence, if a system stores—on the average—equal amounts of magnetic and electric energies, it does not consume any reactive power. In this case, the imaginary part of the complex Poynting vector **S** corresponds to instantaneous power that time averages to zero.

Example 1.3 To comprehend the significance of the reactive power, consider a simple LC circuit as shown in Fig. 1.4 driven by a time harmonic voltage source. The current I_g is given by

$$I_g = V_g(j\omega C + \frac{1}{j\omega L}) = j\omega C V_g(1 - \frac{1}{\omega^2 LC})$$

At the resonance frequency of the tank circuit, $\omega = 1/\sqrt{LC}$, its input impedance is infinite, and hence $I_g = 0$. Therefore, there is no power delivered from the generator, be it real or reactive. However, $I_1 = -I_2 \neq 0$ at resonance, and as the tank circuit is resonating, the electric field energy stored in the capacitor C is being converted

Fig. 1.4 A resonant circuit

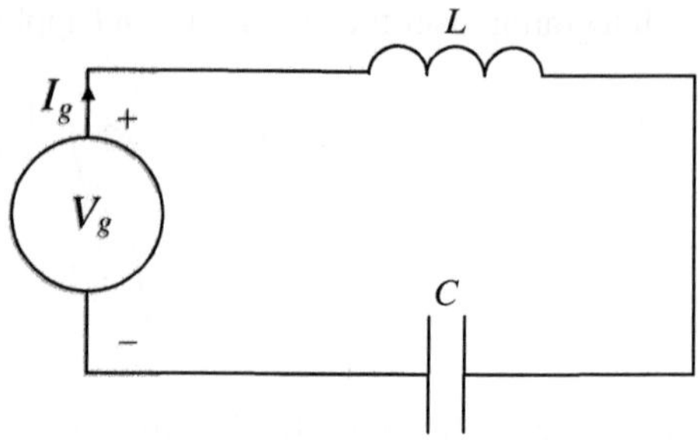

into the magnetic field energy stored in the self L. Therefore, $\frac{1}{2}L|I|^2 = \frac{1}{2}C|V|^2$ can be easily verified for a resonating tank circuit. This is precisely the case mentioned above.

Away from the resonance, I_g is 90° out-of-phase with V_g, and the complex power, $V_g I_g^\star$ is purely imaginary. This implies that there is no time average power delivered by the source, but it delivers nonzero reactive power. The magnetic and electric stored energies are not in perfect balance with respect to each other, and the system needs to be augmented with external reactive power. □

For an antenna, the reactive power is always present in the near field and must be provided by the antenna even though it remains stored in the adjacent medium. The reactive power, therefore, diminishes the efficiency of the antenna as a radiator.

Notice that in the presence of dielectric or magnetic losses (in addition to conduction losses), ϵ and μ may be complex. In this case, the imaginary parts of these parameters contribute to the real power dissipated in the medium. Thus,

$$w_e = \frac{\epsilon'}{4}|\mathbf{E}|^2 \tag{1.176}$$

$$w_m = \frac{\mu'}{4}|\mathbf{H}|^2 \tag{1.177}$$

$$p_\ell = \frac{\omega\epsilon''}{2}|\mathbf{E}|^2 + \frac{\omega\mu''}{2}|\mathbf{H}|^2 + \frac{1}{2}\sigma|\mathbf{E}|^2. \tag{1.178}$$

1.10 Specific Absorption Rate

At radio frequencies, biological media are lossy and they dissipate incident electromagnetic power to heat through a thermodynamic mechanism. Specific Absorption Rate (SAR) is used to represent the power loss per unit mass of biological media, when the incident power flux density is 1 mW/cm^2. If the density of the biological medium is denoted by ρ, we have

$$\text{SAR} = \frac{p_\ell}{\rho}, \text{[W/kg]} \tag{1.179}$$

or equivalently,

$$\text{SAR} = \frac{\sigma |\mathbf{E}|^2}{2\rho} \tag{1.180}$$

where σ is the conductivity of the medium and we assumed $\epsilon'' = \mu'' = 0$. The density ρ is usually taken to be approximately equal to that of water ($\rho = 100\,\text{kg/m}^3$).

1.11 Green's Function Method

A Green's function is the solution of a given differential equation satisfying prescribed boundary conditions, with the source function being an impulse of unit strength at a point in space.

In the case of the inhomogeneous Helmholtz equation, the Green's function satisfies

$$\nabla^2 G(\mathbf{r};\mathbf{r}') + k^2 G(\mathbf{r};\mathbf{r}') = -\delta(\mathbf{r}-\mathbf{r}') \tag{1.181}$$

where the right hand side represents a unit (spherically symmetric) source applied at $\mathbf{r} = 0$.

The Green's function is uniquely specified by the following properties:

1. The Green's function satisfies the homogeneous equation outside the source point,
2. The green's function satisfies the homogeneous boundary conditions,
3. The Green's function is symmetrical with respect to $\mathbf{r}$ and $\mathbf{r}'$. That is

$$G(\mathbf{r};\mathbf{r}') = G(\mathbf{r}';\mathbf{r}) \tag{1.182}$$

   This property is known as *reciprocity*.
4. The Green's function is continuous in $\mathbf{r}$ and $\mathbf{r}'$ for $\mathbf{r} \neq \mathbf{r}'$, and
5. The gradient of the Green's function is discontinuous in the vicinity of the source point

$$\oint_S \nabla G \cdot \mathbf{ds} = -1 \tag{1.183}$$

It can be shown that a function constructed such that the above characteristics are satisfied, is in fact the Green's function of the problem.

1.11.1 Green's Identities

The importance of the Green's function stems from the Green's identities.

1.11.1.1 Green's First Identity

Let f and g be scalar functions of position. Then applying the divergence theorem to the vector $g\nabla f$ over a volume V enclosed by a surface S, we have

$$\int_V \nabla \cdot (g\nabla f)dv = \oint_S (g\nabla f) \cdot \widehat{n}ds \tag{1.184}$$

where $\widehat{n}$ is the unit normal to the closed surface s. Expanding the integrand on the left hand side, we obtain

$$\int_V (\nabla g \cdot \nabla f + g\nabla^2 f)dv = \oint_S (g\nabla f) \cdot \widehat{n}ds \tag{1.185}$$

and therefore

$$\int_V g\nabla^2 f dv = \oint_S (g\nabla f) \cdot \widehat{n}ds - \int_V \nabla g \cdot \nabla f dv \tag{1.186}$$

This is the Green's first identity.

1.11.1.2 Green's Second Identity

If we interchange f and g in Green's first identity and subtract the results, we obtain

$$\int_V (g\nabla^2 f - f\nabla^2 g)dv = \oint_S (g\nabla f - f\nabla g) \cdot \widehat{n}ds \tag{1.187}$$

which is Green's second identity, also referred to as the Green's formula. Green's formula is sometimes written in the following form

$$\int_V (g\nabla^2 f - f\nabla^2 g)dv = \oint_S (g\frac{\partial f}{\partial n} - f\frac{\partial g}{\partial n})ds. \tag{1.188}$$

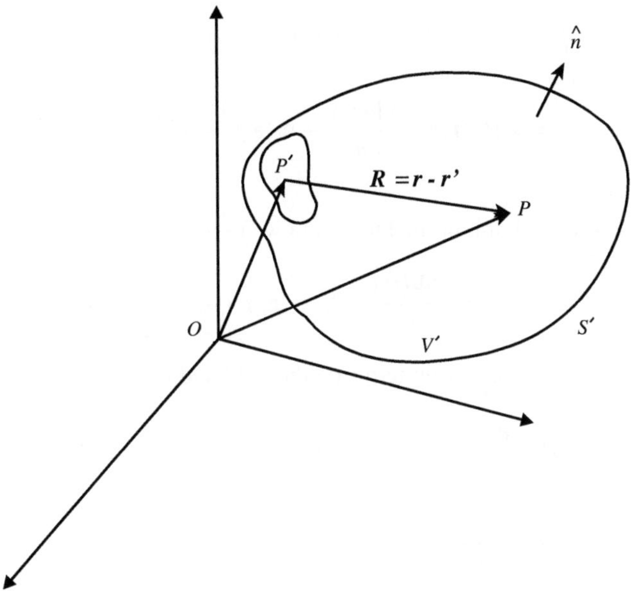

Fig. 1.5 The Green's function

1.11.2 Inhomogeneous Scalar Helmholtz Equation

It is assumed that we know the Green's function $G(\mathbf{r}; \mathbf{r}')$ as a solution of

$$\nabla^2 G(\mathbf{r}; \mathbf{r}') + k^2 G(\mathbf{r}; \mathbf{r}') = -\delta(\mathbf{r} - \mathbf{r}') \tag{1.189}$$

subject to appropriate boundary conditions. It is desired to obtain the solution of the scalar inhomogeneous Helmholtz equation

$$\nabla^2 \Phi + k^2 \Phi = -\rho/\epsilon_0 \tag{1.190}$$

where $\rho(\mathbf{r})$ is the source distribution (Fig. 1.5).

We shall apply Green's theorem (1.188) to find the solution Φ. It should be noted that all derivatives in the integrand must be with respect to the variable coordinates, that is, the variable of integration. Choose a region V' in space bounded by a closed surface S' containing all the source distribution $\rho(\mathbf{r}')$. Let the field point $P(\mathbf{r})$ be fixed and the source point $P'(\mathbf{r}')$ be variable with respect to an arbitrary origin O. From Green's second identity, we have

$$\int_{V'} [G(\mathbf{r};\mathbf{r}')\nabla'^2\Phi(\mathbf{r}') - \Phi(\mathbf{r}')\nabla'^2 G(\mathbf{r};\mathbf{r}')]dv' = \oint_{S'} [G(\mathbf{r};\mathbf{r}')\frac{\partial\Phi(\mathbf{r}')}{\partial n} - \Phi(\mathbf{r}')\frac{\partial G(\mathbf{r};\mathbf{r}')}{\partial n}]\cdot\widehat{n}ds' \tag{1.191}$$

where $\widehat{n}$ is the outward unit normal to S' and we used

$$\nabla'\Phi(\mathbf{r}')\cdot\widehat{n} = \frac{\partial\Phi(\mathbf{r}')}{\partial n}, \quad \nabla' G(\mathbf{r};\mathbf{r}')\cdot\widehat{n} = \frac{\partial G(\mathbf{r};\mathbf{r}')}{\partial n} \tag{1.192}$$

Substituting for the Laplacian terms on the left hand integral, we obtain

$$\int_{V'} \{G[-\frac{\rho}{\epsilon_0} - k^2\Phi] - \Phi[-\delta(\mathbf{r}-\mathbf{r}') - k^2 G(\mathbf{r};\mathbf{r}')]\}dv' = \oint_{S'} (G\frac{\partial\Phi}{\partial n} - \Phi\frac{\partial G}{\partial n})\cdot\widehat{n}ds' \tag{1.193}$$

or

$$\int_{V'} \Phi(\mathbf{r}')\delta(\mathbf{r}-\mathbf{r}')dv' = \int_{V'} \frac{G(\mathbf{r};\mathbf{r}')\rho(\mathbf{r}')}{\epsilon_0}dv' + I \tag{1.194}$$

where for simplicity we have used I to represent the surface integral on the right hand side. By definition

$$\begin{aligned}\int_{V'} \Phi(\mathbf{r}')\delta(\mathbf{r}-\mathbf{r}')dv' &= \Phi(\mathbf{r}), \quad \text{if} \quad \mathbf{r}\in V' \\ &= 0, \quad \text{if} \quad \mathbf{r}\notin V'\end{aligned} \tag{1.195}$$

Assuming $\mathbf{r}\in V'$, we obtain the desired solution as

$$\Phi(\mathbf{r}) = \int_{V'} \frac{G(\mathbf{r};\mathbf{r}')\rho(\mathbf{r}')}{\epsilon_0}dv' + \oint_{S'} [G(\mathbf{r};\mathbf{r}')\frac{\partial\Phi(\mathbf{r}')}{\partial n} - \Phi(\mathbf{r}')\frac{\partial G(\mathbf{r};\mathbf{r}')}{\partial n}]ds'. \tag{1.196}$$

1.11.3 Green's Function of the First Kind

Assume that Φ is given on S. Then the Green's function of the problem satisfies

$$\nabla^2 G_1(\mathbf{r};\mathbf{r}') + k^2 G_1(\mathbf{r};\mathbf{r}') = -\delta(\mathbf{r}-\mathbf{r}') \tag{1.197}$$

subject to the homogeneous Dirichlet boundary condition

$$G_1(\mathbf{r};\mathbf{r}') = 0 \quad \text{on } S \tag{1.198}$$

The function G_1 is called the Green's function of the first kind. With G_1 so defined, we now have from (1.196)

$$\Phi(\mathbf{r}) = \int_{V'} \frac{G_1(\mathbf{r};\mathbf{r}')\rho(\mathbf{r}')}{\epsilon_0} dv' - \oint_S \Phi(\mathbf{r}') \frac{\partial G_1(\mathbf{r};\mathbf{r}')}{\partial n} ds' \tag{1.199}$$

Thus, the source distribution plus a knowledge of Φ on S is sufficient to determine Φ everywhere inside S.

1.11.4 Green's Function of the Second Kind

Assume now that the normal derivative of Φ is given on S. Then the Green's function of the problem satisfies

$$\nabla^2 G_2(\mathbf{r};\mathbf{r}') + k^2 G_2(\mathbf{r};\mathbf{r}') = -\delta(\mathbf{r}-\mathbf{r}') \tag{1.200}$$

subject to the homogeneous Newman condition

$$\frac{\partial G_2(\mathbf{r};\mathbf{r}')}{\partial n} = 0 \quad \text{on } S \tag{1.201}$$

The function G_2 is called the Green's function of the second kind. In this case, (1.196) gives

$$\Phi(\mathbf{r}) = \int_{V'} \frac{G_2(\mathbf{r};\mathbf{r}')\rho(\mathbf{r}')}{\epsilon_0} dv' + \oint_S G_2(\mathbf{r};\mathbf{r}') \frac{\partial \Phi}{\partial n} ds' \tag{1.202}$$

Thus, the source distribution plus a knowledge of the normal derivative of Φ on S is sufficient to determine Φ everywhere inside S.

1.11.5 The Free Space Green's Function

For $\rho(\mathbf{r}')$ confined in space, we can take S' as farther away as we please, provided that there is no other boundary present. Let $\mathbf{r}$ be finite, r' be very large compared to both r and the source dimension. By moving S' to infinity, the volume integral on the right hand side of (1.196) remains unchanged. We are interested in finding out the behavior of the integral

$$I = \oint_{S'} [G_0(\mathbf{r}; \mathbf{r}') \frac{\partial \Phi(\mathbf{r}')}{\partial n} - \Phi(\mathbf{r}') \frac{\partial G_0(\mathbf{r}'; \mathbf{r}')}{\partial n}] ds' \tag{1.203}$$

as $r' \to \infty$. Let us first examine the solution of (1.181) in an unbounded free space.

We are looking for the function $\mathcal{G}_0$ satisfying

$$\nabla^2 \mathcal{G}_0 - \frac{1}{c^2} \frac{\partial^2 \mathcal{G}_0}{\partial t^2} = -\delta(t)\delta(\mathbf{r}) \tag{1.204}$$

subject to the causality condition

$$\mathcal{G}_0(\mathbf{r}, t) = 0, \quad t < 0 \tag{1.205}$$

Notice that we have used the subscript 0 to denote the unbounded free space Green's function. In order to solve this problem, we first take the Fourier transform of both sides and obtain

$$\nabla^2 G_0(\mathbf{r}, \omega) + k^2 G_0(\mathbf{r}, \omega) = -\delta(\mathbf{r}) \tag{1.206}$$

where $G_0(\mathbf{r}, \omega)$ is the Fourier transform of $\mathcal{G}_0(\mathbf{r}, t)$ and $k^2 = \omega^2/c^2$. Away from the source, G_0 satisfies the homogeneous wave equation

$$\nabla^2 G_0(\mathbf{r}, \omega) + k^2 G_0(\mathbf{r}, \omega) = 0 \tag{1.207}$$

which due to the spherical symmetry of the problem takes the form

$$\frac{1}{r^2} \frac{\partial}{\partial r} (r^2 \frac{\partial G_0}{\partial r}) + k^2 G_0 = 0 \tag{1.208}$$

The latter may be written as

$$\frac{\partial^2}{\partial r^2} (r G_0) + k^2 (r G_0) = 0 \tag{1.209}$$

The general solution of this equation is

$$G_0 = A_1 e^{jkr}/r + A_2 e^{-jkr}/r \tag{1.210}$$

Taking the inverse Fourier transform of G_0, we obtain

$$\mathcal{G}_0(\mathbf{r}, t) = \frac{A_1 \delta(t + r/c)}{r} + \frac{A_2 \delta(t - r/c)}{r} \tag{1.211}$$

It is clear from causality (1.205) that A_1 is identically zero.

$$A_1 \equiv 0 \tag{1.212}$$

To determine A_2, we integrate (1.206) in a spherical volume V with center at $\mathbf{r} = 0$. That is

$$\int\limits_V \nabla \cdot \nabla G_0 dv + k^2 \int\limits_V G_0 dv = -\int\limits_V \delta(\mathbf{r}) dv \tag{1.213}$$

The integral on the right hand side is by definition equal to unity. Applying the divergence theorem, we have

$$\oint\limits_S \nabla G_0 \cdot \mathbf{ds} + k^2 \int\limits_V G_0 dv = -1 \tag{1.214}$$

where S is the spherical surface of radius r enclosing the volume V. Substituting G_0 from (1.210) and taking the limit as $r \to 0$, we obtain

$$A_2 = 1/4\pi \tag{1.215}$$

Thus,

$$\mathcal{G}_0(\mathbf{r}, t) = \frac{\delta(t - r/c)}{4\pi r} \tag{1.216}$$

and

$$G_0(\mathbf{r}, \omega) = \frac{e^{-jkr}}{4\pi r} \tag{1.217}$$

If the source is located at $\mathbf{r} = \mathbf{r}'$, instead of at the origin, we have

$$G_0(\mathbf{r}; \mathbf{r}') = \frac{e^{-jkR}}{4\pi R} \tag{1.218}$$

where $R = |\mathbf{r} - \mathbf{r}'|$ and $\mathbf{r}$ is the field (observation) point.

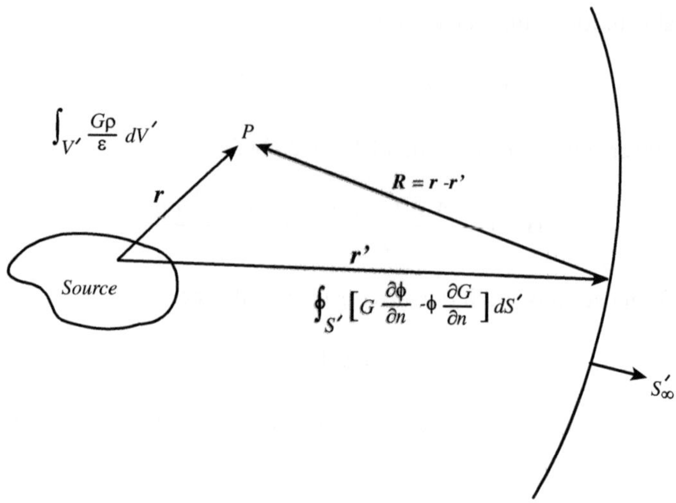

Fig. 1.6 The radiation condition

1.11.5.1 Radiation Condition

We now examine the behavior of the surface integral I in (1.203) at large distances (Fig. 1.6). In particular, we have

$$G_0 \simeq \frac{e^{-jkr'}}{4\pi r'} \tag{1.219}$$

and

$$\frac{\partial G_0}{\partial n} = \nabla' G_0 \cdot \widehat{n} = \frac{\partial G_0}{\partial r'}(\widehat{r'} \cdot \widehat{r}) \simeq \frac{\partial G_0}{\partial r'} \tag{1.220}$$

Or, equivalently

$$\frac{\partial G_0}{\partial n} \simeq \frac{\partial}{\partial r'}\left(\frac{e^{-jkr'}}{4\pi r'}\right) = -(\frac{jk}{r'} + \frac{1}{r'^2})\frac{e^{-jkr'}}{4\pi} \tag{1.221}$$

Thus, we have as $r' \to \infty$

$$\begin{aligned} I &= \oint_{S'} [\tfrac{e^{-jkr'}}{4\pi r'} \tfrac{\partial \Phi}{\partial r'} + \Phi(\mathbf{r}')(\tfrac{jk}{r'} + \tfrac{1}{r'^2})\tfrac{e^{-jkr'}}{4\pi}]ds' \\ &\simeq \tfrac{1}{4\pi} \oint_{S'} \left(r'[\tfrac{\partial \Phi}{\partial r'} + jk\Phi]e^{-jkr'} + \Phi e^{-jkr'}\right) d\Omega \end{aligned} \tag{1.222}$$

where

$$d\Omega = \frac{ds'(\widehat{n} \cdot \widehat{r'})}{r'^2} \simeq \frac{ds'}{r'^2} \tag{1.223}$$

is the element of solid angle.

In the static case, for any confined source, the potential goes to zero at infinity. Mathematically, this implies that the potential function is bounded

$$\lim_{r' \to \infty} r'\Phi(\mathbf{r}') < K \tag{1.224}$$

where K is a constant. If we impose the above condition on the potential function in the electrodynamic case, we would have

$$\lim_{r' \to \infty} \oint_{S'} \Phi(\mathbf{r}')e^{-jkr'} d\Omega = 0 \tag{1.225}$$

However, this is not sufficient to ensure a vanishing contribution of the integral I. Sommerfeld[5] assumed that

$$\lim_{r' \to \infty} r'(\frac{\partial \Phi}{\partial r'} + jk\Phi) = 0 \tag{1.226}$$

This is the radiation condition to be satisfied by Φ. We now remove the $'$ and restate *the Sommerfeld radiation condition* as

$$\lim_{r \to \infty} r \left(\frac{\partial \Phi(\mathbf{r})}{\partial r} + jk\Phi(\mathbf{r}) \right) = 0 \tag{1.227}$$

The free space Green's function satisfies the radiation condition.

For a given source distribution in an unbounded free space, the desired solution to the inhomogeneous Helmholtz equation reduces to

$$\Phi(\mathbf{r}) = \int_{V'} \frac{G_0(\mathbf{r}; \mathbf{r}')\rho(\mathbf{r}')}{\epsilon_0} dv' \tag{1.228}$$

with the free space Green's function given by

$$G_0(\mathbf{r}; \mathbf{r}') = \frac{e^{-jk|\mathbf{r}-\mathbf{r}'|}}{4\pi|\mathbf{r} - \mathbf{r}'|}. \tag{1.229}$$

---

[5] Arnold J. W. Sommerfeld (1868–1951), German mathematician and physicist born in Russia.

1.11.6 The Modified Green's Function

Let there be some conducting body (bodies) in the region. Denote the surface of the conducting body by S. Note the definition of the unit normal $\widehat{n}$ on S. We then have

$$\Phi(\mathbf{r}) = \int_{V'} \frac{G(\mathbf{r};\mathbf{r}')\rho(\mathbf{r}')}{\epsilon_0} dv' + \oint_S [G\frac{\partial \Phi}{\partial n} - \Phi\frac{\partial G}{\partial n}]ds' + \oint_{S'} [G\frac{\partial \Phi}{\partial n} - \Phi\frac{\partial G}{\partial n}]ds' \quad (1.230)$$

If we move S' to infinity, then the radiation condition makes the contribution of the third integral on the right-hand side of the above equation vanish. Hence

$$\Phi(\mathbf{r}) = \int_{V'} \frac{G(\mathbf{r};\mathbf{r}')\rho(\mathbf{r}')}{\epsilon_0} dv' + \oint_S [G(\mathbf{r};\mathbf{r}')\frac{\partial \Phi(\mathbf{r}')}{\partial n} - \Phi(\mathbf{r}')\frac{\partial G(\mathbf{r};\mathbf{r}')}{\partial n}]ds' \quad (1.231)$$

We may now offer a physical interpretation for the above result (Fig. 1.7). The volume integral on the right-hand side is the primary (incident) field due to the confined sources, while the surface integral represents the induced (scattered) field due to the presence of the conducting body S. The total (diffracted) field Φ can therefore be written as

$$\Phi(\mathbf{r}) = \Phi^i(\mathbf{r}) + \Phi^s(\mathbf{r}) \quad (1.232)$$

The above analysis constitutes the foundation for integral equation formulation developed in scattering problems involving conducting bodies. In such problems, a similar expression for one of the field components is combined by an appropriate boundary condition on the surface of the conductor to establish an integral equation for the field.

1.11.7 Eigenfunction Presentation

In this section, we consider the expansion of the Green's function in terms of the eigenfunctions in a closed region V surrounded by the surface S. The Green's function satisfies

$$(\nabla^2 + k^2)G = -\delta(\mathbf{r} - \mathbf{r}') \quad \text{in V} \quad (1.233)$$

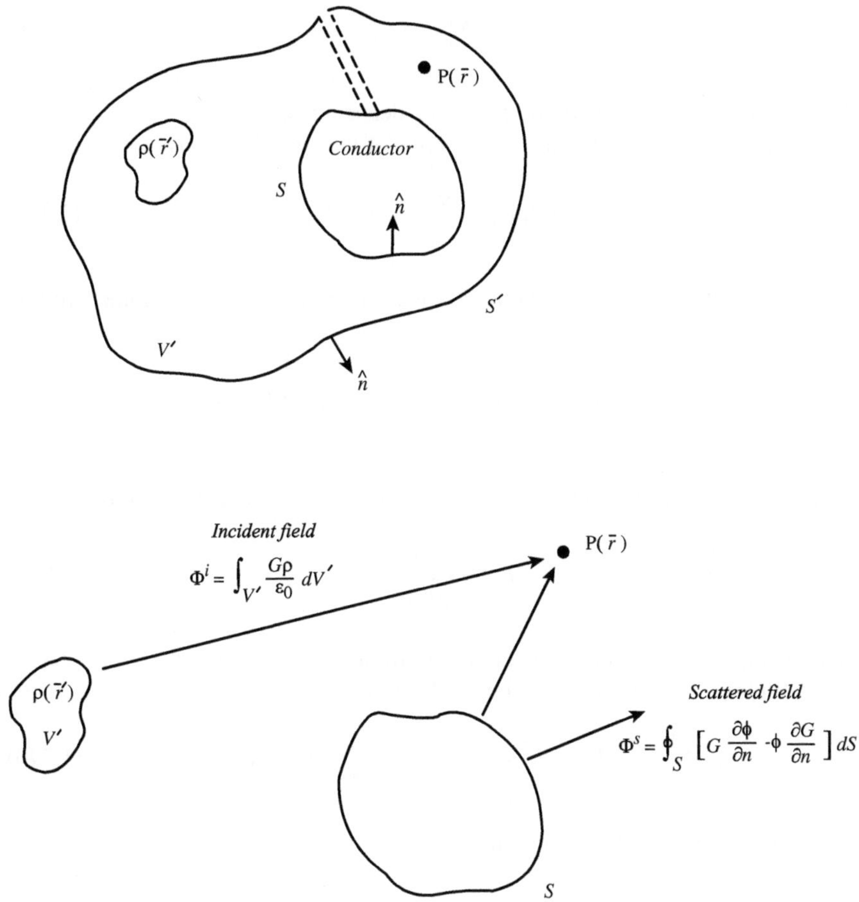

Fig. 1.7 The modified Green's function and the scattering problem

subject to the homogeneous boundary conditions on S. On the other hand the eigenfunctions $\{\psi_n\}$ satisfy

$$(\nabla^2 + k_n^2)\psi_n(\mathbf{r}) = 0 \tag{1.234}$$

subject to the same boundary conditions. Since the eigenfunctions are orthogonal

$$\int_V \psi_n(\mathbf{r})\psi_m(\mathbf{r}')dv = 0, \quad \mathrm{n} \neq \mathrm{m} \tag{1.235}$$

we represent G in the form

$$G = \sum_n a_n \psi_n(\mathbf{r}) \tag{1.236}$$

where a_n are the unknown coefficients to be determined.

We also expand the delta function in terms of the eigenfunctions

$$\delta(\mathbf{r}, \mathbf{r}') = \sum_n b_n \psi_n(\mathbf{r}) \tag{1.237}$$

Multiplying both sides of (1.237) by ψ_m and integrating over the volume, we find that

$$\int_V \delta(\mathbf{r} - \mathbf{r}')\psi_m(\mathbf{r})dv = \sum_n b_n \int_V \psi_n(\mathbf{r})\psi_m(\mathbf{r})dv \tag{1.238}$$

Using orthogonality of the eigenfunctions, the coefficients b_n are found to be

$$b_n = \frac{\psi_n(\mathbf{r}')}{||\psi_n||^2} \tag{1.239}$$

where $||\psi_n||^2 = \int_V |\psi_n(\mathbf{r})|^2 dv$ is the L^2 norm of the eigenfunction.

Substituting (1.236) and (1.239) in (1.233), we obtain

$$\sum_n (\nabla^2 + k^2) a_n \psi_n(\mathbf{r}) = -\sum_n b_n \psi_n(\mathbf{r}) \tag{1.240}$$

Substituting from (1.234), we find that

$$\sum_n (k^2 - k_n^2) a_n \psi_n(\mathbf{r}) = -\sum_n b_n \psi_n(\mathbf{r}) \tag{1.241}$$

Equating the coefficients of the eigenfunctions on both sides, we find

$$a_n = \frac{b_n}{(k_n^2 - k^2)} \tag{1.242}$$

The Green's function is thus given by

$$G = \sum_n \frac{\widehat{\psi}_n(\mathbf{r})\widehat{\psi}_n(\mathbf{r}')}{(k_n^2 - k^2)}, \quad k_n \neq k \tag{1.243}$$

where $\{\widehat{\psi}_n\}$ are the normalized eigenfunctions.

Example 1.4 Let $x \subset [0, 1] | f(x) = f(-x)$. Then expanding $f(x)$ in cosine Fourier series, we have

$$f(x) = \sum_{k=0}^{\infty} a_k \cos k\pi x$$

where

$$a_k = 2 \int_0^1 f(\xi) \cos k\pi \xi d\xi$$

Thus,

$$\begin{aligned} f(x) &= 2 \sum_{k=0}^{\infty} \cos k\pi x \left(\int_0^1 f(\xi) \cos k\pi \xi d\xi \right) \\ &= 2 \int_0^1 f(\xi) \left(\sum_{k=0}^{\infty} \cos k\pi x \cos k\pi \xi \right) d\xi \\ &= \int_0^1 f(\xi) \delta(x - \xi) d\xi \end{aligned}$$

where the last equality is deduced from distribution theory and we have

$$2 \sum_{k=0}^{\infty} \cos k\pi x \cos k\pi \xi = \delta(x - \xi)$$

Now let $g(x) \subset \mathcal{C}[0, 1] | g(x) = -g(-x)$ and expand $g(x)$ in sine Fourier series

$$g(x) = \sum_{k=1}^{\infty} \beta_k \sin k\pi x$$

where

$$\beta_k = 2 \int_0^1 g(\xi) \sin k\pi \xi d\xi$$

Hence

$$g(x) = 2\sum_{k=1}^{\infty} \sin k\pi x \left(\int_0^1 g(\xi) \sin k\pi \xi d\xi \right)$$
$$= 2 \int_0^1 g(\xi) \left(\sum_{k=1}^{\infty} \sin k\pi x \sin k\pi \xi \right) d\xi$$
$$= \int_0^1 g(\xi)\delta(x - \xi) d\xi$$

and therefore,

$$2\sum_{k=1}^{\infty} \sin k\pi x \sin k\pi \xi = \delta(x - \xi)$$

In addition we have from distribution theory

$$\delta(\frac{x}{a}) = a\delta(x)$$

and we may also write

$$2\sum_{k=0}^{\infty} \cos \frac{k\pi}{a} x \cos \frac{k\pi}{a} \xi = a\delta(x - \xi)$$
$$2\sum_{k=1}^{\infty} \sin \frac{k\pi}{a} x \sin \frac{k\pi}{a} \xi = a\delta(x - \xi)$$

□

Example 1.5 Consider the equation

$$(\frac{d^2}{dx^2} + k^2)G(x, x') = -\delta(x - x')$$

subject to the boundary conditions

$$G(0, x') = G(d, x') = 0$$

The eigenfunctions and eigenvalues are given by

$$\psi_n(x) = \sin k_n x, \quad k_n = \frac{n\pi}{d}, \quad n = 1, 2, \ldots$$

The Green's function is therefore given by

$$G(x, x') = \sum_{n=1}^{\infty} \frac{2}{d} \sin n\pi x/d \frac{\sin n\pi x'/d}{(n\pi/d)^2 - k^2}.$$

□

1.12 Inhomogeneous Vector Helmholtz Equation

In many applications, we require the solution to the inhomogeneous vector Helmholtz equation. For example, the vector magnetic potential **A** satisfies the equation

$$\nabla^2 \mathbf{A} + k^2 \mathbf{A} = -\mu_0 \mathbf{J} \tag{1.244}$$

Writing the magnetic potential and the current density in terms of their orthogonal components, we may easily see that each component may be expressed as

$$A_i = \frac{\mu_0}{4\pi} \int_V J_i(\mathbf{r}') \frac{e^{-jk|\mathbf{r}-\mathbf{r}'|}}{|\mathbf{r}-\mathbf{r}'|} dv' \tag{1.245}$$

Thus,

$$\mathbf{A}(\mathbf{r}) = \frac{\mu_0}{4\pi} \int_V \mathbf{J}(\mathbf{r}') \frac{e^{-jk|\mathbf{r}-\mathbf{r}'|}}{|\mathbf{r}-\mathbf{r}'|} dv' \tag{1.246}$$

A similar procedure may be applied to obtain the solution for the vector Hertz potential

$$\nabla^2 \pi + k^2 \pi = -\frac{\mathbf{J}}{j\omega\epsilon_0} \tag{1.247}$$

Thus

$$\pi(\mathbf{r}) = -j\frac{Z_0}{k} \int_V \mathbf{J}(\mathbf{r}') \frac{e^{-jk|\mathbf{r}-\mathbf{r}'|}}{4\pi|\mathbf{r}-\mathbf{r}'|} dv' \tag{1.248}$$

For a given source $\rho(\mathbf{r}, \omega)$ in free space, we may obtain the time-dependent potential $\Phi(\mathbf{r}, t)$ by taking the inverse Fourier transform of (1.228). We have

$$\begin{aligned} \Phi(\mathbf{r}, t) &= \tfrac{1}{2\pi} \int_{-\infty}^{\infty} [\tfrac{1}{4\pi\epsilon_0} \int_V e^{-j\frac{\omega}{c}R} \tfrac{\rho(\mathbf{r}')}{R} dv'] e^{j\omega t} d\omega \\ &= \tfrac{1}{4\pi\epsilon_0} \int_V [\tfrac{1}{2\pi} \int_{-\infty}^{\infty} \tfrac{\rho(\mathbf{r}')}{R} e^{j\omega(t-R/c)} d\omega] dv' \end{aligned} \tag{1.249}$$

and equivalently

$$\Phi(\mathbf{r}, t) = \frac{1}{4\pi\epsilon_0} \int\limits_V \frac{\rho(\mathbf{r}', t - R/c)}{R} dv' \tag{1.250}$$

where $R = |\mathbf{r} - \mathbf{r}'|$. Because of the retarded time involved on the right-hand side of (1.250), Φ so obtained is called the *retarded potential*. This implies that the field at a point $\mathbf{r}$ in space at time t is related to the state of the source at time $t - R/c$ integrated over the entire source.

The conventional notation with the time dependence $e^{j\omega t}$ understood and suppressed is as follows

$$\rho(t - R/c) \Longleftrightarrow \rho(\mathbf{r})e^{-jkR}$$

A similar analysis shows that the retarded magnetic potential is given by

$$\mathcal{A}(\mathbf{r}, t) = \frac{\mu_0}{4\pi} \int\limits_V \frac{\mathbf{J}(\mathbf{r}', t - R/c)}{R} dv' \tag{1.251}$$

It should be noted that the concept of retarded potentials only applies to wave propagation in nondispersive media. It may not, therefore, apply to lossy media.

Exercises

1.1: Consider the following solutions for the electric and magnetic fields each satisfying the source-free homogeneous wave equation

$$\mathbf{E} = \widehat{x} f_1(z - vt), \quad \mathbf{H} = \widehat{y} f_2(z - vt)$$

This represents a plane wave propagating in the z-direction. Show that the intrinsic impedance E_x/H_y is given by

$$Z_0 = E_x/H_y = \sqrt{\mu_0/\epsilon_0} = 377\Omega$$

1.2: The electric field in a source free conducting region of conductivity σ has only one component $E_y(x, t)$ which is independent of y and z. Find the scalar differential equation for E_y from Maxwell's equations and give its general solution.

1.3: Consider the vector fields

$$\mathbf{E} = a\omega \sin(ky - \omega t)\widehat{z}$$
$$\mathbf{B} = b\omega \sin(ky - \omega t)\widehat{x}$$

(a) Derive conditions on the constants a, b, k and ω that must be fulfilled in order for the fields above to be physically possible electromagnetic fields in vacuum.

(b) Calculate the potentials ϕ and A for the electromagnetic field above in the Lorenz gauge.

1.4: Show that in an isotropic but inhomogeneous pure dielectric ($\sigma = 0$), **E** satisfies the wave equation

$$(\nabla^2 - \frac{1}{c^2}\frac{\partial^2}{\partial t^2})\mathbf{E} + \nabla(\log\mu) \times \nabla \times \mathbf{E} + \nabla[\mathbf{E} \cdot \nabla(\log\epsilon)] = 0$$

and deduce the analogous equation for **H**.

1.5: The constitutive relations in a chiral medium are given by

$$\mathbf{D} = \epsilon\mathbf{E} - j\gamma\mathbf{B}, \quad \mathbf{H} = -j\gamma\mathbf{E} + \frac{\mathbf{B}}{\mu}$$

where γ is the chirality parameter. Show that in a source-free chiral medium all field quantities satisfy the vector wave equation

$$\nabla \times \nabla \times \mathbf{A} - k^2\mathbf{A} + \mathbf{F} = 0$$

and find **F**.

1.6: Show that the Lorentz gauge is compatible with the equation of continuity.

1.7: Show that the real Poynting vector **S** can be written as

$$\mathcal{S} = \Re\mathrm{e}\mathbf{S} + \mathbf{F}e^{2j\omega t}$$

where **S** is the complex Poynting vector and find **F**.

1.8: Consider a lossy cylindrical material of conductivity σ, length ℓ, and radius a. The direct current I passes through this "resistor". Calculating the Poynting vector and "radiated" power from the cylindrical surface of the resistor on one hand, and the power loss inside the resistor on the other hand, verify the Poynting theorem.

1.9: Consider a circular parallel-plate capacitor filled with a pure dielectric material. A dc current I_0 is switched on at $t = 0$.

(a) Find the fields **E**, **H**, and the Poynting vector **S** as functions of position **r** and time inside the capacitor.
(b) Find the electromagnetic energy density W.
(c) Verify the Poynting theorem.

(Assume that the fringe field is negligible and that the fields are confined within the capacitor).

1.10: Let Φ be the electrostatic potential and **J** the current density. Show that for static electric and magnetic fields, the vector $\mathbf{S}' = \Phi\mathbf{J}$ is physically equivalent to the Poynting vector.

1.11: Determine the instantaneous values for the energy density and Poynting vector associated with the electromagnetic fields for the following cases:

$$\text{(a)}\begin{cases} \mathbf{E} = \hat{y}E_0 e^{-j(k_x x - \omega t)} & k_x = \omega\sqrt{\mu\epsilon} \\ \mathbf{H} = Y\hat{k} \times \mathbf{E} & k_y = k_z = 0 \end{cases}$$

(b) same as in (a) but for $k_x = -j\alpha$

$$\text{(c)}\begin{cases} B_r = \frac{\mu m}{2\pi}\frac{\cos\theta}{r^3} \\ B_\theta = \frac{\mu m}{4\pi}\frac{\sin\theta}{r^3} \end{cases}$$

Also, identify these cases.

1.12: The electromagnetic field of a certain radiator is given by

$$E_\theta = \frac{V_0\cos(\omega t - kr)}{r}, \quad E_r = E_\phi = 0$$
$$H_\phi = \frac{V_0\cos(\omega t - kr)}{Z_0 r}, \quad H_r = H_\theta = 0$$

(a) Find the instantaneous power radiated through a spherical surface of radius R.
(b) Find the average power flowing through the spherical surface.

1.13: An elliptically polarized wave has an electric field intensity of the form

$$\mathbf{E}(z,t) = 3\cos(\omega t - k_0 z)\hat{x} + 4\sin(\omega t - k_0 z)\hat{y} \quad (\mathrm{V/m})$$

(a) Express the above field in phasor form.
(b) Find the instantaneous and time-averaged power flux densities.

1.14: A microwave polarized in the $\hat{x}$ direction is propagating in a lossy biological medium towards the $\hat{z}$ direction. The power flux density at $z = 0$ is $1\,\mathrm{mW/cm^2}$.

(a) Give the expressions for the electric and magnetic field intensities.
(b) Calculate the stored electric and magnetic energy densities, w_e and w_m, and the power loss density, p_ℓ as functions of z.
(c) Show that they satisfy the Poynting theorem by considering the real and imaginary power densities separately.
(d) What is the Specific Absorption Rate (SAR).

(Assume $f = 915\,\mathrm{MHz}$, $\epsilon_r = 31$, $\sigma = 1.6$ S/m).

1.15: What is the Lorenz gauge in a lossy medium of conductivity σ?

1.16: Investigate the *temporal gauge* $\phi = 0$ which enables one to describe **E** and **H** with the single vector **A**.

1.17: Consider the magnetic field given in cylindrical coordinates,

$$\mathbf{B}(\rho < \rho_0, \phi, z) = \hat{z}B$$
$$\mathbf{B}(\rho > \rho_0, \phi, z) = 0$$

Determine the vector potential **A** that "generated" this magnetic field.

1.18: The electric Hertz vector π for a small wire antenna in free space is given by

$$\pi = \frac{Ae^{-jkr}}{4\pi r}\hat{z}$$

where A is a constant. Find **E** and **H** and express all fields components in the spherical coordinate system.

1.19: A sheet of uniform current $I_0(A/m)$ flowing in the x-direction is located in free space at $z = 0$. This is represented by the source current density $\mathbf{K} = \hat{x}I_0\delta(z)$.

(a) Find the Hertz vector.
(b) Find **E** and **H** at z.
(c) Consider a rectangular volume enclosed by the planes at $x = 1\,\text{m}$, $y = 1\,\text{m}$, $z = 2\,\text{m}$ and show that the real power flowing out of the volume is equal to the power supplied by the current source.
(d) Calculate the reactive power.

1.20: Consider the homogeneous vector Helmholtz equation

$$\nabla \times \nabla \times \mathbf{F} - k^2\mathbf{F} = 0$$

with $\nabla \cdot \mathbf{F} = 0$. Show that

$$\mathbf{F} = \nabla \times (\mathbf{a}\psi)$$

is a solution where **a** is a constant vector, and ψ satisfies the scalar wave equation

$$\nabla^2\psi + k^2\psi = 0.$$

1.21: Given the Hertz vector

$$\pi_e(\mathbf{r}) = \frac{1}{j\omega\epsilon}\int_V \mathbf{J}(\mathbf{r}')G_0(\mathbf{r};\mathbf{r}')dv'$$

show that

$$\mathbf{E}(\mathbf{r}) = \int_V (-j\omega\mu\mathbf{J} + \frac{1}{\epsilon}\rho\nabla')G_0(\mathbf{r};\mathbf{r}')dv'$$

1.22: Using the fact that $\delta(x) = \frac{dU(x)}{dx}$ where $U(x)$ is the unit step function,

(a) Show that $\delta(-x) = \delta(x)$.
(b) Verify by direct substitution that the function $h(x) = \frac{j}{2k}e^{-jk|x|}$ satisfies

$$(\frac{d^2}{dx^2} + k^2)h(x) = \delta(x)$$

1.23: Show that $\nabla \cdot \mathbf{r} = 3$ and $\nabla \cdot \frac{\hat{r}}{r^2} = 4\pi\delta(\mathbf{r})$.

1.24: Verify that $f(\mathbf{r}) = -\frac{1}{4\pi|\mathbf{r}-\mathbf{r}'|}$ satisfies $\nabla^2 f(\mathbf{r}) = \delta(\mathbf{r} - \mathbf{r}')$.

1.25: Using the divergence theorem, prove (1.183).

1.26: Using Green's second identity, show that reciprocity holds for the Green's function

$$G(\mathbf{r}; \mathbf{r}') = G(\mathbf{r}'; \mathbf{r}).$$

1.27: The electrostatic potential Ψ satisfies Poisson's equation

$$\nabla^2\Psi = -\rho/\epsilon_0$$

where ρ is the volume charge density. By means of Green's theorem and with the aid of the free space Green's function

$$G_0(\mathbf{r}; \mathbf{r}') = \frac{1}{4\pi|\mathbf{r} - \mathbf{r}'|}$$

find the solution for Ψ in a closed region.

1.28: Show that the one-dimensional Green's function for the inhomogeneous Helmholtz equation in free space, subject to radiation condition is given by

$$G(x, x') = \frac{1}{2jk_0}e^{-jk_0|x-x'|}.$$

1.29: Find the two-dimensional Green's function in cylindrical coordinate system that satisfies

$$\left[\frac{1}{\rho}\frac{\partial}{\partial\rho}(\rho\frac{\partial}{\partial\rho}) + \frac{1}{\rho^2}\frac{\partial^2}{\partial\phi^2} + k^2\right]G = -\frac{\delta(\rho - \rho')\delta(\phi - \phi')}{\rho}$$

in the region $0 \leq \rho' < a$ subject to the Dirichlet's boundary condition at $\rho = a$.

1.30: In the rectangular region $x \in [0, a]$ and $y \in [0, b]$, determine the static Green's function $G(\boldsymbol{\rho}; \boldsymbol{\rho}')$ such that

$$\nabla^2 G(\boldsymbol{\rho}; \boldsymbol{\rho}') = -\delta(\boldsymbol{\rho} - \boldsymbol{\rho}')$$

with Dirichlet's boundary conditions on the walls $x = 0, x = a, y = 0$ and $y = b$.

1.31: A shorted parallel plate region has perfectly conducting walls at $x = 0, a$ $(0 \le z < \infty, -\infty < y < \infty)$ and $z = 0$ $(0 \le x < a, -\infty < y < \infty)$ and contains a pure dielectric. The time harmonic Green's function satisfies

$$\nabla^2 G + k^2 G = -\delta(x - x')\nabla(z - z')$$

where $0 < x' < a, z' > 0$ with $G = 0$ on the walls. There is no y dependence.

(a) Find G in both regions $z < 0$ and $z > 0$.
(b) Are the Green's functions in the two regions symmetric in (x, z) and (x', z')?

1.32: In a parallel plate waveguide with perfectly conducting walls at $x = 0$ and $x = a$, the region $z \ge 0$ is filled with a homogeneous nonmagnetic dielectric (ϵ_1, μ_0), while the region $z < 0$ is free space (ϵ_0, μ_0). For an electric line source at $x = x'$ with $0 < x' < a$ and $z = z' = 0$, find the Green's function.

1.33: Fourier transform Maxwell's equations in the spatial domain.

1.34: Use the Fourier version of Maxwell's equations to investigate the possibility of waves that do not propagate energy; such waves are called *static waves*.

Chapter 2
Radiation

In this chapter, we discuss the basic principles of electromagnetic radiation due to simple and distributed sources.

2.1 General Considerations

An antenna is a device used to propagate or to capture electromagnetic waves. When an antenna is used for transmission (propagation) of radio waves, electric currents are made to oscillate over the antenna. Energy from this oscillating charge is emitted into space as electromagnetic radio waves. When an antenna is used for reception, these waves induce a weak electric current in the antenna. This current is amplified by the radio receiver. An antenna can generally be used for reception and transmission on the same wavelength.

Electric energy is fed to an antenna by means of a transmission line, or a coaxial cable. In reflector antennas, microwave energy is reflected from a metallic paraboloid that shapes it into a narrow beam.

The dimensions of an antenna usually depend on the wavelength, or frequency, of the radio wave for which the antenna is designed. The length of an antenna must be such that it resonates electrically at the desired wavelength. The basic antenna length must be at least half the wavelength of the radio waves it is designed to transmit or receive. It can also be an integral multiple of the one-half wavelength. Antennas with such dimensions are called resonant antennas. A resonant antenna is an efficient propagator and receptor of electromagnetic energy at its design wavelength.

Let a source distribution $(\mathbf{J}, \rho)$ be confined in a region V in free space. We have seen that the Hertz vector potential at $\mathbf{r}$ due to $\mathbf{J}$ satisfies

$$\nabla^2 \pi + k_0^2 \pi = -\frac{\mathbf{J}}{j\omega\epsilon_0} \tag{2.1}$$

© Springer International Publishing Switzerland 2015

K. Barkeshli, *Advanced Electromagnetics and Scattering Theory*,
DOI 10.1007/978-3-319-11547-4_2

so that

$$\pi(\mathbf{r}) = -j\frac{Z_0}{k_0}\int_V \mathbf{J}(\mathbf{r}')\frac{e^{-jk_0|\mathbf{r}-\mathbf{r}'|}}{4\pi|\mathbf{r}-\mathbf{r}'|}dv' \tag{2.2}$$

where k_0 is the free space wavenumber. The electromagnetic fields are then given by

$$\mathbf{E} = \nabla\nabla\cdot\pi + k_0^2\pi \tag{2.3}$$

$$\mathbf{H} = j\omega\epsilon_0\nabla\times\pi \tag{2.4}$$

We begin our discussion of the radiating systems by elementary sources, and then extend the concepts to distributed sources.

2.2 Elementary Sources

In this section, we are concerned with simple sources which are *small* in extent. For these sources, the maximum dimension of the source is much less than free space wavelength. We shall assume that the point of observation or the field point P is located at a large distance from the source. This implies that

$$k_0 r' \ll \lambda \tag{2.5}$$

$$r' \ll r \tag{2.6}$$

Note that no restriction is placed on the order of magnitude of r in comparison with the wavelength. Later on, we define various field regions based on the magnitude of r with respect to the wavelength.

We now modify (2.2) on the basis of the approximations represented by (2.5) and (2.6). Referring to Fig. 2.1, we have

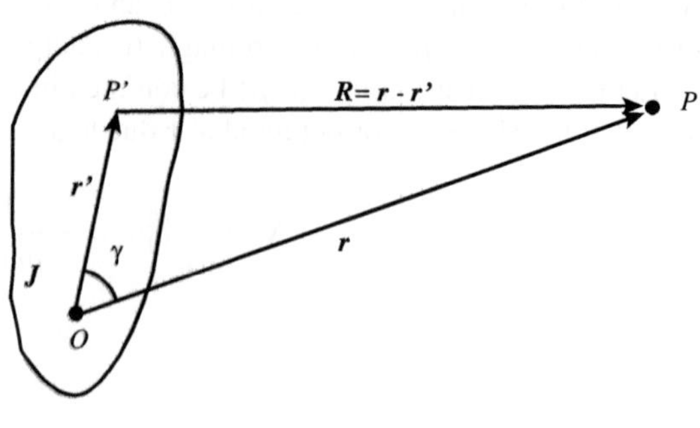

Fig. 2.1 An elementary current source radiating in free space

$$|\mathbf{r} - \mathbf{r}'| = R \simeq r - r' \cos\gamma \tag{2.7}$$

where γ is the included angle between $\mathbf{r}$ and $\mathbf{r}'$. That is

$$\cos\gamma = \widehat{r} \cdot \widehat{r'} \tag{2.8}$$

Also, since $k_0 r' \cos\gamma \ll 1$, we have

$$e^{-jk_0|\mathbf{r}-\mathbf{r}'|} \simeq e^{-jk_0 r}(1 + jk_0 r' \cos\gamma) \tag{2.9}$$

and

$$\frac{1}{|\mathbf{r} - \mathbf{r}'|} \simeq \frac{1}{r}(1 + \frac{r'}{r} \cos\gamma) \tag{2.10}$$

Using the above, (2.2) can now be written as

$$\pi(\mathbf{r}) \simeq -j\frac{Z_0}{k_0}\frac{e^{-jk_0 r}}{4\pi r}\int_V \mathbf{J}(\mathbf{r}')[1 + (jk_0 + \frac{1}{r})r' \cos\gamma]dv' \tag{2.11}$$

Note that although the second term within the bracket is small compared with unity, we have retained it in (2.11), because in some cases the integral $\int_V \mathbf{J}(\mathbf{r}')dv'$ is zero in which case $\int_V \mathbf{J}(\mathbf{r}')(jk_0 + \frac{1}{r}r' \cos\gamma dv')$ would be the leading term. If $\int_V \mathbf{J}(\mathbf{r}')dv'$ is not zero, then we generally neglect the term involving r' in (2.11).

2.2.1 The Short Electric Dipole

Consider a small linear current element of length ℓ carrying a constant current I_0 (actually $I_0 e^{j\omega t}$), and oriented along the $\widehat{z}$ direction (Fig. 2.2). For such a current element

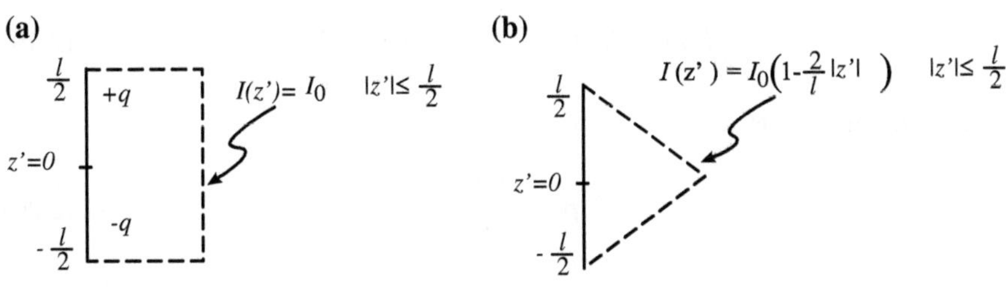

Fig. 2.2 $\widehat{z}$-directed dipoles located at the origin **a** a short Hertzian dipole, **b** an Abraham dipole

$$I(z) = I_0, \quad |z| \leq \ell/2 \tag{2.12}$$

We define the total current moment as

$$\mathbf{p}_i = \int_v \mathbf{J}(\mathbf{r}')dv' = \widehat{z} \int_{-\ell/2}^{\ell/2} I(z')dz' = I_0 \ell \widehat{z} \tag{2.13}$$

Consider now two time-varying charges $\pm q$ separated by a distance ℓ and placed along the z-axis. For small ℓ, this represents a time varying dipole with moment $\mathbf{p}$ defined as

$$\mathbf{p} = q\ell\widehat{z} \tag{2.14}$$

Substituting $I_0 = j\omega q$, we have

$$\mathbf{p} = \frac{I_0 \ell}{j\omega}\widehat{z} \tag{2.15}$$

which indicates that the current moment is related to the dipole moment by

$$\mathbf{p}_i = j\omega \mathbf{p} \tag{2.16}$$

We previously defined the polarization vector $\mathbf{P}$ in connection with the Hertz potential as

$$\mathbf{P} = \frac{\mathbf{J}}{j\omega} \tag{2.17}$$

Thus

$$\int_v \mathbf{P}(\mathbf{r}')dv' = \frac{1}{j\omega}\int \mathbf{J}(\mathbf{r}')dv' = \frac{I_0\ell}{j\omega}\widehat{z} = \mathbf{p} \tag{2.18}$$

Hence, $\mathbf{P}$ is indeed related to the dipole concept. The time varying dipole is referred to as the *Hertzian* dipole. For an elementary current source having the distribution

$$I(z) = I_0(1 - \frac{2}{\ell}|z|), \quad |z| \leq \ell/2 \tag{2.19}$$

the current moment is

$$\mathbf{p}_i = \frac{I_0\ell}{2}\widehat{z} \tag{2.20}$$

which is half the current moment of a Hertzian dipole of the same length. This is called the *Abraham* dipole.

Using the approximation similar to (2.11), it can be shown that the potential for a Hertzian dipole of moment $\mathbf{p}_i = j\omega\mathbf{p}$ is

$$\pi(\mathbf{r}) \simeq \frac{e^{-jk_0 r}}{j\omega 4\pi\epsilon_0 r}\mathbf{p}_i = \mathbf{p}\frac{e^{-jk_0 r}}{4\pi\epsilon_0 r} \tag{2.21}$$

The magnetic field is given by

$$\begin{aligned}\mathbf{H} &= j\omega\epsilon_0 \nabla \times \pi \\ &= \frac{j\omega}{4\pi}[\frac{1}{r^2} + \frac{jk_0}{r}]e^{-jk_0 r}(\mathbf{p} \times \widehat{r})\end{aligned} \tag{2.22}$$

where $\widehat{r}$ is the unit vector in the direction of the field point. The electric field $\mathbf{E}$ can be obtained from

$$\begin{aligned}\mathbf{E} &= \nabla\nabla \cdot \pi + k_0^2 \pi \\ &= \frac{e^{-jk_0 r}}{4\pi\epsilon_0}[(\frac{1}{r^3} + \frac{jk_0}{r^2})\{3(\mathbf{p}\cdot\widehat{r})\widehat{r} - \mathbf{p}\} - \frac{k_0^2}{r}\{\widehat{r} \times (\widehat{r} \times \mathbf{p})\}]\end{aligned} \tag{2.23}$$

In the far zone ($k_0 r \gg 1$), only the terms depending on $1/r$ dominate, and we have

$$\mathbf{E} = -\frac{k_0^2}{4\pi\epsilon_0}\frac{e^{-jk_0 r}}{r}[\widehat{r} \times (\widehat{r} \times \mathbf{p})] \tag{2.24}$$

$$\mathbf{H} = \frac{\omega k_0}{4\pi}\frac{e^{-jk_0 r}}{r}(\widehat{r} \times \mathbf{p}) \tag{2.25}$$

For a current element $I_0 \mathbf{d}\ell$, the above expressions can be used provided that $\mathbf{p}$ is replaced by $I_0 \mathbf{d}\ell / j\omega$.

The above expressions for the electromagnetic fields are valid for any orientation of the dipole. For a $\widehat{z}$-oriented dipole, we have

$$\mathbf{p} = p\cos\theta\,\widehat{r} - p\sin\theta\,\widehat{\theta} \tag{2.26}$$

The spherical components of the electromagnetic fields are, therefore, given by

$$\begin{aligned}E_r &= p\frac{e^{-jk_0 r}}{2\pi\epsilon_0}[\frac{1}{r^3} + \frac{jk_0}{r^2}]\cos\theta \\ E_\theta &= p\frac{e^{-jk_0 r}}{4\pi\epsilon_0}[\frac{1}{r^3} + \frac{jk_0}{r^2} - \frac{k_0^2}{r}]\sin\theta \\ H_\phi &= j\omega p\frac{e^{-jk_0 r}}{2\pi}[\frac{1}{r^2} + \frac{jk_0}{r}]\end{aligned} \tag{2.27}$$

while

$$E_\phi = H_r = H_\theta = 0 \tag{2.28}$$

The radiation fields of a $\hat{z}$-directed Hertzian dipole located at the origin are given in component form as

$$E_\theta = j\frac{k_0^2}{4\pi\omega\epsilon_0}I_0 dz\frac{e^{-jk_0r}}{r}\sin\theta \tag{2.29}$$
$$H_\phi = j\frac{k_0}{4\pi}I_0 dz\frac{e^{-jk_0r}}{r}\sin\theta$$

with $E_r = E_\phi = H_r = H_\theta = 0$. Note that in the far field, the ratio of the transverse field components is

$$\frac{|E_\theta|}{|H_\phi|} = \frac{k_0}{\omega\epsilon_0} = Z_0 \tag{2.30}$$

where Z_0 is the free space intrinsic impedance. The radiation fields of an electric dipole behave as a transverse electromagnetic (TEM) wave with their amplitude decreasing as $1/r$. Also, the Poynting vector is given by

$$\mathbf{S} = \frac{1}{2}\mathbf{E}\times\mathbf{H}^\star = \frac{p^2\omega k_0^3}{32\pi^2\epsilon_0 r^2}\sin^2\theta\,\hat{r} \tag{2.31}$$

representing a real power flow density in the $\hat{r}$-direction.

If a $\hat{z}$-directed dipole of current $I(z)$ is located at a position $\mathbf{r}'$ as shown in Fig. 2.3, then the far fields are given by

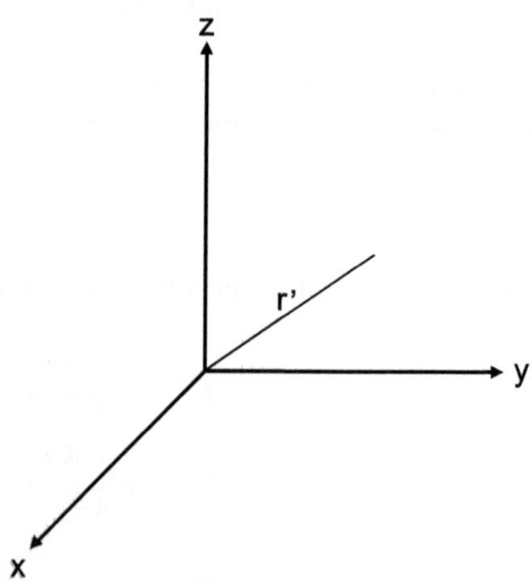

Fig. 2.3 A short $\hat{z}$-directed Hertzian dipole positioned at $\mathbf{r}'$

$$E_\theta = j\frac{k_0^2}{4\pi\omega\epsilon_0}\sin\theta\frac{e^{-jk_0(r-\mathbf{r}'\cdot\widehat{r})}}{r}I(z)dz \tag{2.32}$$

$$H_\phi = j\frac{k_0}{4\pi}\sin\theta\frac{e^{-jk_0(r-\mathbf{r}'\cdot\widehat{r})}}{r}I(z)dz \tag{2.33}$$

2.2.2 The Small Magnetic Dipole

The half-wavelength dimensional rule applies to all antennas except wire loop antennas. Small loop antennas used in transistor radios are resonant at the long, 300 m wavelengths of the broadcast (AM) band because they contain a core of magnetic material called ferrite. Ferrite loop antennas are used in ultracompact transistor radios.

Consider a small loop of area Δ carrying a current I_0 as shown in Fig. 2.4. Such a current loop is called a magnetic dipole. For this loop, we note that

$$\int_V \mathbf{J}(\mathbf{r}')dv' = \oint_C I_0\mathbf{d}\ell = I_0\oint_C \mathbf{d}\ell = 0 \tag{2.34}$$

However, this does not imply that the current moment is zero. We define the magnetic dipole moment as

$$\mathbf{m} = \frac{1}{2}\int_V \widehat{r'} \times \mathbf{J}dv' \tag{2.35}$$

Applying this definition to the present case, we have

$$\mathbf{m} = \frac{1}{2}\oint_C \widehat{r'} \times I_0\mathbf{d}\ell' = I_0\oint_C \frac{1}{2}(\widehat{r'} \times \mathbf{d}\ell') \tag{2.36}$$

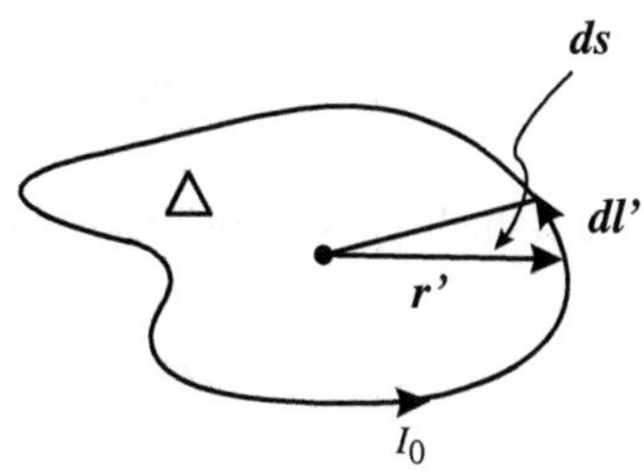

Fig. 2.4 A small magnetic dipole

Therefore,

$$\mathbf{m} = \widehat{n} I_0 \int_S ds' = \widehat{n} I_0 \Delta \tag{2.37}$$

which is clearly independent of the shape of the loop. In order to find the Hertz potential, we invoke (2.11). Thus

$$\begin{aligned}
\pi(\mathbf{r}) &\simeq -j\frac{Z_0}{k_0}\frac{e^{-jk_0 r}}{4\pi r}\int_V \mathbf{J}(\mathbf{r}')(jk_0 + \frac{1}{r})r' \cos\gamma dv' \\
&= -j\frac{Z_0}{k_0}\frac{e^{-jk_0 r}}{4\pi r}(jk_0 + \frac{1}{r})\int_V \mathbf{J}(\mathbf{r}')r'\widehat{r'} \cdot \widehat{r} dv' \\
&= -j\frac{Z_0}{k_0}\frac{e^{-jk_0 r}}{4\pi r}(jk_0 + \frac{1}{r})I_0 \oint_C \frac{\mathbf{r}' \cdot \mathbf{r}}{r}\mathbf{d\ell}'
\end{aligned} \tag{2.38}$$

Employing the vector theorem

$$\oint_C \psi \mathbf{d\ell}' = -\int_S \nabla'\psi \times \widehat{n} ds' \tag{2.39}$$

we obtain

$$\begin{aligned}
\pi(\mathbf{r}) &= j\frac{Z_0 I_0}{k}\frac{e^{-jk_0 r}}{4\pi r}(jk_0 + \frac{1}{r})\int_S \frac{\nabla'(\mathbf{r}' \cdot \mathbf{r}) \times \widehat{n} ds'}{r} \\
&= j\frac{Z_0 I_0}{k_0}\frac{e^{-jk_0 r}}{4\pi r}(jk_0 + \frac{1}{r})\int_S \widehat{r} \times \widehat{n} ds'
\end{aligned} \tag{2.40}$$

Substituting for the magnetic current moment $\mathbf{m}$, we have

$$\pi(\mathbf{r}) = -j\frac{Z_0}{k_0}\frac{e^{-jk_0 r}}{4\pi r}(\mathbf{m} \times \widehat{r})(jk_0 + \frac{1}{r}) \tag{2.41}$$

and if the loop is oriented so that $\widehat{n} = \widehat{z}$, we find

$$\pi(\mathbf{r}) = -j\frac{Z_0}{k_0}\frac{e^{-jk_0 r}}{4\pi r}m(jk_0 + \frac{1}{r}) \sin\theta\widehat{\phi} \tag{2.42}$$

It is noted that the Hertz potential is directed entirely in the $\widehat{\phi}$ direction, regardless of the size of the loop. Using (2.22) and (2.23), it can be shown that the complete electromagnetic fields for a $\widehat{z}$ oriented magnetic dipole are given by

$$\mathbf{E} = m\frac{e^{-jk_0 r}}{4\pi}k_0^2 Z_0\left(-\frac{jk_0}{r^2}+\frac{1}{r}\right)\sin\theta\widehat{\phi} \tag{2.43}$$

$$\mathbf{H} = m\frac{e^{-jk_0 r}}{4\pi}[2\left(\frac{1}{r^3}+\frac{jk_0}{r^2}\right)\cos\theta\widehat{r}+\left(\frac{1}{r^3}+\frac{jk_0}{r^2}-\frac{k_0^2}{r}\right)\sin\theta\widehat{\theta}] \tag{2.44}$$

For such a magnetic dipole, $E_r = E_\theta = h_\phi \equiv 0$.

In the far zone, as r goes to infinity, we obtain from the above

$$\begin{aligned}\mathbf{E} &\simeq m\frac{e^{-jk_0 r}}{4\pi r}k_0^2 Z_0\sin\theta\widehat{\phi}\\ \mathbf{H} &\simeq -m\frac{e^{-jk_0 r}}{4\pi r}k_0^2\sin\theta\widehat{\theta}\end{aligned} \tag{2.45}$$

Also, the Poynting vector is given by

$$\mathbf{S} = \frac{1}{2}\mathbf{E}\times\mathbf{H}^\star = \frac{m^2 k_0^4 Z_0}{32\pi^2 r^2}\sin^2\theta\widehat{r} \tag{2.46}$$

again representing real power flow in the radial direction.

2.3 Wire Antennas

Consider a straight wire antenna driven by a current distribution $\Re e[I(z)e^{j\omega t}]$ similar to half of what is shown in Fig. 2.5. Using expressions (2.32) and (2.33), we may find the fields due to a wire antenna by the superposition integral. Thus, the electric field is given by

$$E_\theta \simeq \frac{jk_0^2}{4\pi\omega\epsilon_0}\sin\theta\frac{e^{-jk_0 r}}{r}\int I(z')e^{jk_0\mathbf{r}'\cdot\widehat{r}}dz' \tag{2.47}$$

We may write the above expression as

$$E_\theta \simeq \frac{jk_0^2\ell}{4\pi\omega\epsilon_0}\frac{e^{-jk_0 r}}{r}I_0 e^{j\phi_0}\psi(\theta) \tag{2.48}$$

where ψ is defined as

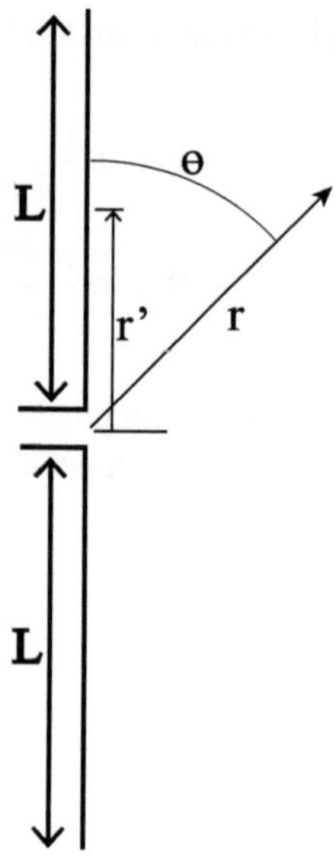

Fig. 2.5 A center-fed wire antenna of length $2L$

$$\psi(\theta) \equiv \frac{\sin\theta}{\ell} \int_{-\ell/2}^{\ell/2} I(z') e^{jk_0 \mathbf{r}' \cdot \hat{r}} dz' \tag{2.49}$$

and ℓ is the length of the antenna. The quantity ψ is dimensionless and is independent of I_0 and ϕ_0. It displays the angular dependence of the radiated field and is sometimes referred to as the *field radiation pattern.* In order to evaluate ψ, the current distribution should be known.

Example 2.1 Let the current distribution on a center-fed wire antenna of length $2L$ shown in Fig. 2.5 be given by

$$I(z) = I_m \frac{\sin k_0(L - |z|)}{\sin(k_0 L)}$$

Using (2.49), we obtain

$$\psi(\theta) = \frac{\sin\theta}{2L} \int_{-L}^{L} \frac{\sin k_0(L - |z'|)}{\sin(k_0 L)} e^{jk_0 z' \cos\theta} dz'$$

Evaluating the integral, we find that

$$\psi(\theta) = \frac{1}{k_0 L \sin(k_0 L)} \left[\frac{\cos(k_0 L) - \cos(k_0 L \cos\theta)}{\sin\theta} \right]$$

The radiated electric field is given by (2.48). □

2.4 Field Regions

The vector potential at a point P due to a source distribution $\mathbf{J}$ in free space is given by (2.2). Depending on the location of the observation point, the maximum linear dimension of the source and the wavelength, various approximations are made to evaluate (2.2), and hence the fields.

We consider point sources and extended sources separately.

2.4.1 Point Sources

If the maximum linear dimension of the source is small compared to the wavelength, that is, $r' \ll \lambda$ or $kr' \ll 1$, then the exponential term in the integrand of (2.2) may be approximated by the first two terms of its Taylor series expansion. This will yield (2.11) applicable for point or small sources. For such cases, the nature of the fields produced are different for $kr \ll 1$ and $kr \gg 1$, with the source assumed to be located at the origin. The two regions, so defined, are called the (reactive) near field and (radiating) far field regions of the source (Fig. 2.6). The common boundary of these two regions are arbitrarily chosen to be at $kr = 1$ or, equivalently, $r = \lambda/2\pi$.

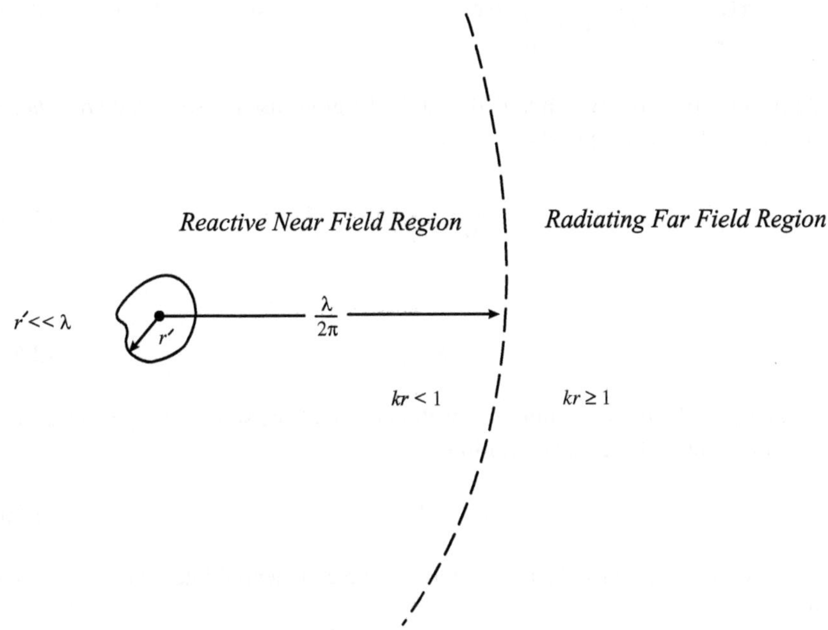

Fig. 2.6 Field regions for point sources in the near field region, the reactive energy is dominant, while in the far field, the radiating energy is dominant

2.4.2 Extended Sources

In many cases, the maximum dimension of the source may be much larger than λ. Due to the extended nature of the source, the far field region can no longer be assumed to start at $\frac{\lambda}{2\pi}$.

Under the assumption

$$r \gg r', \quad k_0 r \gg 1 \tag{2.50}$$

we may write

$$|\mathbf{r} - \mathbf{r}'| \simeq r - (\widehat{r} \cdot \mathbf{r}') + \frac{1}{2r}[r'^2 - (\widehat{r} \cdot \mathbf{r}')^2] + \mathcal{O}\left(\frac{r'}{r}\right)^3 \tag{2.51}$$

when making phase calculations and

$$|\mathbf{r} - \mathbf{r}'| \simeq r \tag{2.52}$$

for amplitude considerations. Under these assumptions, (2.2) reduces to

$$\pi(\mathbf{r}) \simeq -j\frac{Z_0}{k_0}\frac{e^{-jk_0 r}}{4\pi r}\int_V \mathbf{J}(\mathbf{r}')e^{jk_0[(\widehat{r} \cdot \mathbf{r}') + \frac{(\widehat{r} \cdot \mathbf{r}')^2}{2r} - \frac{r'^2}{2r}]}dv' \tag{2.53}$$

According to the IEEE Standard, the far field region, also known as the *Fraunhofer* region starts at a distance r where

$$\frac{r'^2_{max}}{2r} = \lambda/16 \tag{2.54}$$

that is

$$r = 8r'^2_{max}/\lambda \tag{2.55}$$

Assuming D to be the maximum linear dimension of the source, $r'_{max} = D/2$, and we obtain the far field region definition as

$$r \geq 2D^2/\lambda \tag{2.56}$$

Under this condition, only the $(\widehat{r} \cdot \mathbf{r}')$ term in the exponential integrand of (2.53) is retained.

$$\pi(\mathbf{r}) \simeq -j\frac{Z_0}{k_0}\frac{e^{-jk_0 r}}{4\pi r}\int_V \mathbf{J}(\mathbf{r}')e^{jk_0 \mathbf{r}' \cdot \widehat{r}}dv' \tag{2.57}$$

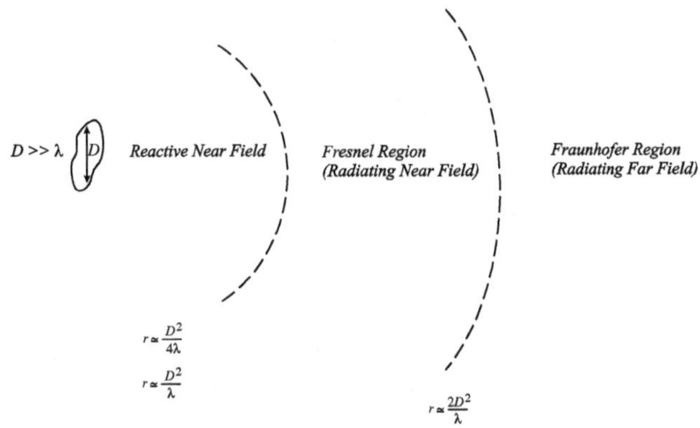

Fig. 2.7 Field regions for extended sources. In the Fresnel region, the angular field distribution depends on the distance from the antenna, while in the Franuhofer region, it is independent of the distance

This is the usual approximation used when determining the radiation fields of antennas. The neglect of lower order terms introduces a maximum phase error of $k_0 r'^2/2r$ implying an error of $\pi/8$ at the far field boundary $r = 2D^2/\lambda$. For larger distances, the error will be less.

The far field region is dominated by radiating energy and is very important in antenna analysis and design.

The region in which the second order term $k_0 r'^2/2r$ must be retained for field calculations is referred to as the *Fresnel region*. This is also known as the quasi-far field or the radiating near field region. There is no clearly marked boundary for the specification of the Fresnel region. However, for electrically large sources ($D \gg \lambda$), the region may be defined as

$$D^2/4\lambda \leq r < 2D^2/\lambda \tag{2.58}$$

as shown in Fig. 2.7.

The near field region extends from the source up to the lower boundary of the Fresnel region. For this region in which the reactive energy dominates, no general approximation is made in the evaluation of the potential and the fields.

2.5 Far Field Calculation for General Antennas

The far field expression (2.57) for the Hertz potential may be written as

$$\pi(\mathbf{r}) = -j\frac{Z_0}{k_0}\frac{e^{-jk_0 r}}{4\pi r}\mathbf{N}(\theta, \phi) \tag{2.59}$$

where

$$\mathbf{N} = \int_V \mathbf{J}(\mathbf{r}')e^{jk_0\mathbf{r}'\cdot\widehat{r}}dv' \tag{2.60}$$

The magnetic field is given by

$$\mathbf{H}(\mathbf{r}) = j\omega\epsilon_0\nabla \times \pi(\mathbf{r}) \tag{2.61}$$

Since we are interested in the far fields, we wish to express the result only retaining those terms behaving like e^{-jk_0r}/r as $r \to \infty$. Carrying out the differentiation, we have

$$\begin{aligned}\nabla \times \{\frac{e^{-jk_0r}}{r}\mathbf{N}(\theta,\phi)\} &= \nabla(\frac{e^{-jk_0r}}{r}) \times \mathbf{N} + \frac{e^{-jk_0r}}{r}\nabla \times \mathbf{N} \\ &= -(\frac{jk_0}{r} + \frac{1}{r^2})e^{-jk_0r}(\widehat{r} \times \mathbf{N}) + \frac{e^{-jk_0r}}{r}\nabla \times \mathbf{N} \\ &= (1 + \frac{j}{k_0r})\frac{e^{-jk_0r}}{r}(-jk_0)\widehat{r} \times \mathbf{N} + \frac{e^{-jk_0r}}{r}\nabla \times \mathbf{N}\end{aligned} \tag{2.62}$$

Note that the curl of the vector **N** in the above equation is

$$\nabla \times \mathbf{N} = \mathcal{O}(1/r) \tag{2.63}$$

Therefore

$$\nabla \times \{\frac{e^{-jk_0r}}{r}\mathbf{N}(\theta,\phi)\} \simeq \frac{e^{-jk_0r}}{r}(-jk_0)(\widehat{r} \times \mathbf{N}) + \mathcal{O}(\frac{1}{r^2}) \tag{2.64}$$

Hence, the operation $(\nabla\times)$ can be replaced by $(-jk_0\widehat{r}\times)$ in the far field calculations. We may, therefore, write

$$\mathbf{H}(\mathbf{r}) = -jk_0\frac{e^{-jk_0r}}{4\pi r}(\widehat{r} \times \mathbf{N}) + \mathcal{O}(\frac{1}{r^2}) \tag{2.65}$$

Noting that $\widehat{r} \times \mathbf{N} = \widehat{r} \times \mathbf{N}_t$, we have

$$\mathbf{H}(\mathbf{r}) = -jk_0\frac{e^{-jk_0r}}{4\pi r}(\widehat{r} \times \mathbf{N}_t) \tag{2.66}$$

where

$$\mathbf{N}_t = \mathbf{N} - \widehat{r}N_r = -\widehat{r} \times \widehat{r} \times \mathbf{N} \tag{2.67}$$

is referred to as the radiation vector of the current distribution[1] and the terms neglected are of the order $1/k_0r^2$. The radiation vector has the dimension of $[A.m]$. We may write

$$\mathbf{N}_t = I_i\widehat{h}(\theta, \phi) \tag{2.68}$$

where I_i is a reference current, usually taken as the input current to the antenna. Then $\widehat{h}(\theta, \phi)$ is called the *vector effective height function*.

The electric field in the region away from the sources is given by

$$\mathbf{E}(\mathbf{r}) = \frac{1}{j\omega\epsilon_0}\nabla \times \mathbf{H}(\mathbf{r}) \tag{2.69}$$

Thus, we have

$$\begin{aligned}\mathbf{E}(\mathbf{r}) &= \frac{1}{j\omega\epsilon_0}(-jk_0\widehat{r}) \times \mathbf{H}(\mathbf{r}) \\ &= jk_0Z_0\frac{e^{-jk_0r}}{4\pi r}[\widehat{r} \times (\widehat{r} \times \mathbf{N}_t)]\end{aligned} \tag{2.70}$$

Therefore, the transverse components of the fields dominate in the far field.

Summarizing the above results, we use the following procedure to find the far fields of any antenna.

$$\begin{aligned}\mathbf{N} &= \int_V \mathbf{J}(\mathbf{r}')e^{jk_0\mathbf{r}'\cdot\widehat{r}}dv' \\ \mathbf{N}_t &= N_\theta\widehat{\theta} + N_\phi\widehat{\phi} \\ &= -\widehat{r} \times \widehat{r} \times \mathbf{N}\end{aligned} \tag{2.71}$$

$$\begin{aligned}\mathbf{E} &= E_\theta\widehat{\theta} + E_\phi\widehat{\phi} \\ &= -jk_0Z_0\frac{e^{-jk_0r}}{4\pi r}\mathbf{N}_t \\ \mathbf{H} &= \frac{1}{Z_0}\widehat{r} \times \mathbf{E}\end{aligned}$$

It is noted that the direction of the Poynting vector is that of $\mathbf{N}_t \times (\widehat{r} \times \mathbf{N}_t)$, that is $\widehat{r}$ and $\mathbf{E}$, $\mathbf{H}$ and $\widehat{r}$ form a right handed perpendicular system of vectors.

The above prescription is widely used to obtain the far fields of various antennas, provided the current distribution is known.

---

[1] This vector is due to Schelkunoff.

2.6 Antenna Parameters

In this section, we discuss various antenna parameters. Some of these parameters such as the radiation intensity and the directive gain pertain to the far field behavior of the antenna, while the others like the antenna impedance are near field quantities.

2.6.1 Antenna Patterns and Radiation Intensity

A plot of $|\mathbf{E}|$ with constant r as a function of (θ, ϕ) is called the field radiation pattern of the antenna.

The power radiated from an antenna per unit solid angle is called the *radiation intensity* of the antenna. If $\mathbf{S}$ is the Poynting vector, then

$$U = r^2 S_r(\theta, \phi) \tag{2.72}$$

is the radiation intensity in Watts per unit solid angle. In the far field, the Poynting vector is given by $\mathbf{S} = \frac{1}{2Z_0}|\mathbf{E}|^2\hat{r}$ and

$$U(\theta, \phi) = \frac{r^2}{2Z_0}|\mathbf{E}|^2 \tag{2.73}$$

A plot of U as function of (θ, ϕ) is called the antenna power pattern. These patterns are usually plotted in the far-field and are directly related to the magnitude of the vector effective height function. The normalized power pattern is defined as

$$U_n(\theta, \phi) = \frac{U(\theta, \phi)}{U(\theta, \phi)_{max}} \tag{2.74}$$

For a short electric dipole, The normalized field radiation and power patterns are shown in Fig. 2.8.

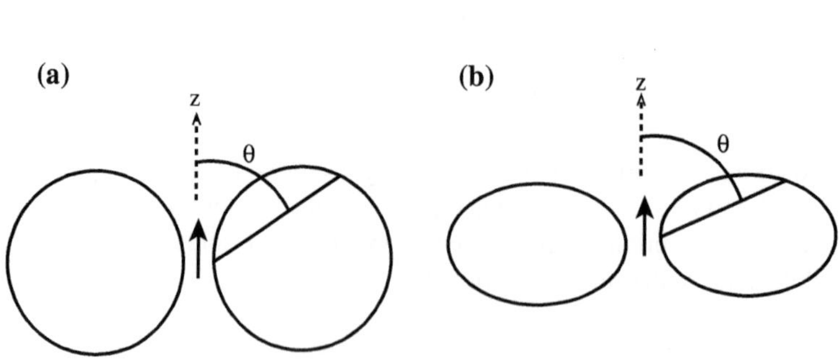

Fig. 2.8 The normalized **a** field radiation pattern, and **b** power pattern pf a short Hertzian dipole

2.6.2 Directive Gain

The *directive gain* of an antenna in a given direction is the ratio of the radiation intensity in that direction to the radiation intensity of an equivalent isotropic antenna radiating the same total average power. Thus

$$D_g = \frac{U}{U_0} = \frac{U}{P_{rad}/4\pi} \tag{2.75}$$

where P_{rad} can be written in terms of the radiated power density S_r

$$P_{rad} = \oint_S r^2 S_r d\Omega \tag{2.76}$$

where $d\Omega$ is the element of solid angle. Thus, using the above expression and (2.72), the directive gain (2.75) can also be expressed as

$$D_g(\theta, \phi) = \frac{4\pi\, S_r(\theta, \phi)}{\int_{4\pi} S_r(\theta, \phi) d\Omega} \tag{2.77}$$

If the direction is not specified, it is implied that D_g is specified in the direction of maximum gain. This is referred to as *directivity*. Directivity is denoted by D_0 and is given by

$$D_0 = \frac{U_{max}}{P_{rad}/4\pi} = \frac{4\pi}{\int_{4\pi} U_n(\theta, \phi) d\Omega} \tag{2.78}$$

The directive gain can be expressed in terms of the directivity and the normalized power pattern as

$$D(\theta, \phi) = D_0 U_n(\theta, \phi) \tag{2.79}$$

Example 2.2 A short Hertzian dipole transmits or receives most of its energy at right angles to the wire; little energy is transferred along the length of the wire. Such directivity is one of the most important electric qualities of an antenna. It allows transmission or reception to be beamed in a particular direction, to the exclusion of signals in other directions.

The complex Poynting vector for a short electric dipole is given by (2.31)

$$S_r = \frac{p^2 \omega k^3}{32\pi^2 \epsilon_0 r^2} \sin^2\theta$$

The directive gain is expressed as

$$D_g(\theta) = \frac{\sin^2\theta}{\frac{1}{2}\int_0^{\pi}\sin^3\theta d\theta} = \frac{3}{2}\sin^2\theta$$

and the directivity is $D_0 = 3/2$. □

The directive gain may also be written in terms of the vector effective height function $\widehat{h}(\theta,\phi)$ as

$$D_g = \frac{|\widehat{h}(\theta,\phi)|^2}{\frac{1}{4\pi}\int_{\Omega}|\widehat{h}(\theta,\phi)|^2 d\Omega} \tag{2.80}$$

2.6.3 Gain

The *gain* of an antenna in a specified direction is defined as the ratio of the power density radiated by the antenna, $S_r(\theta,\phi)$, to the power density radiated by a lossless isotropic antenna, S_{ri}, provided both antennas are supplied with the same amount of power, P_t

$$G(\theta,\phi) = \frac{S_r(\theta,\phi)}{S_{ri}} \tag{2.81}$$

The total power radiated by the antenna is given by

$$P_{rad} = \oint_S S_r(\theta,\phi)ds \tag{2.82}$$

while the total power radiated by the lossless isotropic antenna is given by

$$P_{rad}^i = 4\pi r^2 S_{ri} \tag{2.83}$$

This is equal to the total power delivered to the antenna P_t. However, due to the losses in the antenna system, part of the power is dissipated in the antenna structure. Designating this power loss as P_ℓ, the *radiation efficiency* is defined as

$$\eta_\ell = \frac{P_{rad}}{P_t} \tag{2.84}$$

Combining (2.82) to (2.84), we find that

$$S_{ri} = \frac{1}{4\pi\eta_\ell}\int_{4\pi} S_r(\theta,\phi)d\Omega \tag{2.85}$$

and substituting in (2.81), we obtain

$$G(\theta, \phi) = \frac{4\pi \eta_\ell S_r(\theta, \phi)}{\int_{4\pi} S_r(\theta, \phi) d\Omega} \tag{2.86}$$

In view of (2.77), the gain can be written in terms of the directive gain as

$$G(\theta, \phi) = \eta_\ell D(\theta, \phi) \tag{2.87}$$

The antenna gain accounts for ohmic losses in the antenna structure.

2.6.4 Effective Aperture

The ability of an antenna to capture energy from an incident wave and to convert it to an intercepted power for delivering to a matched load is characterized by the *effective aperture* A_e. If the incident power density at the position of the receiving antenna is S_i, then the intercepted power is given by

$$P_{int} = A_e S_i \tag{2.88}$$

The effective aperture is also known as *effective area* and *receiving cross section*.

The effective aperture can be written in terms of directivity of the antenna as

$$A_e = \frac{\lambda^2 D_0}{4\pi} \tag{2.89}$$

2.6.5 Antenna Impedance

Consider the antenna shown in Fig. 2.9. Enclose the antenna by a closed surface S consisting of three surfaces: Surface S_a in the far-field region, surface S_b enclosing the antenna and the generator, and surface S_c a tube-shaped surface connecting S_a and S_b. Thus, we may write $S = S_a + S_b + S_c$. We now write the Poynting theorem for the surface S enclosing the volume V

$$-\frac{1}{2}\oint_S \mathbf{S} \cdot d\mathbf{S} = j2\omega \int_V \frac{1}{4}[\mu_0 |\mathbf{H}|^2 - \epsilon_0 |\mathbf{E}|^2] dV + \int_V \frac{1}{2}\mathbf{E} \cdot \mathbf{J}_a dV \tag{2.90}$$

We may reduce the diameter of the tube S_c as far as we are pleased. Therefore, the contribution of the surface integral S_c is negligible. Hence, we have

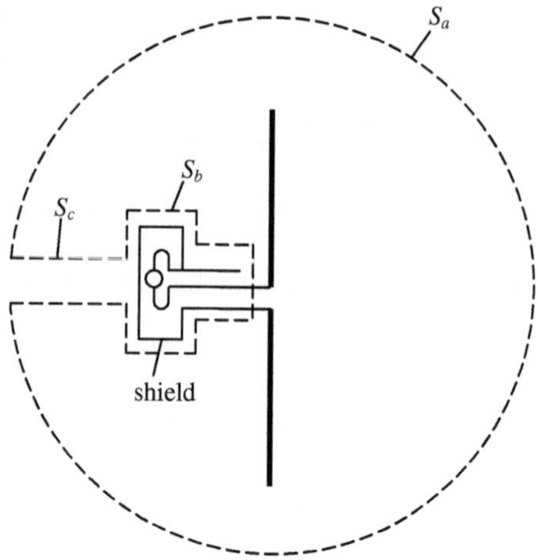

Fig. 2.9 The Poynting theorem for radiating antennas

$$-\oint_{S_a} \frac{1}{2}\frac{|\mathbf{E}|^2}{Z_0} dS - \oint_{S_b} \frac{1}{2}(\mathbf{E} \times \mathbf{H}^\star) \cdot \mathbf{dS} = j2\omega \int_V \frac{1}{4}[\mu_0|\mathbf{H}|^2 - \epsilon_0|\mathbf{E}|^2] dV \quad (2.91)$$

where we used the expression for **S** in the far-field region. We now give an interpretation for the second integral on the left. If, in accordance with the concepts of circuit theory, we ignore the displacement current and magnetic induction effects, we may write

$$\mathbf{E} = -\nabla\Phi \quad (2.92)$$

so that

$$-\oint_{S_b} \frac{1}{2}(\mathbf{E} \times \mathbf{H}^\star) \cdot \mathbf{dS} = -\oint_{S_b} \frac{1}{2}(-\nabla\Phi \times \mathbf{H}^\star) \cdot \mathbf{dS} \quad (2.93)$$

Using the vector identity

$$\nabla \times (\Phi\mathbf{H}^\star) \equiv \nabla\Phi \times \mathbf{H}^\star + \Phi\nabla \times \mathbf{H}^\star \quad (2.94)$$

we have

$$-\oint_{S_b} \frac{1}{2}(\mathbf{E} \times \mathbf{H}^\star) \cdot \mathbf{dS} = \frac{1}{2}\oint_{S_b} \nabla \times (\Phi\mathbf{H}^\star) \cdot \mathbf{dS} - \frac{1}{2}\oint_{S_b} (\Phi\nabla \times \mathbf{H}^\star) \cdot \mathbf{dS} \quad (2.95)$$

But since $\nabla \times \mathbf{H} = \mathbf{J}$,

$$-\oint_{S_b} \frac{1}{2}(\mathbf{E} \times \mathbf{H}^{\star}) \cdot \mathbf{dS} = -\frac{1}{2}\oint_{S_b} \Phi \mathbf{J}^{\star} \cdot \mathbf{dS} \tag{2.96}$$

Clearly, the right hand side of the above equation is the power generated by the source

$$-\oint_{S_b} \frac{1}{2}(\mathbf{E} \times \mathbf{H}^{\star}) \cdot \mathbf{dS} = \frac{1}{2}\Phi_i I_i^{\star} \tag{2.97}$$

The complex voltage of the antenna is proportional to the complex terminal current

$$\Phi = Z_{ant} I \tag{2.98}$$

where Z_{ant} is the antenna impedance. The real part of the antenna can be defined as

$$\Re e Z_{ant} \equiv R_{rad} = \frac{1}{Z_0}\oint_S \frac{|\mathbf{E}|^2}{|I_0|^2} dS \tag{2.99}$$

where R_{rad} is the *radiation resistance* and the imaginary part is related to the reactive power supplied to the antenna

$$\Im m Z_{ant} = \frac{\omega \int_V (\mu_0 |\mathbf{H}|^2 - \epsilon_0 |\mathbf{E}|^2) dV}{|I_0|^2} \tag{2.100}$$

Example 2.3 The radiation resistance of a short electric dipole of length ℓ can be found by the total power radiated by the dipole in the far-field

$$P = \oint_S \mathbf{S} \cdot \mathbf{ds}$$

where $\mathbf{S}$ is the Poynting vector. Using (2.96), we have

$$\begin{aligned} P &= \int_0^{2\pi}\int_0^{\pi} (I_0\ell)^2 \frac{k_0^2 Z_0}{2(4\pi)^2} \sin^2\theta ds \\ &= (I_0\ell)^2 \frac{k_0^2 Z_0}{12\pi} = (\frac{I_0\ell}{\lambda_0})^2 \frac{\pi}{3} Z_0 \end{aligned}$$

The radiation resistance is given by

$$R_{rad} = 2P/I_0^2$$

Thus

$$R_{rad} = Z_0(\frac{2\pi}{3})(\ell/\lambda_0)^2$$

This expression is valid for short dipoles ($\ell \ll \lambda_0$), but it is a good approximation for dipoles of length $\ell \leq \lambda_0/4$. □

2.6.6 Friis Transmission Formula

Consider a transmitting and a receiving antenna positioned in the direction of their maximum gain in free space separated by a distance R, as shown in Fig. 2.10 . If the power transmitted by the transmitting antenna is P_t, the power density at the receiver is given by

$$S_r = G_t \frac{P_t}{4\pi R^2} \tag{2.101}$$

where G_t is the gain of the transmitting antenna. The intercepted power at the receiving antenna is expressed as

$$P_{int} = A_r S_r = A_r G_t \frac{P_t}{4\pi R^2} \tag{2.102}$$

where A_r is the effective aperture of the receiving antenna. The received power can be written in terms of the intercepted power as

$$P_{rec} = \eta_r P_{int} \tag{2.103}$$

where η_r is the receiving antenna efficiency. Substituting from (2.102), we get

$$P_{rec} = \eta_r A_r G_t \frac{P_t}{4\pi R^2} = G_t G_r P_t \left(\frac{\lambda}{4\pi R}\right)^2 \tag{2.104}$$

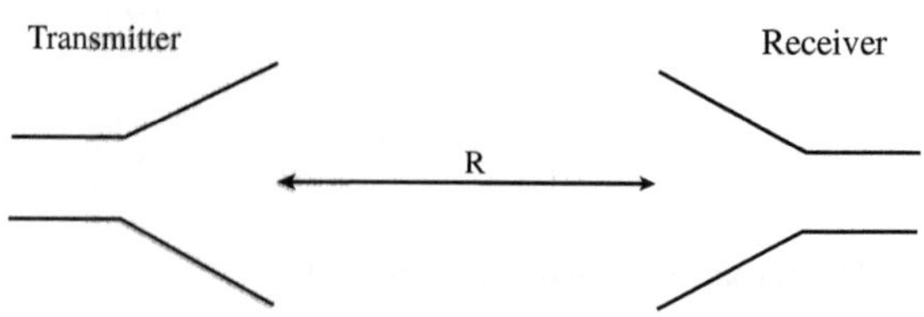

Fig. 2.10 The configuration for the derivation of the Friis transmission formula

where use has been made of (2.89). The power transfer ratio is given by

$$\frac{P_{rec}}{P_t} = G_t G_r \left(\frac{\lambda}{4\pi R}\right)^2 \tag{2.105}$$

If the antennas are positioned arbitrarily, then this ratio is given by

$$\frac{P_{rec}}{P_t} = G_t G_r \left(\frac{\lambda}{4\pi R}\right)^2 U_t(\theta_t, \phi_t) U_r(\theta_r, \phi_r) \tag{2.106}$$

which is known as *Friis transmission formula.*

Exercises

2.1: If $\mathbf{F} = \mathbf{A} f(r)$ where $\mathbf{A}$ is a constant vector, f is a function of r only, and $\mathbf{r} = r\widehat{r} = x\widehat{x} + y\widehat{y} + z\widehat{z}$ is the position vector,
(a) show

$$\nabla \times \mathbf{F} = \widehat{r} \times \mathbf{A}\frac{df}{dr}$$

(b) In particular, if $f(r) = \frac{e^{-jkr}}{r}$, show that for $kr \gg 1$ (corresponding to the far field)

$$\nabla \times \mathbf{F} \simeq -jk\widehat{r} \times \mathbf{F}$$

(c) More generally, if $\mathbf{F} = \mathbf{F}(r, \theta, \phi) = \mathbf{A}\frac{e^{-jkr}}{r}$ where $\mathbf{A}$ is any vector *independent of* r but not necessarily constant, show that in the far field

$$\nabla \times \mathbf{F} \simeq -jk\widehat{r} \times \mathbf{F}$$
$$\nabla \times \nabla \times \mathbf{F} \simeq -k^2\widehat{r} \times (\widehat{r}\mathbf{F})$$

In other words, in the far field the operator $\nabla(.)$ is equivalent to $-jk\widehat{r} \times (.)$. Thus, given a Hertz vector, the resulting $\mathbf{E}$ and $\mathbf{H}$ in the *far field region* can be obtained *without any differentiation.*

2.2: A straight wire of length L carrying the current $\widehat{z}I_0 e^{-j\beta z}$ lies on the z-axis ($0 \le z \le L$). With β a real constant, this represents a travelling wave antenna.
(a) State the Hertz vector(s) associated with this source.
(b) If the point of observation $\mathbf{r}$ is such that $r \gg L$, show that

$$|\mathbf{r} - \mathbf{r}'| \simeq r - z' \cos\theta$$

(c) Under the assumption that

$$\frac{e^{-jk|\mathbf{r}-\mathbf{r}'|}}{|\mathbf{r}-\mathbf{r}'|} \simeq \frac{e^{-jkr}}{r} e^{jkz'} \cos\theta$$

evaluate the Hertz vector(s) in the far field.
(d) Determine **E** in the far field, showing that $\mathbf{E} = \widehat{\theta} E_\theta$

2.3: Find the radiation resistance of a short Hertzian dipole.

Chapter 3
Fundamental Theorems

In this chapter, we discuss the fundamental theorems of electromagnetics.

3.1 Uniqueness Theorem

When working with electromagnetic fields, it is assumed that a field which satisfies the wave equation and appropriate boundary conditions, is unique. The question of extreme importance is what are necessary and sufficient boundary conditions to ensure uniqueness.

Consider a homogeneous medium with the electric source **J** and bounded by a surface S. The medium is assumed to be lossy with constitutive parameters

$$\begin{aligned}\epsilon &= \epsilon' - j\epsilon'' - j\frac{\sigma}{\omega}\\ \mu &= \mu' - j\mu''\end{aligned} \tag{3.1}$$

Then in the volume V interior to S, we have

$$\begin{aligned}\nabla \times \mathbf{E} &= -j\omega\mu\mathbf{H}\\ \nabla \times \mathbf{H} &= j\omega\epsilon\mathbf{E} + \mathbf{J}\end{aligned} \tag{3.2}$$

Suppose there are two different fields $(\mathbf{E}^1, \mathbf{H}^1)$ and $(\mathbf{E}^2, \mathbf{H}^2)$ that can exist in volume V generated by different sources outside. Then the difference fields $(\delta\mathbf{E}, \delta\mathbf{H})$ must satisfy

$$\begin{aligned}\nabla \times \delta\mathbf{E} &= -j\omega\mu\delta\mathbf{H}\\ \nabla \times \delta\mathbf{H} &= j\omega\epsilon\delta\mathbf{E}\end{aligned} \tag{3.3}$$

implying

$$\nabla \times \delta\mathbf{H}^\star = -j\omega\epsilon^\star\delta\mathbf{E}^\star \tag{3.4}$$

© Springer International Publishing Switzerland 2015

K. Barkeshli, *Advanced Electromagnetics and Scattering Theory*,
DOI 10.1007/978-3-319-11547-4_3

Consider now the quantity

$$\nabla \cdot (\delta \mathbf{E} \times \delta \mathbf{H}^\star) = \delta \mathbf{H}^\star \cdot \nabla \times \delta \mathbf{E} - \delta \mathbf{E} \cdot \nabla \times \delta \mathbf{H}^\star$$
$$= -j\omega\mu|\delta \mathbf{H}|^2 + j\omega\epsilon^\star|\delta \mathbf{E}|^2 \tag{3.5}$$

Hence, form divergence theorem

$$\oint_S (\delta \mathbf{E} \times \delta \mathbf{H}^\star) \cdot \widehat{n} ds = -j\omega \int_V (\mu|\delta \mathbf{H}|^2 - \epsilon^\star|\delta \mathbf{E}|^2) dv$$
$$= -\omega \int_V \left[\mu''|\delta \mathbf{H}|^2 + (\epsilon'' + \frac{\sigma}{\omega})|\delta \mathbf{E}|^2\right] dv$$
$$- j\omega \int_V \left[\mu'|\delta \mathbf{H}|^2 - \epsilon'|\delta \mathbf{E}|^2\right] dv \tag{3.6}$$

If the surface integral vanishes, the volume integral must do so. In particular, the real part must vanish.[1] Thus, if μ'' or either of ϵ'' and σ is nonzero in V, no matter how small it is, $\delta\mathbf{E}$ and $\delta\mathbf{H}$ must vanish throughout V, i.e. the solution is unique.

But what is (are) the condition(s) under which the surface integral is zero? We note that

$$\oint_S \widehat{n} \cdot (\delta \mathbf{E} \times \delta \mathbf{H}^\star) ds = \oint_S (\widehat{n} \times \delta \mathbf{E}) \cdot \delta \mathbf{H}^\star ds$$
$$= \oint_S (\delta \mathbf{H}^\star \times \widehat{n}) \cdot \delta \mathbf{E} ds \tag{3.7}$$

Thus, the field is unique if

$\widehat{n} \times \delta\mathbf{E} = 0$ on S, i.e. $\widehat{n} \times \mathbf{E}$ specified, or
$\widehat{n} \times \delta\mathbf{H} = 0$ on S, i.e. $\widehat{n} \times \mathbf{H}$ specified.

The uniqueness principle gives the necessary and sufficient boundary conditions which should be imposed on a surface *completely* enclosing the region under study to ensure unique solution to the wave equation. It states that the solution of the wave equation in a given lossy region is uniquely determined if and only if

- the tangential electric field $\widehat{n} \times \mathbf{E}$ is specified on the surrounding surface, or
- the tangential magnetic field $\widehat{n} \times \mathbf{H}$ is specified on the surrounding surface, or
- the tangential electric field $\widehat{n} \times \mathbf{E}$ is specified on part of the surface and the tangential magnetic field $\widehat{n} \times \mathbf{H}$ is specified on the remainder.

[1] The requirement that the imaginary part must vanish does not ensure uniqueness, because ϵ' and μ' are never zero.

The fact that $\widehat{n} \times \mathbf{E}$ and/or $\widehat{n} \times \mathbf{H}$ are the surface quantities to be specified shows why the consideration of surface electric and magnetic currents is so convenient.

The uniqueness was established for general lossy media. The fields in a lossless medium are considered to be the limit of the corresponding fields in a lossy medium as the losses go to zero. We can also extend the uniqueness principle to non-homogeneous media by separate application to small volumes each assumed homogeneous. In fact, the ultimate limitation to the method is to *linear* media.

In the case where the region extends to infinity, we require an equivalent condition that is imposed at large distances that will serve to ensure a unique solution. This is, of course, the radiation condition.

The radiation condition imposes a particular rate of decay for the outgoing fields in an unbounded lossy medium. In particular, the *Sommerfeld radiation condition* (1949) for a scalar field ψ is given by

$$\lim_{r\to\infty} r(\frac{\partial\psi}{\partial r} + jk\psi) = 0 \tag{3.8}$$

It can be shown that the Sommerfeld radiation condition is unnecessarily strong; a milder sufficient condition

$$\lim_{r\to\infty} \int_S |\frac{\partial\psi}{\partial r} + jk\psi|^2 ds = 0 \tag{3.9}$$

proposed by Wilcox (1956) is sufficient to ensure a unique solution that is outgoing.

For an outward-travelling spherical electromagnetic wave at infinity, the magnetic field is related to the electric field by the equation $\mathbf{H} = \widehat{r} \times \mathbf{E}/Z_0$. Thus, the radiation conditions for the electromagnetic fields can be expressed as

$$\lim_{r\to\infty} r[\nabla \times \mathbf{E} + jk_0\widehat{r} \times \mathbf{E}] = 0 \tag{3.10}$$

$$\lim_{r\to\infty} r[\mathbf{H} - \frac{\widehat{r} \times \mathbf{E}}{Z_0}] = 0 \tag{3.11}$$

where Z_0 is the free space intrinsic impedance. The radiation condition implies that the fields are outgoing and that their radial components decrease faster than $1/r$. In other words, the Poynting vector is pointed outward and decreases as $1/r^2$.

3.2 Duality

Duality is a consequence of the mathematical similarity of the equations governing electric and magnetic phenomena as reflected by Maxwell's equations, and expressions are said to be the dual of one another if they display this similarity. It may be desirable to preserve this symmetry whenever possible, even to the extent of introducing the concepts of fictitious magnetic currents and charges. The symmetric form of the equations are as follows

$$\nabla \times \mathbf{E} = -\mu \frac{\partial \mathbf{H}}{\partial t} - \mathbf{M} \tag{3.12}$$

$$\nabla \times \mathbf{H} = \epsilon \frac{\partial \mathbf{E}}{\partial t} + \mathbf{J} \tag{3.13}$$

$$\nabla \cdot \mathbf{D} = \rho_e \tag{3.14}$$

$$\nabla \cdot \mathbf{B} = \rho_m \tag{3.15}$$

where $\mathbf{M}$ and ρ_m denote the magnetic current and charge, respectively.

It is noted that the above equations are invariant to the following transformations in field quantities and material properties.

$$\mathbf{E} \to \mathbf{H} \qquad \mathbf{H} \to -\mathbf{E}$$

$$\mathbf{J} \to \mathbf{M} \qquad \mathbf{M} \to -\mathbf{J} \tag{3.16}$$

$$\rho_e \to \rho_m \qquad \rho_m \to -\rho_e \tag{3.17}$$

$$\epsilon \to \mu \qquad \mu \to \epsilon \tag{3.18}$$

This form of the duality principle requires the use of dual fields in a dual medium. In many cases, it is more convenient to maintain the same medium when applying duality. In such cases, we use the following transformations:

$$\begin{aligned} \mathbf{E} &\to Z\mathbf{H} \qquad & \mathbf{H} &\to -\mathbf{E}/Z \\ \mathbf{J} &\to \mathbf{M}/Z \qquad & \mathbf{M} &\to -Z\mathbf{J} \\ \rho_e &\to \rho_m/Z \qquad & \rho_m &\to -Z\rho_e \end{aligned} \tag{3.19}$$

where $Z = \sqrt{\frac{\mu}{\epsilon}}$ is the intrinsic impedance of the medium.

Example 3.1 Using duality, we can find the radiation fields of a magnetic dipole from those of an electric dipole.

The radiation fields of a short $\hat{z}$-directed Hertzian dipole of length ℓ located at the origin are given by (2.29).

$$E_\theta = j \frac{k_0^2}{4\pi\omega\epsilon_0} I_0 \ell \frac{e^{-jk_0 r}}{r} \sin\theta$$

$$H_\phi = j \frac{k_0}{4\pi} I_0 \ell \frac{e^{-jk_0 r}}{r} \sin\theta$$

Assume that the magnetic dipole is located in the xy-plane and centered at the origin. By duality, we have

$$H_\theta = j\frac{k_0^2}{4\pi\omega\epsilon_0} I_0\ell \frac{e^{-jk_0r}}{r}\sin\theta$$
$$E_\phi = j\frac{k_0}{4\pi} I_0\ell \frac{e^{-jk_0r}}{r}\sin\theta$$
$$\mathbf{E} = m\frac{e^{-jk_0r}}{4\pi r}k_0^2 Z_0 \sin\theta\widehat{\phi}$$
$$\mathbf{H} = -m\frac{e^{-jk_0r}}{4\pi r}k_0^2 \sin\theta\widehat{\theta}$$

This is the same result obtained in Sect. 2.2 (Eq. (2.45)). □

We previously defined the electric Hertz potential π_e in terms of which the electromagnetic fields where given by (1.157) and (1.158). We may define the *magnetic* Hertz potential π_m in terms of fictitious magnetic sources. Thus, we write

$$\mathbf{P}_m = \frac{\mathbf{M}}{j\omega} \quad \text{and} \quad \nabla \cdot \mathbf{P}_m = -\rho_m \tag{3.20}$$

Therefore, we have

$$\nabla^2 \pi_m + k^2 \pi_m = -\frac{\mathbf{P}_m}{\mu} \tag{3.21}$$

The electromagnetic fields due to magnetic sources are then given by

$$\mathbf{E} = -j\omega\mu\nabla \times \pi_m \tag{3.22}$$
$$\mathbf{H} = \nabla(\nabla \cdot \pi_m) + k^2\pi_m \tag{3.23}$$

Using superposition, the most general solution in presence of both electric and magnetic current densities is expressed as follows

$$\mathbf{E} = \nabla(\nabla \cdot \pi_e) + k^2\pi_e - j\omega\mu\nabla \times \partial\pi_m \tag{3.24}$$
$$\mathbf{H} = j\omega\epsilon\nabla \times \partial\pi_e + \nabla(\nabla \cdot \pi_m) + k^2\pi_m \tag{3.25}$$

In view of (3.21) the magnetic Hertz potentials in free space is given by

$$\pi_m = -jkY\int_V \mathbf{M}(\mathbf{r}')G(\mathbf{r};\mathbf{r}')dv' \tag{3.26}$$

Example 3.2 Consider a narrow slot of width δ excited by a transverse uniform electric field of intensity E_0 as shown in Fig. 3.1. Using the concept of the magnetic current, we write

$$\mathbf{K}_m = \mathbf{E} \times \widehat{n} = \widehat{s}E_0 \tag{3.27}$$

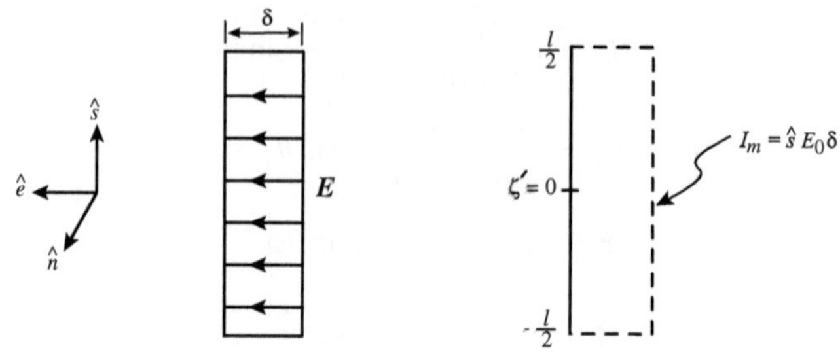

Fig. 3.1 A radiating elementary slot

The total magnetic current is therefore

$$\mathbf{I}_m = \widehat{s} E_0 \delta \tag{3.28}$$

The moment of the magnetic current is thus

$$\mathbf{p}_m = I_m \ell \widehat{s} = E_0 \delta \ell \widehat{s} \tag{3.29}$$

The above result shows that the elementary slot of length ℓ oriented in the $\widehat{n}$ direction and excited by a transverse electric field $\mathbf{E} = \widehat{t} E_0$ may be considered as a magnetic current $\mathbf{I}_m = \widehat{t} \times \widehat{n} E_0 \delta$ (Volts) of the same length.

In view of the above analogy, the fields of an elementary slot can be directly obtained from those of an electric dipole using duality. □

Duality can increase our understanding of the physical phenomena involved by displaying the equivalence of different sources. Therefore, it allows us to deduce a solution from a knowledge of its dual. In some occasions, duality can simplify an analysis by allowing us to split a problem into two parts, one dual of the other, e.g. one due to magnetic sources and the other due to electric ones.

3.3 Image Theory

If sources are present in the vicinity of perfectly conducting boundaries and if the boundary is infinite in extent, we may be able to use image theory to simplify the problem.

For example, suppose we have a wire element bearing an x-directed current above the perfectly conducting plane $z = 0$ as shown in Fig. 3.2. We seek the field in $z \geq 0$.

The boundary condition at $z = 0$ is

$$\widehat{n} \times \mathbf{E} = 0 \tag{3.30}$$

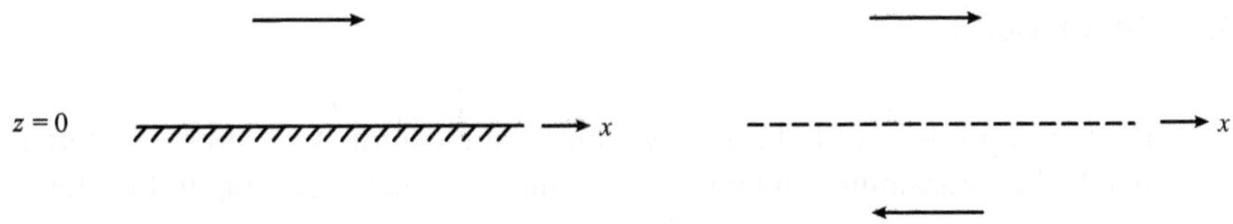

Fig. 3.2 A wire element above a perfectly conducting ground plane

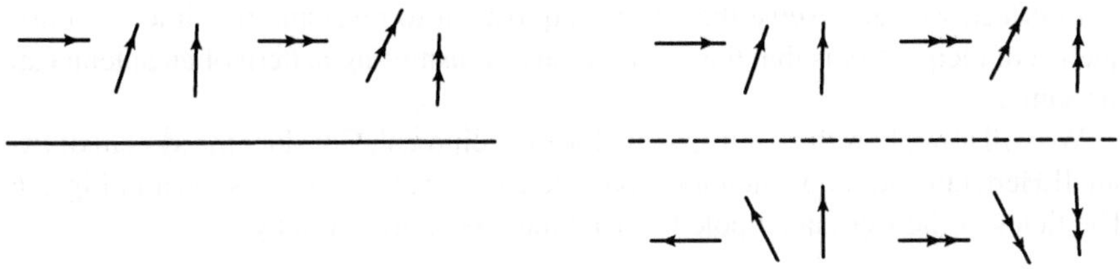

Fig. 3.3 Electric and magnetic dipoles above a perfectly conducting ground plane

This boundary condition can be satisfied by removing the perfectly conducting plane and inserting an oppositely-oriented current $-\mathbf{J}_e$ at the image point. In $z \geq 0$ the field is due to the two sources radiating in infinitely-extended space.

Since the necessary boundary condition is satisfied, the resulting field is the one and the only correct one by the uniqueness theorem. Note that image theory does not give the field in $z < 0$ (it is obviously zero there), but only in the half space of interest.

For any type of source above a perfectly conducting plane, the equivalent problem is as shown in Fig. 3.3.

It is noted that analogous images exist for perfect magnetic planes, but they are of less interest practically. Also, the theory (with localized images) is exact only for perfect boundaries.

Example 3.3 Consider the magnetic dipole examined in Example 3.1 located in front of a perfectly conducting ground plane. Find the radiated fields in the region $z > 0$.

The geometry of the problem is shown in Fig. 3.4. □

Fig. 3.4 **a** A magnetic dipole located above a perfectly conducting ground plane. **b** The dipole with its image

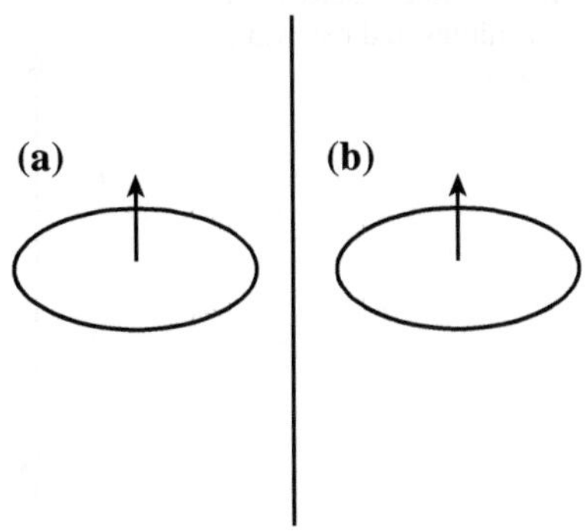

3.4 Reciprocity

Reciprocity is a property of fields in any reciprocal medium. A reciprocal system is unchanged when transmitter and receiver are interchanged. Referring to Fig. 3.5, if a voltage V_1 applied to the terminals of a transmitting antenna T produces a voltage V_2 at the terminals of a receiving antenna R, then V_2 applied at the terminals of R regarded as a transmitter will producc V_1 at T acting as a receiver.

In effect, we can reverse the arrows, equivalent to reversing the time. A consequence of reciprocity is that the receiving and transmitting pattern of an antenna are the same.

We will now show that isotropic media are reciprocal. Consider the system of two small Hertzian and loop antennas separated a distance d apart as shown in Fig. 3.6. The fields of the Hertzian dipole far from the dipole are given by

$$E_\theta^d = jZk(I_d d\ell)\frac{e^{-jkr}}{4\pi r}\sin\theta \tag{3.31}$$

$$H_\phi^d = jk(I_d d\ell)\frac{e^{-jkr}}{4\pi r}\sin\theta \tag{3.32}$$

and the fields of the loop antenna are given by duality as

$$H_\theta^\ell = -k^2 I_\ell(\pi a^2)\frac{e^{-jkr}}{4\pi r}\sin\theta \tag{3.33}$$

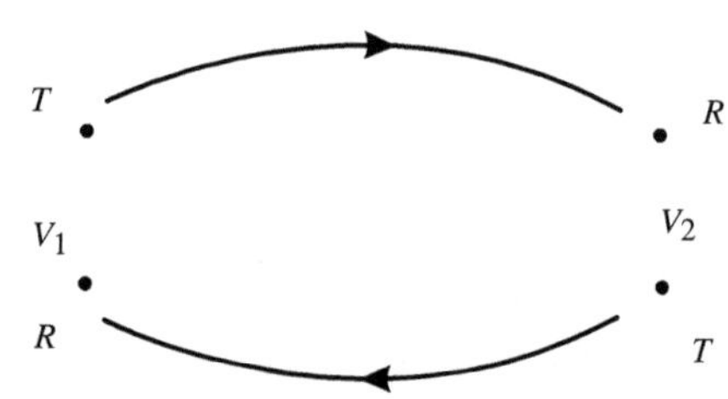

Fig. 3.5 Reciprocity for transmitting and receiving antennas

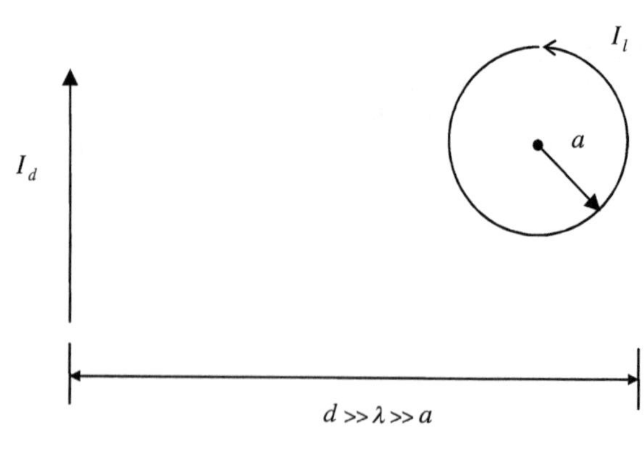

Fig. 3.6 Reciprocity for a transmitting and receiving dipoles

$$E_\phi^\ell = Zk^2 I_\ell(\pi a^2)\frac{e^{-jkr}}{4\pi r}\sin\theta \tag{3.34}$$

where θ and ϕ in the latter equations are measured with respect to the loop axis.

The induced open circuit voltage in the loop due to radiation from the dipole is given by

$$\mathcal{V}_{\text{emf}}^\ell = -d\Phi/dt \tag{3.35}$$

Therefore

$$\begin{aligned}\mathcal{V}_{\text{emf}}^\ell = V_\ell &= -j\omega\mu \oint_S \mathbf{H}\cdot\widehat{n}ds \simeq -j\omega\mu H_\phi^d S \\ &= k\omega\mu(\pi a^2) I_d d\ell \frac{e^{-jkd}}{4\pi d}\end{aligned} \tag{3.36}$$

The induced open circuit voltage on the dipole due to radiation from the loop antenna is

$$\mathcal{V}_{\text{emf}}^d = V_d = \int \mathbf{E}\cdot\mathbf{d}\ell \simeq Zk^2 I_\ell(\pi a^2)\frac{e^{-jkd}}{4\pi d}d\ell \tag{3.37}$$

It is now easy to observe that

$$V_d I_d = V_\ell I_\ell \tag{3.38}$$

which is a statement of reciprocity: the interaction of the induced voltage in the dipole (due to the field of the loop) with the dipole current is equal to the interaction of the induced voltage in the loop antenna (due to the field of the dipole) with the loop current. This is sometimes written simply as

$$\langle 2, 1\rangle = \langle 1, 2\rangle \tag{3.39}$$

In order to give a formal presentation of the reciprocity, consider two independent sets of electromagnetic fields with the same frequency of oscillations in free space due to sources $\mathbf{J}_1$ and $\mathbf{J}_2$

$$(\mathbf{J}_1, \mathbf{E}_1, \mathbf{H}_1) \quad (\mathbf{J}_2, \mathbf{E}_2, \mathbf{H}_2)$$

as shown in Fig. 3.7.

From Maxwell's equations, we have

$$\nabla\times\mathbf{E}_1 = -j\omega\mu\mathbf{H}_1 \tag{3.40}$$

$$\nabla\times\mathbf{H}_1 = j\omega\epsilon\mathbf{E}_1 + \mathbf{J}_1 \tag{3.41}$$

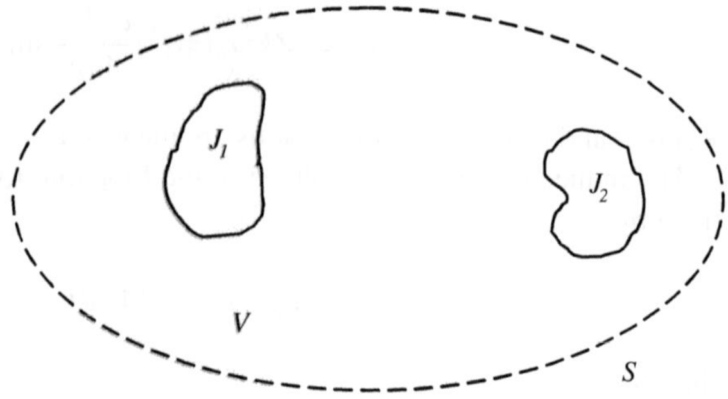

Fig. 3.7 Reciprocity for two radiating electric current distributions

and

$$\nabla \times \mathbf{E}_2 = -j\omega\mu\mathbf{H}_2 \quad (3.42)$$
$$\nabla \times \mathbf{H}_2 = j\omega\epsilon\mathbf{E}_2 + \mathbf{J}_2 \quad (3.43)$$

We multiply (3.40) by $\mathbf{H}_2$ and (3.42) by $\mathbf{H}_1$ and subtract the two resulting equations to get

$$\mathbf{H}_2 \cdot \nabla \times \mathbf{E}_1 - \mathbf{H}_1 \cdot \nabla \times \mathbf{E}_2 = 0 \quad (3.44)$$

Similarly, we multiply (3.41) by $\mathbf{E}_2$ and (3.43) by $\mathbf{E}_1$ and subtract the two resulting equations and obtain

$$\mathbf{E}_2 \cdot \nabla \times \mathbf{H}_1 - \mathbf{E}_1 \cdot \nabla \times \mathbf{H}_2 = \mathbf{J}_1 \cdot \mathbf{E}_2 - \mathbf{J}_2 \cdot \mathbf{E}_1 \quad (3.45)$$

Adding now the last two equations, we have

$$\begin{aligned}\mathbf{J}_1 \cdot \mathbf{E}_2 - \mathbf{J}_2 \cdot \mathbf{E}_1 &= \mathbf{E}_2 \cdot \nabla \times \mathbf{H}_1 - \mathbf{H}_1 \cdot \nabla \times \mathbf{E}_2 \\ &\quad - (\mathbf{E}_1 \cdot \nabla \times \mathbf{H}_2 - \mathbf{H}_2 \cdot \nabla \times \mathbf{E}_1) \\ &= \nabla \cdot (\mathbf{E}_1 \times \mathbf{H}_2 - \mathbf{E}_2 \times \mathbf{H}_1) \quad (3.46)\end{aligned}$$

Integrating over a closed volume V, we obtain *Lorentz* reciprocity theorem

$$\int_V (\mathbf{J}_1 \cdot \mathbf{E}_2 - \mathbf{J}_2 \cdot \mathbf{E}_1) dv = \oint_S (\mathbf{E}_1 \times \mathbf{H}_2 - \mathbf{E}_2 \times \mathbf{H}_1) \cdot \widehat{n} ds \quad (3.47)$$

For a source-free region, (3.47) reduces to

$$\oint_S (\mathbf{E}_1 \times \mathbf{H}_2 - \mathbf{E}_2 \times \mathbf{H}_1) \cdot \widehat{n} ds = 0 \quad (3.48)$$

As an example, suppose that $(\mathbf{E}_1, \mathbf{H}_1)$ and $(\mathbf{E}_2, \mathbf{H}_2)$ represent two different modes present in a section of a hollow waveguide. Then these two pairs of electromagnetic fields must satisfy (3.48).

If the volume V extends to infinity, the surface integral on the right hand side of (3.47) vanishes due to radiation conditions (3.10). Therefore, the relation

$$\int_V (\mathbf{J}_1 \cdot \mathbf{E}_2 - \mathbf{J}_2 \cdot \mathbf{E}_1) dv = 0 \tag{3.49}$$

holds true. This is the statement of *Rayleigh-Carson* reciprocity theorem. The volume integral $\int_V \mathbf{J}_1 \cdot \mathbf{E}_2 dv$ is referred to as the reaction of the field $\mathbf{E}_2$ to the source $\mathbf{J}_1$, while $\int_V \mathbf{J}_2 \cdot \mathbf{E}_1 dv$ is the reaction of the field $\mathbf{E}_1$ to the source $\mathbf{J}_2$.

Example 3.4 Using reciprocity, prove that a current sheet $\mathbf{K}$ denoted as source 1 impressed on the surface of a perfect conductor produces null fields everywhere.

If the surface of the conductor is planar, the image theory assures us that no field is produced by the current. However, if the conductor is not planar, then we can use reciprocity to show that the current sheet produces null fields everywhere.

Referring to Fig. 3.8, let there be a dipole denoted as source 2 that measures the field produced by $\mathbf{K}$. The dipole antenna produces no tangential electric field along the surface of the conductor. The reaction of $\mathbf{K}$ and the dipole is, therefore, $\langle 1, 2 \rangle = 0$. By reciprocity we have

$$\langle 2, 1 \rangle = \langle 1, 2 \rangle = 0$$

But $\langle 2, 1 \rangle = 0$ is the field arising from the impressed source 1 as measured by source 2, and 2 can be any arbitrary source (electric or magnetic) of arbitrary orientation.

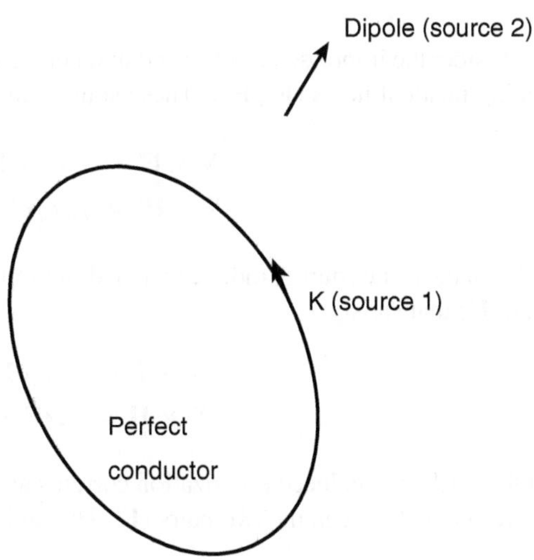

Fig. 3.8 Illustration of a surface current on a perfect conductor; there is no normal surface current component

Thus, source 2 measures no field. Therefore, impressed current sheets on the surface of a perfectly conducting body produce no field. □

Reciprocity is a physical attribute of a system, and when constructing an approximate solution to a problem, for example a numerical solution, it is important to try to ensure that reciprocity is satisfied.

3.5 Equivalence Principles

The equivalence principle is an extremely useful concept in electromagnetic theory. In order to calculate the field in a given region of space, the equivalence principle shows how we can replace the *actual* sources by *fictitious* sources which are *equivalent* in the sense that they yield the same field in that region. The equivalence rests on the uniqueness theorem for the solution of Maxwell's equations.

There are several variations of the principle which are basically divided into two classes, namely

1. volume equivalence principles.
2. surface equivalence principles, and

The volume equivalence principles are based on equivalent volumetric currents and the surface equivalence principles rely on equivalent surface currents. In what follows, we discuss the basic aspects of both types of currents.

3.5.1 Equivalent Volumetric Currents

Consider the impressed electric and magnetic currents $\mathbf{J}_i$ and $\mathbf{M}_i$ in free space generating incident fields $(\mathbf{E}^i, \mathbf{H}^i)$. These sources and fields satisfy Maxwell's equations

$$\begin{aligned} \nabla \times \mathbf{E}^i &= -j\omega\mu_o \mathbf{H}^i - \mathbf{M}_i \\ \nabla \times \mathbf{H}^i &= j\omega\epsilon_o \mathbf{E}^i + \mathbf{J}_i \end{aligned} \tag{3.50}$$

When the same sources radiate in a medium represented by (ϵ, μ), they radiate fields $(\mathbf{E}, \mathbf{H})$ that satisfy

$$\begin{aligned} \nabla \times \mathbf{E} &= -j\omega\mu \mathbf{H} - \mathbf{M}_i \\ \nabla \times \mathbf{H} &= j\omega\epsilon \mathbf{E} + \mathbf{J}_i \end{aligned} \tag{3.51}$$

Obviously, the induced polarization currents are responsible for the difference (scattered) field between the two pairs $(\mathbf{E}^i, \mathbf{H}^i)$ and $(\mathbf{E}, \mathbf{H})$

$$\mathbf{E}^s = \mathbf{E} - \mathbf{E}^i \quad \mathbf{H}^s = \mathbf{H} - \mathbf{H}^i \tag{3.52}$$

By subtracting the two pair of Eqs. (3.50) and (3.51), we find that the scattered fields are source free and satisfy the following equations

$$\begin{aligned} \nabla \times \mathbf{E}^s &= -j\omega(\mu\mathbf{H} - \mu_o\mathbf{H}^i) \\ \nabla \times \mathbf{H}^s &= j\omega(\epsilon\mathbf{E} - \epsilon_o\mathbf{E}^i) \end{aligned} \tag{3.53}$$

Adding and subtracting the terms $j\omega\mu_o\mathbf{H}$ and $j\omega\epsilon_o\mathbf{E}$ to the right hand sides of the first and second equations, respectively yield

$$\begin{aligned} \nabla \times \mathbf{E}^s &= -j\omega(\mu - \mu_o)\mathbf{H} - j\omega\mu_o\mathbf{H}^s \\ \nabla \times \mathbf{H}^s &= j\omega(\epsilon - \epsilon_o)\mathbf{E} + j\omega\epsilon_o\mathbf{E}^s \end{aligned} \tag{3.54}$$

or equivalently

$$\begin{aligned} \nabla \times \mathbf{E}^s &= -\mathbf{M}_e - j\omega\mu_o\mathbf{H}^s \\ \nabla \times \mathbf{H}^s &= \mathbf{J}_e + j\omega\epsilon_o\mathbf{E}^s \end{aligned} \tag{3.55}$$

where

$$\begin{aligned} \mathbf{J}_e &= j\omega\epsilon_o(\epsilon_r - 1)\mathbf{E} \\ \mathbf{M}_e &= j\omega\mu_o(\mu_r - 1)\mathbf{H} \end{aligned} \tag{3.56}$$

and (ϵ_r, μ_r) are the relative permittivity and permeability of the medium. Thus, the presence of the dielectric material is actually equivalent to having the equivalent volumetric currents $\mathbf{J}_e$ and $\mathbf{M}_e$ radiating in the unbounded space.

Note that these volumetric currents are the same as the polarization currents and may be directly used to construct integral equations.

3.5.1.1 Resistive Sheet Boundary Condition

A thin conducting sheet or non-magnetic dielectric layer can be represented by a resistive sheet. In the case of a source-free dielectric layer having thickness τ, and a relative complex premittivity ϵ_r, the equivalent volumetric current density is given by

$$\mathbf{J}_{eq} \equiv j\omega\epsilon_o(\epsilon_r - 1)\mathbf{E} \tag{3.57}$$

When the layer is electrically thin (Fig. 3.9), the normal component of the electric field across the layer is negligible. The dielectric layer can therefore be replaced by a resistive sheet of surface current density

$$\mathbf{K} \equiv \lim_{\tau\to 0} \int_{-\tau/2}^{\tau/2} \mathbf{J}_{eq} d\xi \simeq \lim_{\tau\to 0} \tau \left[\mathbf{J}_{eq}\right]_{\tan} \tag{3.58}$$

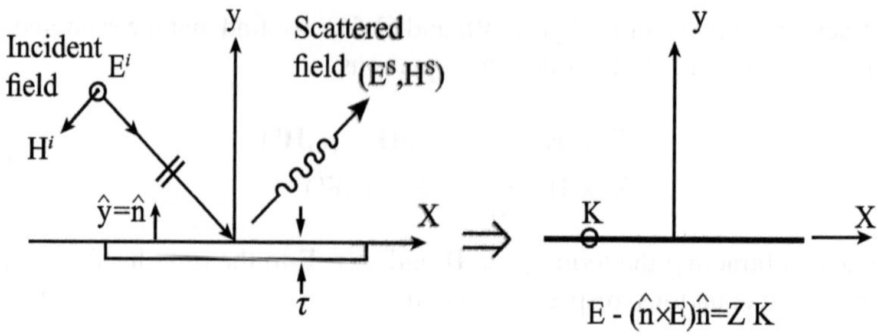

Fig. 3.9 A thin dielectric layer approximated by a resistive sheet

where ξ coordinate is chosen parallel to the direction of the upward unit normal to the layer $\widehat{n}$, and

$$\left[\mathbf{J}_{eq}\right]_{\tan} = \mathbf{J}_{eq} - (\widehat{n} \cdot \mathbf{J}_{eq})\,\widehat{n} \tag{3.59}$$

is the transverse volumetric current flowing inside the layer. In view of (3.57), we may write

$$\mathbf{E} - (\widehat{n} \cdot \mathbf{E})\,\widehat{n} = Z_s \mathbf{K} \tag{3.60}$$

where Z_s is the resistivity of the sheet

$$Z_s = \frac{Z_o}{jk_o\tau(\epsilon_r - 1)} \tag{3.61}$$

Equation (3.60) is known as the resistive sheet boundary condition. Thus, a resistive sheet is an electric current sheet whose strength is proportional to the local tangential electric field. For a thin conducting sheet of conductivity σ, (3.61) reduces to

$$Z_s = \frac{1}{\sigma\tau} \tag{3.62}$$

3.5.2 Equivalent Surface Currents

The equivalent surface currents were first conceived by S.A. Schelkunoff in 1936.

Suppose we have some sources represented by the volume electric and magnetic currents $\mathbf{J}_i$ and $\mathbf{M}_i$. These give rise to a field $(\mathbf{E}, \mathbf{H})$ everywhere. Surround the sources with an imaginary closed surface S. If there is some actual surface, e.g. a perfectly conducting one, present in the problem, it may be convenient to choose S to coincide with all or part of this, but it is not necessary to do so. The task is to define surface currents on S which yield $(\mathbf{E}, \mathbf{H})$ in the region outside.

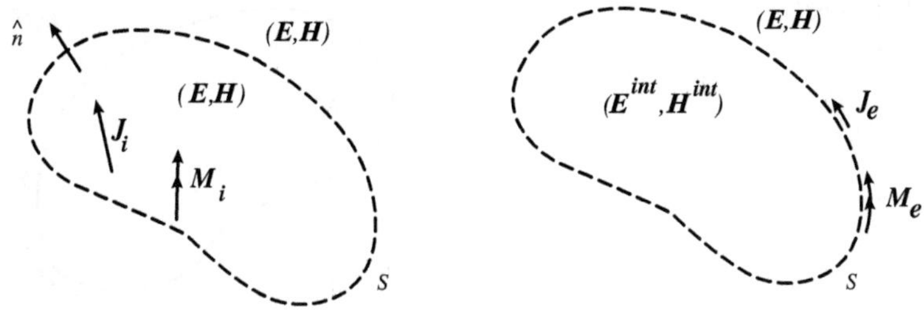

Fig. 3.10 First equivalence

3.5.2.1 First Equivalence

In order to find the equivalent currents, we remove the actual sources and assume that the field within S is $(\mathbf{E}^{int}, \mathbf{H}^{int})$ where the superscript *int* denotes internal fields. Then from the fundamental electromagnetic boundary conditions, S must be a surface supporting surface currents

$$\begin{aligned} \mathbf{J}_e &= \widehat{n} \times (\mathbf{H} - \mathbf{H}^{int}) \\ \mathbf{M}_e &= -\widehat{n} \times (\mathbf{E} - \mathbf{E}^{int}) \end{aligned} \tag{3.63}$$

which radiate in unbounded space — *the same medium everywhere* (Fig. 3.10). They are equivalent sources (surface currents) because they produce the same field outside S as did the original sources, though not the same field inside S. But the latter is of no concern; in fact, we can choose the internal fields *arbitrarily*. Clearly, this means that the choice of $\mathbf{J}_e$ and $\mathbf{M}_e$ is not unique and different combinations of equivalent electric and magnetic currents, corresponding to different choices $(\mathbf{E}^{int}, \mathbf{H}^{int})$ for the internal field, produce the same field outside S.

3.5.2.2 Love's Equivalence Principle

Since the choice of the internal fields is arbitrary, we may as well assume the field inside S to be zero, i.e. $(\mathbf{E}^{int}, \mathbf{H}^{int}) = 0$.

In this case, the equivalent sources are

$$\begin{aligned} \mathbf{J}_e &= \widehat{n} \times \mathbf{H} \\ \mathbf{M}_e &= -\widehat{n} \times \mathbf{E} \end{aligned} \tag{3.64}$$

radiating in unbounded space. This is referred to as Love's equivalence principle.[2]

[2] After A. E. H. Love who developed it in 1941.

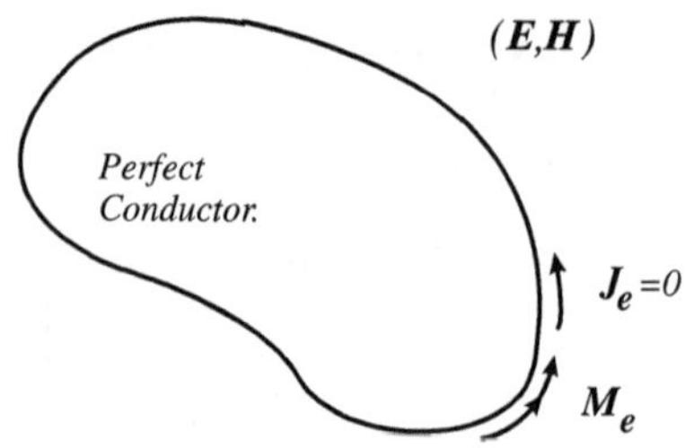

Fig. 3.11 Second equivalence

The Love's principle is widely used in antenna theory (to produce an equivalent aperture for an antenna) and in optics. The sources introduced in Love's theorem are analogous to Stratton-Chu representation.

It should be noted that this and other surface equivalence principles do not necessarily help us to solve a problem; but they do allow us to consider only *surface currents* and we may, for example, be able to construct integral equations for these which can be solved numerically. Once the equivalent surface currents are found, calculation of the external fields will be an easy task.

First equivalence principle requires electric *and* magnetic currents, but the uniqueness theorem indicates that one *or* the other should suffice to specify a unique field. This leads to the second equivalence principle.

3.5.2.3 Second Equivalence

In the above, we chose the field inside S to be zero. Therefore, it will not be disturbed if the properties of the medium inside S are changed. Thus, let us choose the properties to eliminate one or the other current.

Assume S is the surface of a perfect *electric* conductor[3] as shown in Fig. 3.11. The two currents are now flowing just outside a perfectly conducting body. We know that an electric current radiating just outside a perfectly conducting body produces a null field (see Example 3.1). Also, the tangential components of $\mathbf{E}$ are zero on the conductor (just behind $\mathbf{M}_e$), and equal to the original field components just in front of $\mathbf{M}_e$. Thus

$$\begin{aligned} \mathbf{J}_e &= 0 \\ \mathbf{M}_e &= -\widehat{n} \times \mathbf{E} \end{aligned} \tag{3.65}$$

Of course, the current $\mathbf{M}_e$ now radiates in the *presence* of the *conductor* and *not* in unbounded space.

[3] Note that the zero internal field does not *necessitate* the presence of a conducting medium inside S. In fact, we may consider any medium of our choice in the region inside as long as the internal zero field and the equivalent surface currents are not disturbed.

It is noted that the presence of the conductor does not imply that $\widehat{n} \times \mathbf{E}$ and hence $\mathbf{M}_e$ is zero, because the tangential electric field is not identically zero over the surface S.

3.5.2.4 Third Equivalence

We now choose S to be the boundary of a perfect *magnetic* conductor (or Ferrite) as in Fig. 3.12. Then, from reciprocity, $\mathbf{M}_e$ is short-circuited and

$$\begin{aligned} \mathbf{J}_e &= \widehat{n} \times \mathbf{H} \\ \mathbf{M}_e &= 0 \end{aligned} \tag{3.66}$$

with (of course) $\mathbf{J}_e$ radiating in the presence of the magnetic body.

As regards the calculation of the fields outside S, we now have three types of equivalent surface currents.

Example 3.5 (*Fields in a Half Space*) We seek the field in a half space $z \geq 0$ arising from sources lying wholly in $z < 0$ (Fig. 3.13).

By choosing our surface S to be the infinite plane $z = 0$, we are led to a particularly simple representation of the field by combining the second equivalence principle with image theory (Fig. 3.14).

As regards the fields in $z \geq 0$, applying the second equivalence and imaging leads to (c). Hence, in $z \geq 0$ the field can be attributed to the Hertz vector

$$\pi_e = 0 \tag{3.67}$$

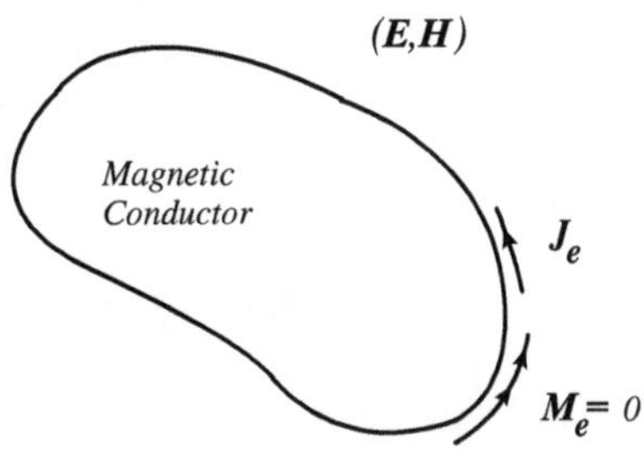

Fig. 3.12 Third equivalence

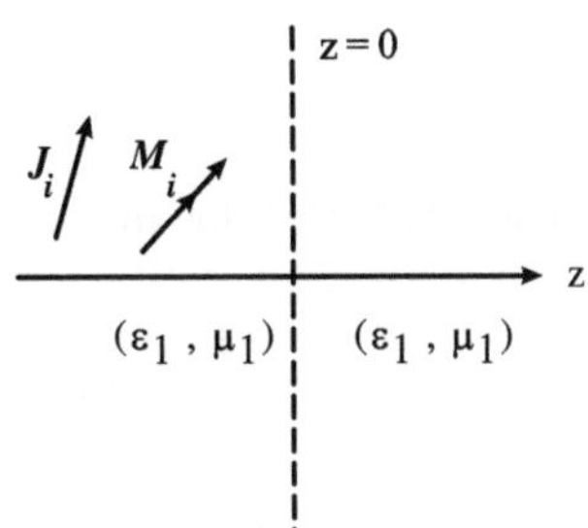

Fig. 3.13 Radiation of sources lying in the region $z < 0$ to the region $z \geq 0$

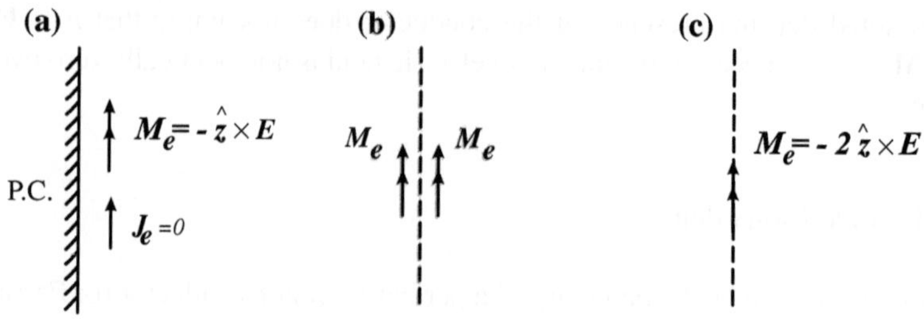

Fig. 3.14 Application of the second equivalence and image theory to the plane at $z = 0$: **a** perfect magnetic conductor across the magnetic current element, **b** replacement of conductor with image, and **c** the current element is placed on the surface

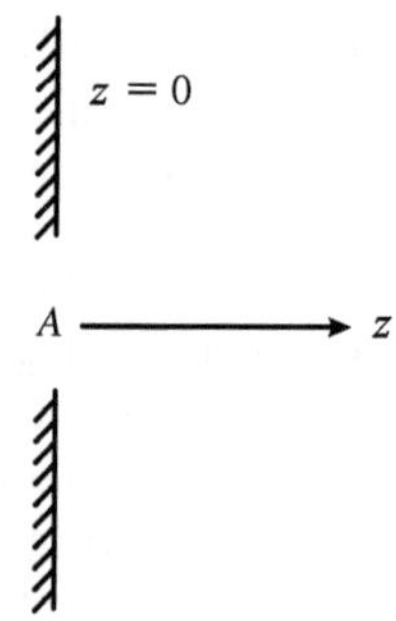

Fig. 3.15 Radiation from an aperture in an infinite ground plane

$$\pi_m = -j\frac{Y}{k}\int_S 2\mathbf{M}(\mathbf{r}')G(\mathbf{r};\mathbf{r}')ds'$$

$$= \frac{jY}{2\pi k}\int_S (\widehat{z} \times \mathbf{E}|_{z=0})\frac{e^{-jk|\mathbf{r}-\mathbf{r}'|}}{|\mathbf{r}-\mathbf{r}'|}dx'dy' \tag{3.68}$$

in terms of which

$$\mathbf{E} = -jkZ\nabla \times \pi_m, \quad \mathbf{H} = \nabla \times \nabla \times \pi_m \tag{3.69}$$

Now consider an aperture A in an otherwise perfectly conducting plane screen $z = 0$ as shown in Fig. 3.15. On the metal we have the condition

$$\widehat{z} \times \mathbf{E} = 0$$

Hence, as regards the field in $z \geq 0$

$$\mathbf{E} = \frac{1}{2\pi}\nabla \times \int_A (\widehat{z} \times \mathbf{E}|_{z=0})\frac{e^{-jk|\mathbf{r}-\mathbf{r}'|}}{|\mathbf{r}-\mathbf{r}'|}dx'dy' \tag{3.70}$$

which is fully determined by a knowledge of $E_{\tan}$ in the aperture.

This is the basis for all transmission problems in optics (where it can generally be assumed that E_{tan} is known in A), and is the vector equivalent of the Rayleigh-Sommerfeld diffraction formula in aperture antennas. It is also related to the angular spectrum representation of radiated fields due to planar sources. □

3.5.2.5 Induction Theorem

The induction theorem (or the induction equivalence) is closely related to Love's equivalence principle. It is, however, used for scattering problems rather than for aperture antenna problems.

Assume that the impressed sources $(\mathbf{J}^i, \mathbf{M}^i)$ are radiating in the presence of an obstacle occupying a region surrounded by the closed surface S (Fig. 3.16). In the presence of the obstacle, the field is $(\mathbf{E}, \mathbf{H})$ and in its absence, it is $(\mathbf{E}^i, \mathbf{H}^i)$ which is assumed known everywhere. The presence of the obstacle has resulted in a difference or disturbance field, referred to as the scattered field

$$\mathbf{E}^s = \mathbf{E} - \mathbf{E}^i \quad \mathbf{H}^s = \mathbf{H} - \mathbf{H}^i \tag{3.71}$$

which can be attributed to the induced conduction and polarization currents on the obstacles.

In order to find the scattered field, we construct the following equivalent problem. We retain the obstacle and define the equivalent surface currents

$$\begin{aligned} \mathbf{J}_e &= \widehat{n} \times (\mathbf{H}^s - \mathbf{H}) \\ \mathbf{M}_e &= -\widehat{n} \times (\mathbf{E}^s - \mathbf{E}) \end{aligned} \tag{3.72}$$

over the surface S. These produce the scattered field $(\mathbf{E}^s, \mathbf{H}^s)$ outside and the original field $(\mathbf{E}, \mathbf{H})$ inside S. Substituting from (3.71), we have

$$\begin{aligned} \mathbf{J}_e &= -\mathbf{n} \times \mathbf{H}^i \\ \mathbf{M}_e &= \mathbf{n} \times \mathbf{E}^i \end{aligned} \tag{3.73}$$

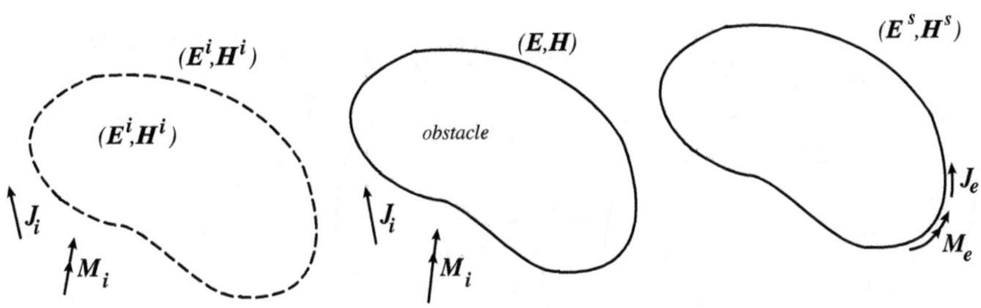

Fig. 3.16 The induction theorem

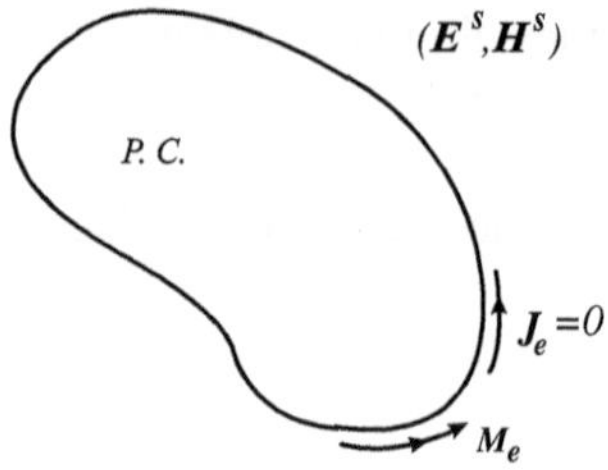

Fig. 3.17 Application of the induction theorem to a perfectly conducting obstacle

radiating in the presence of the obstacle.

If the obstacle is perfectly conducting, the electric current is short circuited and we are left with only the equivalent magnetic current radiating in the presence of the obstacle (Fig. 3.17). Thus,

$$\begin{aligned} \mathbf{J}_e &= 0 \\ \mathbf{M}_e &= \hat{n} \times \mathbf{E}^i \end{aligned} \tag{3.74}$$

3.5.2.6 Physical Equivalence

Physical equivalence is an alternative method for the treatment of scattering from conducting bodies. For scattering problems, the sources are positioned outside S. Physical equivalence is often used to construct integral equations which can be solved analytically or numerically.

Again, referring to Fig. 3.18, assume that the impressed sources ($\mathbf{J}^i$, $\mathbf{M}^i$) are radiating in the presence of a perfect conductor. In the presence of the obstacle, the field is ($\mathbf{E}$, $\mathbf{H}$) in the external region and in its absence, it is ($\mathbf{E}^i$, $\mathbf{H}^i$) everywhere.

In order to find the scattered field, we now construct the following equivalent problem. We retain the obstacle and express the standard boundary conditions for the tangential fields over the conducting surface in terms of *induced* (or physical) surface currents. We have

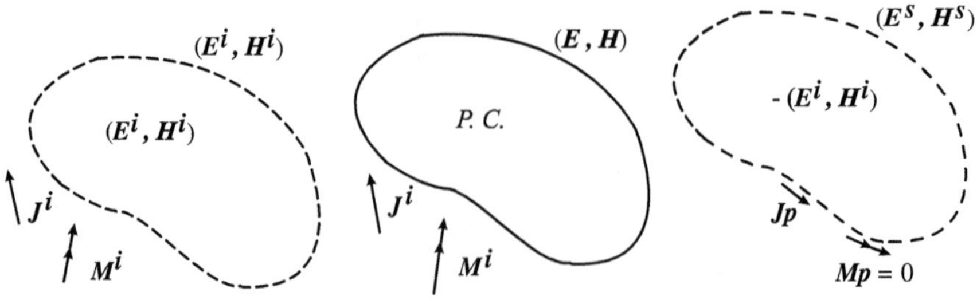

Fig. 3.18 The physical equivalence

$$\begin{aligned}\mathbf{J}_p &= \widehat{n} \times (\mathbf{H} - \mathbf{H}^{int}) \\ \mathbf{M}_p &= -\widehat{n} \times (\mathbf{E} - \mathbf{E}^{int})\end{aligned} \tag{3.75}$$

But the internal field ($\mathbf{E}^{int}$, $\mathbf{H}^{int}$) is zero for a conducting body and we have

$$\begin{aligned}\mathbf{J}_p &= \widehat{n} \times \mathbf{H} = \widehat{n} \times (\mathbf{H}^s + \mathbf{H}^i) \\ \mathbf{M}_p &= -\widehat{n} \times \mathbf{E} = 0\end{aligned} \tag{3.76}$$

where we substituted for the total magnetic field $\mathbf{H} = \mathbf{H}^s + \mathbf{H}^i$ and used the fact that the tangential electric field is zero on the outer surface of a conductor. The second equation above implies that

$$\widehat{n} \times \mathbf{E}^s = -\widehat{n} \times \mathbf{E}^i \tag{3.77}$$

which means that the tangential component of the scattered electric field is known on S.

The equivalent problem is now clear: with the obstacle removed, the physical current $J_p = \widehat{n} \times \mathbf{H}$ radiates in the unbounded space and produces the scattered field ($\mathbf{E}^s$, $\mathbf{H}^s$) outside and negative of the incident field ($-\mathbf{E}^i$, $-\mathbf{H}^i$) inside S.

The advantage of the induction equivalence over the physical equivalence is that the equivalent current produced by the induction principle (3.74) is in terms of a known quantity while that introduced by the physical equivalence (3.76) is expressed in terms of unknown fields. However, the equivalent current $\mathbf{M}_e$ in (3.74) radiates in the presence of the obstacle. Thus, in this case, solving for the fields with a knowledge of the surface current constitutes a boundary value problem. On the other hand, the equivalent current $\mathbf{J}_p$ in (3.76) radiates in the unbounded medium and is thus amenable to free space Green's function. Once the surface current is obtained (either by solving an exact integral equation or by a suitable approximation), the field quantities may be obtained directly.

3.5.2.7 Physical Optics Equivalence

If the conducting scatterer is an infinite, flat, perfect electric conductor (infinite ground plane), then the equivalent problem is as shown in Fig. 3.19.

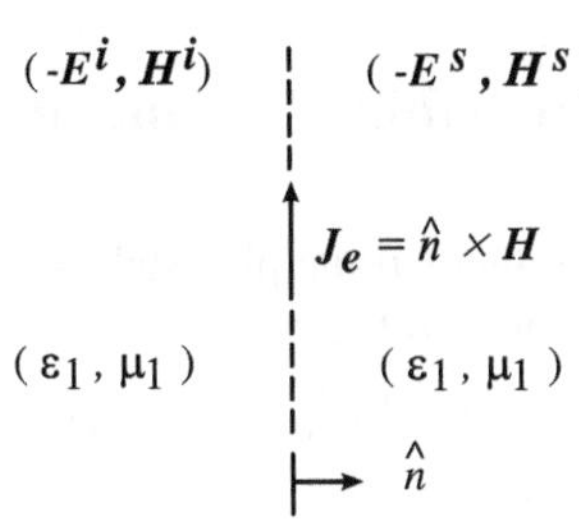

Fig. 3.19 Radiation from an aperture in an infinite ground plane

Since the perfectly conducting plane is infinitely long, the scattered fields are actually the reflected waves. On the surface of the perfect conductor, the tangential electric field is zero, because the reflection coefficient for the electric field is $R = -1$. The reflection coefficient for the magnetic field is unity, and thus

$$\hat{n} \times \mathbf{H}^i = \hat{n} \times \mathbf{H}^s \tag{3.78}$$

Therefore, the equivalent surface electric current is given by

$$\mathbf{J}_{p0} = 2\hat{n} \times \mathbf{H}^i \tag{3.79}$$

which is a known quantity.

In practice, if the extent of S is not infinite, we may still use this equivalence to approximate the fields. Such an approximation is referred to as the *physical optics* approximation. Physical optics approximation is valid when the size and radii of curvature of the conducting body is large compared to the operating wavelength. Radiation from large reflector antennas as well as scattering from large conducting targets can be analyzed using the physical optics approximation. This will be explored in Chap. 7.

3.5.2.8 Equivalent Surface Charges

Based on the surface equivalence principle, the equivalent surface currents are specified by the tangential electric and magnetic fields. Using the continuity relation or the conservation law for the surface charge, we may define equivalent surface charges as follows

$$\begin{aligned} \nabla_s \cdot \mathbf{J}_e &= -j\omega\rho_e \\ \nabla_s \cdot \mathbf{M}_e &= -j\omega m_e \end{aligned} \tag{3.80}$$

Or, equivalently

$$\begin{aligned} \rho_e &= \epsilon\hat{n} \cdot \mathbf{E} \\ m_e &= \mu\hat{n} \cdot \mathbf{H} \end{aligned} \tag{3.81}$$

3.6 Babinet's Principle

Babinet's principle establishes the identity of the fields in two different but closely related problems:

1. the field diffracted by a thin perfectly conducting plate of arbitrary shape $(\mathbf{E}_1, \mathbf{H}_1)$, and

2. the field transmitted through an aperture of the same location and shape as the plate in a perfectly conducting plane screen ($\mathbf{E}_2, \mathbf{H}_2$).

The principle is named after Babinet[4] who compared the fields in the two cases and observed experimentally that

$$\mathbf{E}_1 + \mathbf{E}_2 = \mathbf{E}^i \tag{3.82}$$

The two problems are called *complimentary*.

The result was published in 1837 and certainly seems reasonable (think of the limiting case as either the aperture or screen shrinks to zero).

The principle is trivial in optics: The field at any point behind a plate having a screen, if added to the field at the same point when the complimentary screen is substituted, is equal to the field at the point when no screen is present. This is shown in Fig. 3.20. The source may be a point or a distribution of sources.

But, as shown by H.G. Booker in 1946, the result is mathematically precise only if the fields in the two problems are the duals of one another. This is due to the vector nature of the electromagnetic field.

If one screen is perfectly conducting, the complementary screen must have infinite permeability. Since no infinitely permeable material exists, the equivalent effect may be obtained by making both the original and complementary screens of perfectly conducting material and interchanging electric and magnetic quantities everywhere. In other words, Babinet's principle for electromagnetic fields are mathematically precise only if the fields in the two complementary problems are duals of one another.

A simple proof based on the half-space formula developed from the second equivalence follows.

Problem 3.1 (*plate*) Consider a perfectly conducting plate illuminated by a plane wave as shown in Fig. 3.21a. The plate can only support an electric current $\mathbf{J}_e$; and $\mathbf{M}_e = 0$. We can, therefore, construct the field from the electric Hertz vector alone. The scattered magnetic field is given by

$$\mathbf{H}_1^s = jkY\nabla \times \pi_e \tag{3.83}$$

where

$$\pi_e = -\frac{jZ}{k} \int\!\!\int_A \mathbf{J}_e(\mathbf{r}')G(\mathbf{r};\mathbf{r}')ds' \tag{3.84}$$

$$\mathbf{J}_e = \hat{z} \times \mathbf{H}|_{z=0^-}^{z=0^+} = 2\hat{z} \times \mathbf{H}_1|_{z=0^+} \tag{3.85}$$

[4] Jacques Babinet (1794–1872), French physicist and astronomer.

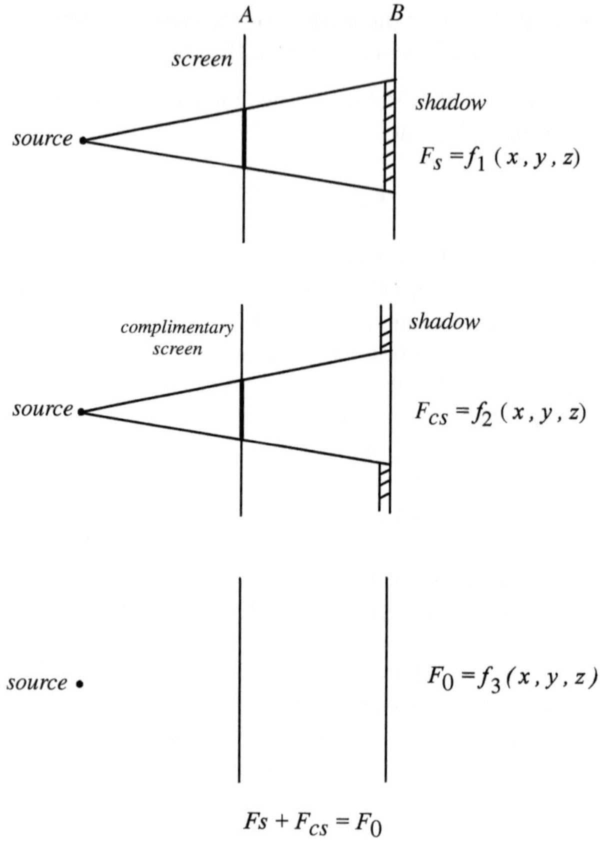

Fig. 3.20 Babinet's principle for optical sources: radiation in the presence of a conducting screen and the complimentary slot

Thus

$$\mathbf{H}_1^s = \nabla \times \iint_A (2\hat{z} \times \mathbf{H}_1|_{z=0^+}) G(\mathbf{r}; \mathbf{r}') ds' \tag{3.86}$$

and the total field is

$$\mathbf{H}_1 = \mathbf{H}_1^i + \mathbf{H}_1^s \tag{3.87}$$

Problem 3.2 (*aperture*) Now consider the complimentary aperture illuminated by the plane wave as depicted in Fig. 3.21b. For this problem we have, from the second equivalence principle,

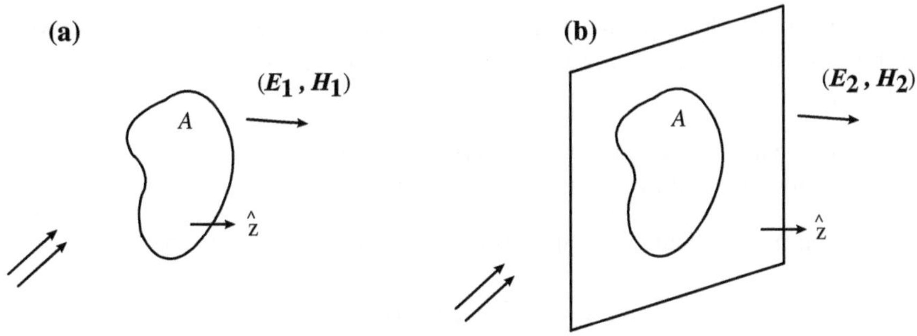

Fig. 3.21 Babinet's principle for electromagnetic fields. **a** Problem I. **b** Problem II

$$\mathbf{E}_2 = \nabla \times \iint_A (2\widehat{z} \times \mathbf{E}_2|_{z=0^+}) G(\mathbf{r}; \mathbf{r}') ds' \tag{3.88}$$

From this relation, we can construct an integral equation for $\widehat{z} \times \mathbf{E}_2$ in A involving only $\mathbf{E}_2^i$—the incident field in the $z \leq 0$ region. From the tangential electric field $\widehat{z} \times \mathbf{E}$ over the aperture, the field $\mathbf{E}_2$ can be found. This can be accomplished by enforcing the continuity of tangential electric field over the aperture.

Comparing the expressions for $\mathbf{H}_1^s$ and $\mathbf{E}_2$, it is clear that the integral equation for $\widehat{z} \times \mathbf{H}_1$ will be the same as that for $\widehat{z} \times \mathbf{E}_2$ in the second problem. Moreover, if $\mathbf{H}_1^i = \mathbf{E}_2^i$, then

$$\mathbf{H}_1^s = \mathbf{E}_2 \tag{3.89}$$

In other words, if the incident fields are duals of one another, then the Babinet's principle ensures that the resulting fields in the two problems are also duals. We also have that

$$\mathbf{H}_1^i = \mathbf{H}_1 - \mathbf{E}_2 = \mathbf{E}_2^i \tag{3.90}$$

By means of Babinet's principle many of the problems of slot antennas can be reduced to situations involving complementary linear antennas for which solutions have already been obtained.

Exercises

3.1: Find the far field radiated field of a small Hertzian dipole vertically located at a distance h above an infinite ground plane. Plot the radiation pattern for $h = 0.46\lambda$ and $h = 2\lambda$.

3.2: A wire of length ℓ is parallel to a perfectly conducting plane a distance h above it and bears a constant current I. At a large distance r in the direction θ $(0 \le \theta \le \pi/2)$ in the far field

(a) derive the appropriate approximation(s) to $|\mathbf{r} - \mathbf{r}'|$.
(b) determine the resulting electric Hertz vector.

3.3: The current moment of an infinitesimal current source is given by

$$\mathbf{p} = p_1\widehat{x} + p_2\widehat{y} + p_3\widehat{z}$$

which is placed at (x', y', z') above a perfectly conducting plane located at $z = 0$. Find the expression for the magnetic field at (x, y, z), and the surface current density on the conducting plane.

3.4: Find the far field in the xy plane of a Hertzian monopole of current moment $\mathbf{p} = I\ell\widehat{y}$ vertically located at the point $(d, 0, 0)$ inside a conducting corner whose vertex coincides with the origin. Express your results in terms of (r, ϕ) variables. (note that in the far field, $r \gg d, \ell$ and $kr \gg 1$).

3.5: Apply the Rayleigh-Carson reciprocity theorem to find the relation between V_1, I_3 and V_3' and I_1'. Verify your result by means of circuit theory.

3.6: The receiving pattern of an antenna is defined as the voltage at the antenna terminals due to a plane wave incident upon the antenna. Using reciprocity, prove that the receiving pattern of any antenna constructed of linear isotropic material is identical to its transmitting pattern.

3.7: Two potential functions V_1 and V_2 due to two separate charge distributions ρ_1 and ρ_2 satisfy the Poisson's equations

$$\nabla^2 V_1 = -\rho_1/\epsilon_0, \quad \nabla^2 V_2 = -\rho_2/\epsilon_0$$

respectively. Develop a reciprocity theorem in free space for V_1, V_2, ρ_1 and ρ_2 similar to the Lorentz's reciprocity theorem.

3.8: Derive the Lorentz and Rayleigh-Carson reciprocity theorems for the case where both electric and magnetic currents are present.

3.9: From Rayleigh-Carson reciprocity theorem, show that if V is bounded by a perfectly conducting surface S and in which two linear current elements $\mathbf{K}_1$ and $\mathbf{K}_2$ exist, then

$$\mathbf{E}_1 \cdot \mathbf{K}_2 = \mathbf{E}_2 \cdot \mathbf{K}_1$$

In the above, $\mathbf{E}_1$ and $\mathbf{E}_2$ are generated by $\mathbf{K}_1$ and $\mathbf{K}_2$, respectively.

3.10: Let $(\mathbf{E}, \mathbf{H})$ be a source-free solution to Maxwell's equations in all of space satisfying radiation condition. Let $(\mathbf{E}', \mathbf{H}')$ be a solution to Maxwell's equations due to an arbitrary current element $\mathbf{J}$ satisfying the radiation condition. By showing that $\mathbf{J} \cdot \mathbf{E} = 0$ for all space, show that the source-free solution $(\mathbf{E}, \mathbf{H})$ must be a null field.

3.11: A resistive sheet is characterized by R Ohms/square. State the boundary conditions for the electromagnetic field vectors, **E**, **H**, **D** and **B** across the sheet. The medium surrounding the sheet is air.

3.12: A point charge q is located at the origin. Find the equivalent surface charges which should be placed at the spherical surface $r = a$, such that the same field $\mathbf{E}$ is maintained in the region $r > a$ with the original point charge q removed.

3.13: A point charge q is located at the origin. Let us postulate the existence of a uniform field $\mathbf{E}_1 = \widehat{x} E_0$ in the region $r < a$. Find the equivalent sources which should be placed at the spherical surface $r = a$ with the original point charge q removed, if the field for $(r > a)$ is to remain unchanged.

3.14: A plane wave is normally incident on a circular hole of radius a on a large conducting screen. Approximating the aperture field with the incident field at the aperture, find the diffracted electric field intensity on the other side far from the screen.

[Hint: Use the integral representation of the Bessel function

$$\int_0^{2\pi} e^{jx\cos(\phi-\phi')} d\phi' = 2\pi J_0(x)$$

and the identity $\int x J_0(x) dx = x J_1(x)$.]

Chapter 4
Wave Harmonics and Guided Waves

Solutions of the homogeneous Helmholtz equation are referred to as wave harmonics. In this chapter, we will discuss the solution of this equation in rectangular, cylindrical and spherical coordinate systems. They are respectively called plane waves, cylindrical waves and spherical waves. We will also introduce sources responsible for generating such waves. We will also discuss wave polarization, wave velocities and plane wave angular spectral representation.

After introducing wave harmonics, we will discuss the planar and cylindrical waveguides.

4.1 Plane Waves

Plane waves are solutions to the source free Maxwell's equations which depend on t and *only* upon distance ζ measured along a single direction in space. This implies that at each instant, **E** and **H** are constant in magnitude and direction over a plane perpendicular to ζ direction.

Consider the Maxwell's equations

$$\nabla \times \mathbf{E} = -\mu \frac{\partial \mathbf{H}}{\partial t} \tag{4.1}$$

$$\nabla \times \mathbf{H} = \sigma \mathbf{E} + \epsilon \frac{\partial \mathbf{E}}{\partial t} \tag{4.2}$$

$$\nabla \cdot \mathbf{E} = 0 \tag{4.3}$$

$$\nabla \cdot \mathbf{H} = 0 \tag{4.4}$$

and define the plane as $\mathbf{r} \cdot \widehat{\zeta} = \zeta$, where $\mathbf{r}$ is any point on the plane, ζ is the distance of the plane from the origin, and $\widehat{\zeta}$ is the unit vector in the increasing ζ-direction. Then, noting that

© Springer International Publishing Switzerland 2015

K. Barkeshli, *Advanced Electromagnetics and Scattering Theory*,
DOI 10.1007/978-3-319-11547-4_4

$$\nabla = \widehat{\zeta}\frac{\partial}{\partial \zeta} \tag{4.5}$$

we have

$$\widehat{\zeta} \times \frac{\partial \mathbf{E}}{\partial \zeta} = -\mu \frac{\partial \mathbf{H}}{\partial t} \tag{4.6}$$

$$\widehat{\zeta} \times \frac{\partial \mathbf{H}}{\partial \zeta} = \sigma \mathbf{E} + \epsilon \frac{\partial \mathbf{E}}{\partial t} \tag{4.7}$$

$$\widehat{\zeta} \cdot \frac{\partial \mathbf{E}}{\partial \zeta} = 0 \tag{4.8}$$

$$\widehat{\zeta} \cdot \frac{\partial \mathbf{H}}{\partial \zeta} = 0 \tag{4.9}$$

Now take $\mathbf{H} = \mathbf{H}(\zeta, t)$ and consider the total differential

$$d\mathbf{H} = \frac{\partial \mathbf{H}}{\partial \zeta} d\zeta + \frac{\partial \mathbf{H}}{\partial t} dt \tag{4.10}$$

We have from (4.6) and (4.9) and above

$$\widehat{\zeta} \cdot d\mathbf{H} = \widehat{\zeta} \cdot [\frac{\partial \mathbf{H}}{\partial \zeta} d\zeta + \frac{\partial \mathbf{H}}{\partial t} dt] \equiv 0 \tag{4.11}$$

implying $dH_\zeta = 0$. This means that H_ζ cannot vary with ζ or t. If we discard the static term, we come to the conclusion that

$$\widehat{\zeta} \cdot \mathbf{H} = 0 \tag{4.12}$$

and, similarly

$$\widehat{\zeta} \cdot \mathbf{E} = 0 \tag{4.13}$$

Thus, **E**, **H** and ζ form a right-handed orthogonal system.

4.1.1 Planar Harmonics

Planar harmonics are the solutions to the homogeneous Helmholtz wave equation

$$\nabla^2 \psi + k^2 \psi = 0 \tag{4.14}$$

in rectangular coordinate system. We shall outline the method of solving this equation by the method of separation of variables (Fig. 4.1).

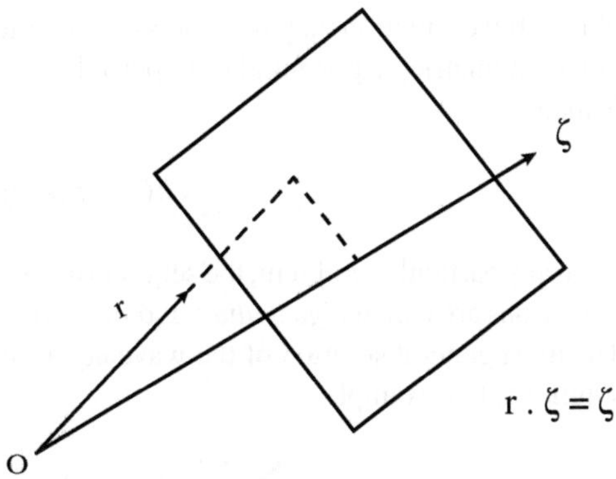

Fig. 4.1 A plane wave with direction ζ

In rectangular coordinates, the operator

$$\nabla^2 = \frac{\partial^2}{\partial x^2} + \frac{\partial^2}{\partial y^2} + \frac{\partial^2}{\partial y^2}$$

applies to any cartesian component ψ of **E** or **H**. Writing

$$\psi = X(x)Y(y)Z(z) \tag{4.15}$$

and substituting in the wave equation, we have after division by ψ

$$\frac{1}{X}\frac{d^2X}{dx^2} + \frac{1}{Y}\frac{d^2Y}{dy^2} + \frac{1}{Z}\frac{d^2Z}{dz^2} + k^2 = 0 \tag{4.16}$$

Since the functions are independent from each other, each of the three terms on the left should be constant. Thus

$$-k_x^2 - k_y^2 - k_z^2 + k^2 = 0$$

where k_x, k_y and k_z are the separation constants. They are related by

$$k^2 = k_x^2 + k_y^2 + k_z^2 \tag{4.17}$$

which is sometimes referred to as the *consistency relation*. Clearly, only two of the constants are independent and the third is obtained from the above relation. Therefore, we have

$$\frac{d^2X}{dx^2} + k_x^2 X = 0 \tag{4.18}$$

and the other two functions Y and Z also satisfy similar equations. Since the separation constants can in general take real, imaginary or complex values, the solution $h(k_x x)$

of the above equation may be expressed in various alternative forms, that is, in terms of trigonometric, exponential or hyperbolic functions. An elementary wave function is then

$$\psi = h(k_x x)h(k_y y)h(k_z z) \tag{4.19}$$

In any particular problem, the separation constants required to satisfy the boundary conditions are called *eigenvalues* and the corresponding ψ are called *eigenfunctions*. The most general solution of the wave equation is a linear combination of the eigen solutions. For example

$$\psi = \sum\sum B(k_x, k_y)h(k_x x)h(k_y y)h(k_z z) \tag{4.20}$$

where $B(k_x, k_y)$ are constants. The choice of the two separation constants to sum over depends on the problem under consideration. The above is a Fourier representation of ψ and would be appropriate when k_x and k_y take only discrete values (as in a waveguide or cavity problem). The eigenvalues then have a discrete spectrum. In other cases, the spectrum is continuous and

$$\psi = \int_C \int_C f(k_x, k_y)h(k_x x)h(k_y y)h(k_z z)dk_x dk_y \tag{4.21}$$

where f is a suitable weighting function and the integrations are over some designated path in the complex k_x and k_y planes.

Consider now the elementary wave function

$$\begin{aligned}\psi &= e^{-jk_x x}e^{-jk_y y}e^{-jk_z z}\\ &= e^{-j\mathbf{k}\cdot\mathbf{r}}\end{aligned} \tag{4.22}$$

where

$$\mathbf{k} = k_x\widehat{x} + k_y\widehat{y} + k_z\widehat{z}, \quad \mathbf{r} = x\widehat{x} + y\widehat{y} + z\widehat{z} \tag{4.23}$$

which can also be written as

$$\psi = e^{-jk\widehat{k}\cdot\mathbf{r}} \tag{4.24}$$

where k is the propagation constant. A valid electromagnetic field is then

$$\mathbf{E} = \mathbf{E}_0\psi \tag{4.25}$$

where $\mathbf{E}_0$ is a constant vector with $\mathbf{E}_0 \cdot \widehat{k} = 0$, that is, $\mathbf{E}_0$ is perpendicular to the direction of propagation, as demanded by the divergence condition $\nabla \cdot \mathbf{E} = 0$, and from (4.5) and (4.6)

$$\mathbf{H} = Y\widehat{k} \times \mathbf{E}_0\psi \tag{4.26}$$

This is a transverse electromagnetic (TEM) wave and is called a plane wave.

4.1.2 The Sheet Current Source

Consider a planar electric current sheet in the xy-plane. We assume linear polarization in the x-direction and no variations in either x or y. We have

$$\mathbf{J}(z) = \widehat{x}J_{so}\delta(z) \tag{4.27}$$

Since the above current source varies only with z, and since there are no scattering objects present, we conclude that $\partial/\partial x = \partial/\partial y = 0$. Thus, from Maxwell's equations,

$$dE_x/dz = -j\omega\mu H_y \tag{4.28}$$

$$-dH_y/dz = J_{so}\delta(z) + j\omega\epsilon E_x \tag{4.29}$$

$$dE_y/dz = j\omega\mu H_x \tag{4.30}$$

$$dH_x/dz = j\omega\epsilon E_y \tag{4.31}$$

The second system of equations is source-free throughout all space and independent from the first set. Thus $E_y = H_x = 0$. The problem is therefore completely described by the first set together with appropriate conditions as $z \to \pm\infty$. Differentiating the first equation, we obtain

$$(\frac{d^2}{dz^2} + k^2)E_x = j\omega\mu J_{s0}\delta(z) \tag{4.32}$$

$$H_y = -\frac{1}{j\omega\mu}dE_x/dz \tag{4.33}$$

where

$$k = k_d\sqrt{1 - j\tan\delta} \tag{4.34}$$

and

$$\tan\delta = \frac{\sigma}{\omega\epsilon_d}, \quad k_d = \omega\sqrt{\mu\epsilon_d} \tag{4.35}$$

for limiting conditions, we demand that for $k \in C$

$$\lim_{z\to\pm\infty} E_x = 0 \tag{4.36}$$

To solve the second order ordinary differential equation, let

$$E_x = -j\omega\mu J_{s0} g \tag{4.37}$$

Therefore

$$\frac{d^2 g}{dz^2} + k^2 g = -\delta(z) \tag{4.38}$$

with

$$\lim_{z\to\pm\infty} g = 0 \tag{4.39}$$

This is a Green's function problem associated with the Strum-Liouville problem of third kind. The solution is given by

$$g(z,0) = \frac{e^{-jk|z|}}{2jk}, \quad \Im m k < 0 \tag{4.40}$$

or, equivalently

$$E_x(z) = -\frac{\omega\mu}{2k} J_{s0} e^{-jk|z|} \tag{4.41}$$

Letting

$$J_{s0} = -\frac{2k}{\omega\mu}$$

so that the normalized electric field is given by

$$E_x(z) = e^{-jk|z|} \tag{4.42}$$

and the accompanying magnetic field

$$\begin{aligned} H_y(z) &= \frac{1}{Z} e^{-jkz}, \quad z > 0 \\ &= -\frac{1}{Z} e^{jkz}, \quad z < 0 \end{aligned} \tag{4.43}$$

where the intrinsic impedance Z is given by $Z = \omega\mu/k$. Note that

$$\widehat{n} \times (\mathbf{H}_1 - \mathbf{H}_2) = \mathbf{J}_s \tag{4.44}$$

as required.

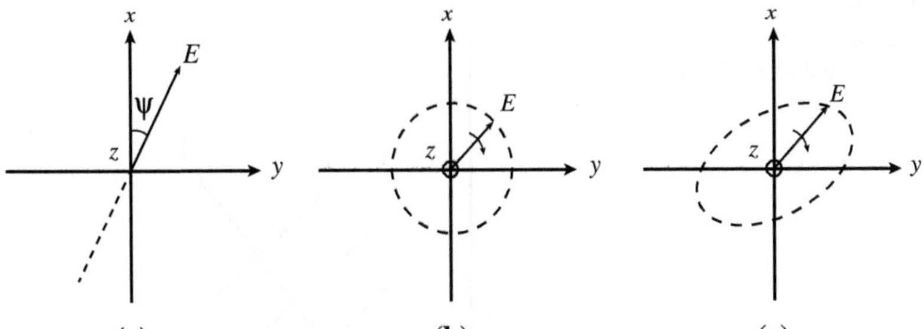

Fig. 4.2 Wave polarizations: **a** linear, **b** circular, and **c** elliptical

4.1.3 Wave Polarization

The polarization of a plane wave is determined by the locus of the "tip" of the electric vector as a function of time. We know that the electric vector must lie in a plane perpendicular to the direction of propagation, but there is no requirement that it maintain a fixed direction in that plane as a function of time (Fig. 4.2).

Consider a plane wave propagating in the z direction. Then **E** must lie in the xy plane. Let

$$\mathbf{E} = (A_x e^{j\delta_x}\widehat{x} + A_y e^{j\delta_y}\widehat{y})e^{-jkz} \tag{4.45}$$

where A_x, A_y, δ_x and δ_y are real. They represent the magnitude and phase of the electric field components in the transverse plane. That is

$$E_x(z,t) = A_x \cos(\omega t - kz + \delta_x) \tag{4.46}$$

$$E_y(z,t) = A_y \cos(\omega t - kz + \delta_y) \tag{4.47}$$

The wave polarization is determined by the relative size of A_x, A_y, δ_x and δ_y.

4.1.3.1 Linear Polarization

If $\delta_x = \delta_y$, then

$$\frac{E_y(z,t)}{E_x(z,t)} = \frac{A_y}{A_x} \tag{4.48}$$

The net electric vector traces a line and remains fixed in direction at an angle $\psi = \tan^{-1}\frac{A_y}{A_x}$ to the x direction (Fig. 4.3).

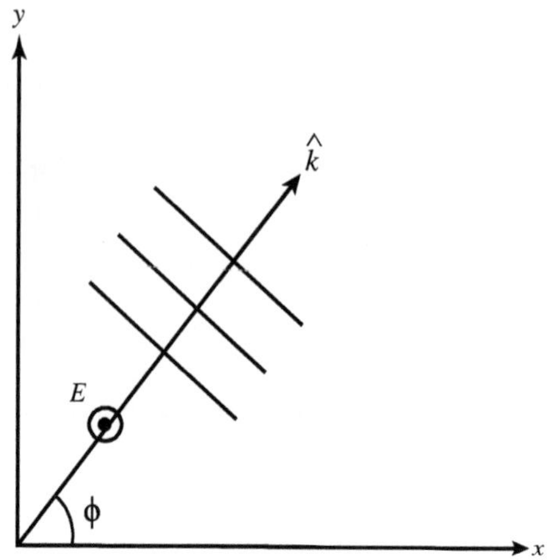

Fig. 4.3 Illustration of propagation direction, wavevector, and phase fronts

4.1.3.2 Circular Polarization

If $\delta_x - \delta_y = \delta = \pm\pi/2$ and $A_x = A_y = A$, then

$$\{E_x(z,t)\}^2 + \{E_y(z,t)\}^2 = A^2 \tag{4.49}$$

The net electric vector traces a circle of radius A in the xy plane. The sense of rotation is right-handed if the direction of rotation is clockwise for an observer looking in the direction of propagation. In this case, the x component of the electric vector leads the y component in phase ($\delta = \pi/2$). On the other hand, the sense of rotation is left-handed if the direction of rotation is counter-clockwise and the x component of the electric vector lags the y component ($\delta = -\pi/2$). The latter is referred to as *left-handed circularly polarized* plane wave, while the former is a *right-handed circularly polarized* plane wave. This is the standard definition in electrical engineering, but is opposite to that used in classical optics.

4.1.3.3 Elliptical Polarization

If $\delta_x - \delta_y \neq \pm\pi/2$ and/or $A_x \neq A_y$, then

$$\{\frac{E_x(z,t)}{A_x}\}^2 + \{\frac{E_y(z,t)}{A_y}\}^2 - 2\frac{E_x(z,t)}{A_x}\frac{E_y(z,t)}{A_y}\cos\delta = \sin^2\delta \tag{4.50}$$

independent of t. The electric vector traces out an ellipse in the xy plane. The sense of rotation (or handedness) is defined as before.

Specific polarizations can be used to enhance target detection and discrimination against unwanted signals and clutters, and this is important in radar and remote

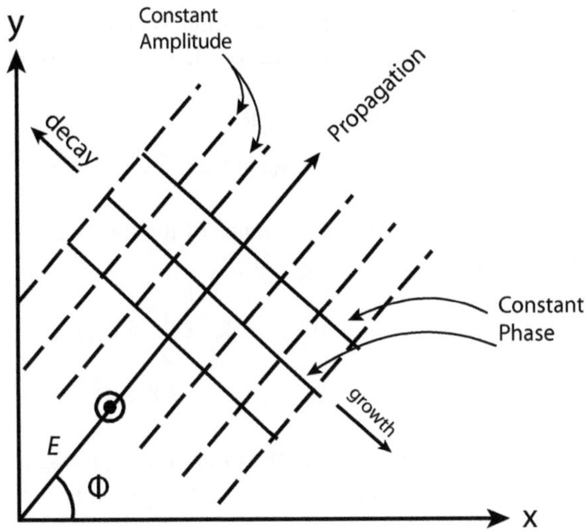

Fig. 4.4 An inhomogeneous plane wave propagating in a lossless medium. The surfaces of constant amplitude are perpendicular to the surfaces of constant phase

sensing. The simplest of all polarizations is linear and since any polarization can be synthesized using two linearly polarized waves with appropriate relative phases and amplitudes, it is sufficient for theoretical purposes to confine attention to linearly polarized plane waves.

4.1.4 Lossless Medium

In a lossless medium, the propagation constant k is real. Assuming that the unit vector $\widehat{k}$ is also real,[1] we observe that the field propagates without any change of form in the direction defined by $\widehat{k}$. The wave amplitude $\mathbf{E}_0$ is constant and the surfaces of constant phase

$$k\widehat{k} \cdot \mathbf{r} = const. \tag{4.51}$$

are planes perpendicular to $\widehat{k}$. Such a field is called a *uniform* or *homogeneous* plane wave.

Example 4.1 A linearly polarized plane wave propagating in the xy plane is given by

$$\begin{aligned}\mathbf{E} &= \widehat{z}E_0 e^{-j(k_x x + k_y y)}\\ &= \widehat{z}E_0 e^{-jk(x\cos\phi + y\sin\phi)}\end{aligned}$$

[1] As we shall see, $\widehat{k}$ does not have to be real.

where ϕ is the angle the propagation vector makes with the x axis. The magnetic field intensity vector lies in the xy-plane and is given by

$$\begin{aligned}\mathbf{H} &= Y\widehat{k} \times \mathbf{E} \\ &= YE_0(\widehat{x}\sin\phi - \widehat{y}\cos\phi)e^{-jk(x\cos\phi + y\sin\phi)}\end{aligned}$$

where Y is the intrinsic admittance of the medium. The complex Poynting vector is obtained from

$$\begin{aligned}\widetilde{\mathbf{S}} &= \frac{1}{2}(\mathbf{E} \times \mathbf{H}^{\star}) \\ &= \widehat{k}Y|E_0|^2/2\end{aligned} \tag{4.52}$$

Clearly, it is a real vector in the direction of propagation. The time-averaged power density flowing in the direction of propagation is $Y|E_0|^2/2$. □

Since all we require is $\widehat{k} \cdot \widehat{k} = 1$, $\widehat{k}$ does not have to be real, even though the medium is lossless. Writing

$$\widehat{k} = \mathbf{k}' - j\mathbf{k}'' \tag{4.53}$$

where $\mathbf{k}'$ and $\mathbf{k}''$ are real vectors, we require that

$$\widehat{k} \cdot \widehat{k} = k'^2 - k''^2 - 2j\mathbf{k}' \cdot \mathbf{k}'' = 1 \tag{4.54}$$

so that either

$$k'' = 0, \quad k' = 1 \tag{4.55}$$

which gives the uniform plane wave, or

$$\mathbf{k}' \cdot \mathbf{k}'' = 0 \tag{4.56}$$

In the latter case, the field is given by

$$\psi = e^{-k\mathbf{k}'' \cdot \mathbf{r}} e^{-jk\mathbf{k}' \cdot \mathbf{r}} \tag{4.57}$$

where the first exponent controls the amplitude, while the second exponent specifies the phase. More specifically, the surfaces of constant phase are given by

$$\mathbf{k}' \cdot \mathbf{r} = const. \tag{4.58}$$

These are planes perpendicular to $\mathbf{k}'$, the direction of propagation. The amplitude is not constant in this case and the surfaces of constant amplitude are

$$\mathbf{k}'' \cdot \mathbf{r} = const. \tag{4.59}$$

which are planes perpendicular to $\mathbf{k}''$ and therefore parallel to $\mathbf{k}'$.

This is an example of *nonuniform* or *inhomogeneous* plane wave (Fig. 4.4). It can also be regarded as a plane wave propagating in a complex direction. Note that in this case, the change in amplitude is not in the direction of propagation.

Example 4.2 Consider the linearly polarized wave we considered in the Example 4.1.

$$\begin{aligned}\mathbf{E} &= \widehat{z}e^{-jk(x\cos\phi + y\sin\phi)}\\ &= \widehat{z}e^{-jk[x(a_1 - ja_2) + y(b_1 - jb_2)]}\\ &= \widehat{z}e^{-k(xa_2 + yb_2)}e^{-jk(xa_1 + yb_1)}\end{aligned}$$

where

$$\begin{aligned}1 &= \cos^2\phi + \sin^2\phi\\ &= (a_1 - ja_2)^2 + (b_1 - jb_2)^2\\ &= a_1^2 + b_1^2 - a_2^2 - b_2^2 - 2j(a_1a_2 + b_1b_2)\end{aligned}$$

with $a_1a_2 + b_1b_2 = 0$. It is seen that the surfaces of constant phase are given by

$$xa_1 + yb_1 = const.$$

and the surfaces of constant amplitude are specified by

$$xa_2 + yb_2 = const.$$

Clearly, surfaces of constant phase and amplitude are perpendicular to each other since the slopes are related by $b_1/a_1 = -a_2/b_2$.

The real and imaginary parts of the complex Poynting vector can be computed as

$$\begin{aligned}\Re\mathrm{e}\widetilde{\mathbf{S}} &= \frac{Y}{2}(a_1\widehat{x} + b_1\widehat{y})e^{-2k(xa_2+yb_2)}\\ \Im\mathrm{m}\widetilde{\mathbf{S}} &= -\frac{Y}{2}(a_2\widehat{x} + b_2\widehat{y})e^{-2k(xa_2+yb_2)}\end{aligned}$$

The first expression is the time-averaged real power flow in the direction of propagation, and the second expression is the time-averaged reactive power flowing perpendicular to the direction of propagation, that is, along the surfaces of constant phase. □

Inhomogeneous plane waves appear in the case of total reflection from a plane dielectric interface, as well as in the analysis of thin film waveguides where it is sometimes referred to as *leaky mode*.

4.1.5 Lossy Medium

For a lossy medium, the permittivity and/or permeability may be complex and hence the wavenumber k is expressed as

$$k = \omega\sqrt{\mu\epsilon} = \omega\sqrt{\mu'\epsilon'(1 - j\tan\delta)} \tag{4.60}$$

where the loss tangent is in general given by

$$\tan\delta = \frac{\epsilon''}{\epsilon'} + \frac{\mu''}{\mu'} + \frac{\sigma}{\omega\epsilon'} \tag{4.61}$$

The wavenumber is, therefore, complex and we have

$$k = \omega\sqrt{\frac{\mu'\epsilon'}{\cos\delta}}e^{-j\delta/2} = k_1 - jk_2 \tag{4.62}$$

In this case, even for propagation in a real direction, $\widehat{k}$, the wave decays in the direction of propagation. Hence

$$\psi = e^{-jk\widehat{k}\cdot\mathbf{r}} = e^{-k_2\widehat{k}\cdot\mathbf{r}}e^{-jk_1\widehat{k}\cdot\mathbf{r}} \tag{4.63}$$

It is customary to classify materials based on their ability to conduct electric currents. Suppose dielectric and magnetic damping losses are negligible and losses are primarily ohmic. Then

$$\tan\delta = \frac{\sigma}{\omega\epsilon'} \tag{4.64}$$

If $\tan\delta \ll 1$, we have a good dielectric and

$$k = \omega\sqrt{\mu'\epsilon'(1 - j\tan\delta)} \simeq \omega\sqrt{\mu'\epsilon'}(1 - j\frac{\tan\delta}{2}) \tag{4.65}$$

Therefore,

$$k_1 = \omega\sqrt{\mu'\epsilon'}, \quad k_2 = \frac{1}{2}\sigma\sqrt{\frac{\mu'}{\epsilon'}} \tag{4.66}$$

On the other hand, if $\tan\delta \gg 1$, the medium is a good conductor and

$$k \simeq \omega\sqrt{-j\mu'\epsilon'\tan\delta} = \frac{1 - j}{\sqrt{2}}\sqrt{\omega\mu'\sigma} \tag{4.67}$$

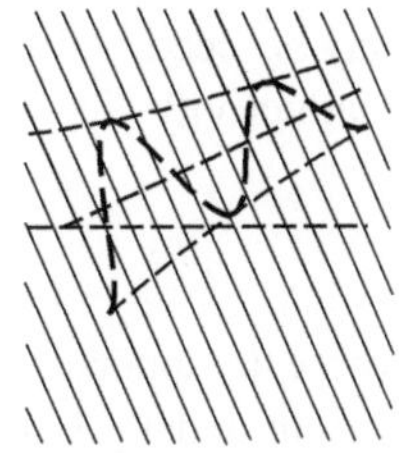

Fig. 4.5 An inhomogeneous plane wave propagating in a lossy medium. The surfaces of constant amplitude are parallel to the surfaces of constant phase

and therefore

$$k_1 \simeq k_2 = \sqrt{\frac{\omega\mu'\sigma}{2}} \tag{4.68}$$

Note that whether a material behaves like a dielectric or a conductor is determined by the operating frequency (assuming all parameters are frequency independent). Thus, a medium like sea water behaves as a good conductor below 100 MHz (and more so as frequency decreases), but looks like a dielectric above 1 GHz (Fig. 4.5).

The distance at which the field amplitude has decreased to $1/e$ of its initial value is known as the *penetration* or *skin depth* and denoted by δ

$$\delta = \sqrt{\frac{2}{\omega\mu'\sigma}} = \sqrt{\frac{1}{\pi f\mu'\sigma}} \tag{4.69}$$

The skin depth decreases with increasing frequencies and increasing conductivities.[2]

4.1.5.1 Impedance Boundary Condition

When a plane wave is incident from the free space upon the surface of a highly conducting half space ($\tan\delta \gg 1$), the angle of refraction θ_t is given by

$$\theta_t \simeq \tan^{-1}\left(\sqrt{\frac{2\omega\epsilon_0}{\sigma}}\sin\theta_i\right) \tag{4.70}$$

where σ is the conductivity of the medium. This is a very small angle indeed and, for all practical purposes, it may be assumed that the refracted wave propagates normally in the conductive medium. In addition, when propagating through an actual conductor, electromagnetic wave amplitude falls off exponentially according to $e^{\frac{-z}{\delta}}$, where z is the distance into the conductor, and δ is the skin depth given by (4.69). Thus, in most cases, the field can be considered as not penetrating into good conductors. However, when it is necessary to account for dissipation losses in conductors, we

[2] Note that the formula (4.69) for penetration depth holds for a good conductor.

may proceed as follows. First, the electromagnetic field is found by assuming perfect conductivity

$$\hat{n} \times \mathbf{E}^{[0]} = 0 \tag{4.71}$$

where the superscript [0] denotes zeroth order approximation for the field. The surface current density **K** is then found from the boundary condition

$$\mathbf{K}^{[0]} = \hat{n} \times \mathbf{H}^{[0]} \tag{4.72}$$

The first order approximation to the tangential component of the electric field is related to this current density by the relationship

$$\hat{n} \times \mathbf{E}^{[1]} = Z_s \hat{n} \times \mathbf{H}^{[0]} \tag{4.73}$$

where Z_s is the surface impedance of the conductor given by

$$Z_s = \frac{1+j}{\sigma\delta} \quad (\Omega) \tag{4.74}$$

and δ is the skin depth (Fig. 4.6).

Example 4.3 The skin depth δ for copper at 1 MHz is

$$\delta = \sqrt{\frac{2}{\omega\mu_0\sigma}}$$

The surface impedance is therefore

$$Z_s = \frac{1+j}{\sigma\delta} = 2.6\times 10^{-4}(1+j)\Omega$$ □

Substituting for **K** in (4.73), we find that

$$-\hat{n} \times \hat{n} \times \mathbf{E} = Z_s \hat{n} \times \mathbf{H} \tag{4.75}$$

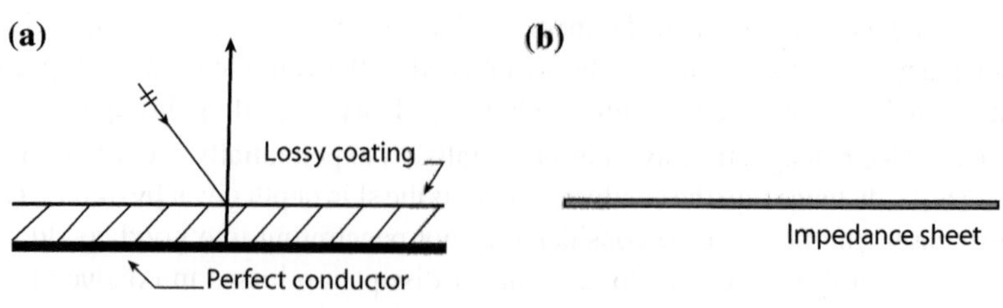

Fig. 4.6 **a** A coated perfectly conducting plate. The coating material has a thickness of τ and is highly lossy. **b** The equivalent impedance sheet at which $-\hat{n} \times \hat{n} \times \mathbf{E} = Z_s \hat{n} \times \mathbf{H}$

This is the *standard impedance boundary condition (SIBC)*, also known as the *Shchukin-Leontovich boundary condition.*[3] The boundary condition gives a relationship between the tangential electric and magnetic fields over the impedance sheet.

For a metal backed thin lossy dielectric coating shown in Fig. 4.6, the surface impedance is given by

$$Z_s = jZ_o \frac{n}{\epsilon_r} \tan(nk_o \tau) \tag{4.76}$$

where τ is the coating thickness and $n = \sqrt{\epsilon_r \mu_r}$ is the index of refraction. The above result is obtained by assuming a transmission line model for the normally transmitted wave into the lossy dielectric.

The validity of the impedance boundary condition has been shown to hold in general if

$$|n| \gg 1, \quad |\Im\mathrm{m}\{n\}| k_o t \gg 1 \tag{4.77}$$

The first condition ensures that within the medium, the field behaves essentially as a plane wave propagating in the direction of the inward normal to the coating. The second condition, on the other hand, imposes the requirement that the inward traveling field suffers enough attenuation so that no outward traveling waves exit at the interface due to reflection. Also, for inhomogeneous materials, the SIBC remains valid if the lateral variations of the impedance in the medium are slow, that is

$$\left| \frac{1}{k\eta_s} \nabla_s \eta_s \right| \ll 1 \tag{4.78}$$

where ∇_s denotes gradient in coordinates transverse to the normal.

Inherently, the standard impedance boundary condition does not permit modeling of polarization currents which are normal to the layer and is thus most suited for near normal incidences.

4.1.6 Reflection from Plane Dielectric Interfaces

In this section, we discuss the reflection and transmission of a plane wave incident on a planar interface between two semi-infinite media. To analyze the reflection and transmission of obliquely incident waves with arbitrary polarization, it is customary to decompose the electric field into its parallel and perpendicular components and analyze each separately. The parallel and perpendicular polarization are defined with respect to the plane of incidence. The plane of incidence is defined as the plane containing the wave vector and the unit vector normal to the interface.

Without loss of generality we consider a boundary in the x-y plane ($z = 0$) and we choose the x-z plane to be the plane of incidence. A plane wave is obliquely incident

[3] Named after Alexandr N. Shchukin (1900–1990) and Mikhail A. Leontovich, Russian scientists.

on the boundary from medium 1 characterized by constitutive parameters ϵ_1 and μ_1 and partially transmitted to medium 2 characterized by constitutive parameters ϵ_2 and μ_2. The angle of incidence θ_i is defined as the angle between the direction of propagation and normal to the interface.

4.1.6.1 Perpendicular Polarization

We start by considering the perpendicular polarization as shown in Fig. 4.7. The wave vectors for the incident, reflected and transmitted waves are denoted by $\mathbf{k}_i$, $\mathbf{k}_r$, and $\mathbf{k}_t$ and can be expressed as

$$\mathbf{k}_i = k_1 \left[\widehat{x} \sin\theta_i + \widehat{z} \cos\theta_i\right] \tag{4.79}$$

$$\mathbf{k}_r = k_1 \left[\widehat{x} \sin\theta_r - \widehat{z} \cos\theta_r\right] \tag{4.80}$$

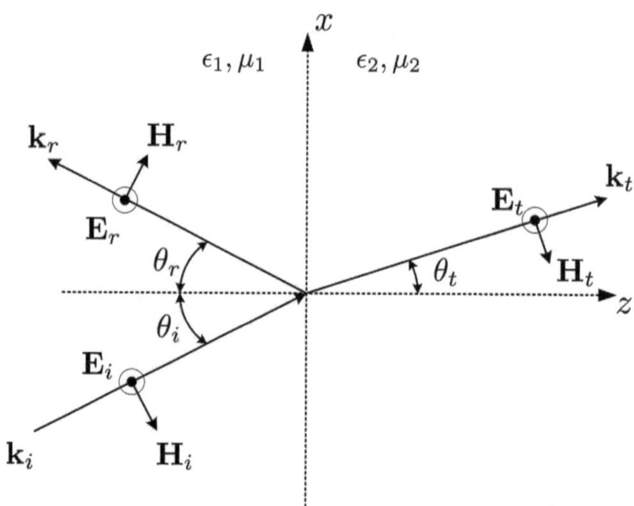

Fig. 4.7 Illustration of refraction and reflection from planar interface of homogeneous dielectrics in TE polarization

$$\mathbf{k}_t = k_2 \left[\widehat{x} \sin\theta_t + \widehat{z} \cos\theta_t\right] \tag{4.81}$$

where $k_1 = \omega\sqrt{\epsilon_1\mu_1}, k_2 = \omega\sqrt{\epsilon_2\mu_2}$, and θ_r and θ_t are the reflection and transmission angles as noted in Fig. 4.7. The incident electric field can be written as

$$\mathbf{E}_i = E_0\widehat{y}e^{-jk_1(z\cos\theta_i + x\sin\theta_i)} \tag{4.82}$$

where E_0 is the magnitude of the incident field. Reflected and transmitted electric fields can be expressed as:

$$\mathbf{E}_r = E_0R_\perp\widehat{y}e^{-jk_1(-z\cos\theta_r + x\sin\theta_r)} \tag{4.83}$$

$$\mathbf{E}_t = E_0T_\perp\widehat{y}e^{-jk_2(z\cos\theta_t + x\sin\theta_t)} \tag{4.84}$$

where $R_\perp$ and $T_\perp$ are respectively the reflection and transmission coefficients for perpendicular polarization. The corresponding magnetic fields are

$$\mathbf{H}_i = \frac{1}{\omega\mu_1}\mathbf{k}_i \times \mathbf{E}_i \tag{4.85}$$

$$\mathbf{H}_r = \frac{1}{\omega\mu_1}\mathbf{k}_r \times \mathbf{E}_r \tag{4.86}$$

$$\mathbf{H}_t = \frac{1}{\omega\mu_2}\mathbf{k}_t \times \mathbf{E}_t \tag{4.87}$$

The values for $R_\perp, T_\perp, \theta_r$, and θ_t are obtained by enforcing boundary conditions at the interface which requires the continuity of tangential electric and magnetic fields:

$$\widehat{z} \times [\mathbf{E}_i + \mathbf{E}_r]|_{z=0} = \widehat{z} \times \mathbf{E}_t|_{z=0} \tag{4.88}$$

$$\widehat{z} \times [\mathbf{H}_i + \mathbf{H}_r]|_{z=0} = \widehat{z} \times \mathbf{H}_t|_{z=0} \tag{4.89}$$

Equations (4.88) and (4.89) can each be decomposed into two equations by separating the real and imaginary parts. Hence a total of four equations are obtained which can then be solved for for $R_\perp$, $T_\perp$, θ_r, and θ_t. Substituting (4.82), (4.83) and (4.84) into (4.88) leads to the law of reflection,

$$\theta_i = \theta_r \tag{4.90}$$

and Snells law of refraction,

$$k_1 \sin\theta_i = k_2 \sin\theta_t \tag{4.91}$$

which can also be expressed in terms of refractive indices as

$$n_1 \sin\theta_i = n_2 \sin\theta_t \tag{4.92}$$

Next, substituting magnetic fields obtained from (4.85), (4.86) and (4.87) into (4.89) and using (4.90) and (4.91) we arrive at expressions for the reflection and transmission coefficients for perpendicular polarization

$$R_{\perp} = \frac{\eta_2 \cos\theta_i - \eta_1 \cos\theta_t}{\eta_2 \cos\theta_i + \eta_1 \cos\theta_t} \tag{4.93}$$

$$T_{\perp} = \frac{2\eta_2 \cos\theta_i}{\eta_2 \cos\theta_i + \eta_1 \cos\theta_t} \tag{4.94}$$

where $\eta_1 = \sqrt{\mu_1/\epsilon_1}$ and $\eta_2 = \sqrt{\mu_2/\epsilon_2}$ are intrinsic impedances.

4.1.6.2 Parallel Polarization

The field configurations for parallel polarization are shown in Fig. 4.8. The wave vectors for the incident, reflected and transmitted waves are identical to perpendicular polarization. The incident electric field lies in the plane of incidence and can be expressed as:

$$\mathbf{E}_i = E_0 \left[\widehat{x} \cos\theta_i - \widehat{z} \sin\theta_i\right] e^{-jk_1(z\cos\theta_i + x\sin\theta_i)} \tag{4.95}$$

Reflected and transmitted electric fields can be written as

$$\mathbf{E}_r = R_{\parallel} E_0 \left[\widehat{x} \cos\theta_r + \widehat{z} \sin\theta_r\right] e^{-jk_1(-z\cos\theta_r + x\sin\theta_r)} \tag{4.96}$$

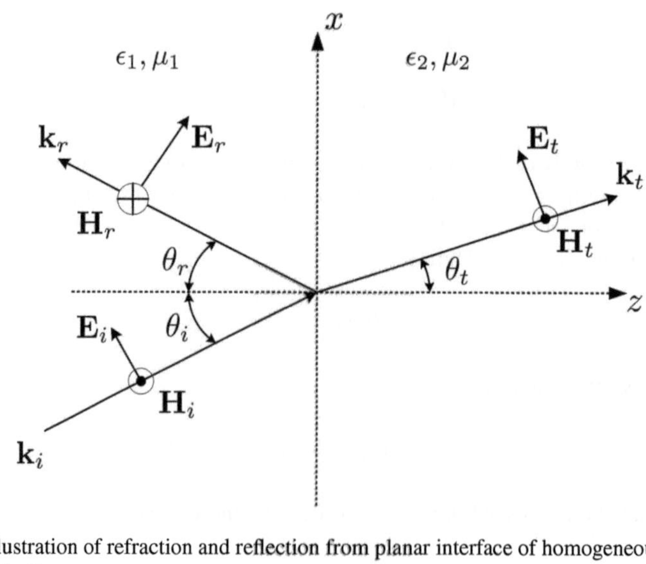

Fig. 4.8 Illustration of refraction and reflection from planar interface of homogeneous dielectrics in TM polarization

$$\mathbf{E}_t = T_{\parallel} E_0 \left[\widehat{x}\cos\theta_t - \widehat{z}\sin\theta_t\right] e^{-jk_2(z\cos\theta_t + x\sin\theta_t)} \tag{4.97}$$

where $R_{\parallel}$ and $T_{\parallel}$ are respectively the reflection and transmission coefficients for parallel polarization. The solution procedure is similar to what was done for perpendicular polarization. Solving the boundary conditions leads to identical relationships between θ_i, θ_r, and θ_t as shown in (4.90) and (4.91). Hence, the law of reflection and Snells law of refraction are valid for both polarizations. The expressions for the reflection and transmission coefficients for parallel polarization are

$$R_{\parallel} = \frac{\eta_2\cos\theta_t - \eta_1\cos\theta_i}{\eta_2\cos\theta_t + \eta_1\cos\theta_i} \tag{4.98}$$

$$T_{\parallel} = \frac{2\eta_2\cos\theta_i}{\eta_2\cos\theta_t + \eta_1\cos\theta_i} \tag{4.99}$$

4.1.6.3 Transmission Line Modeling

Here we make a brief note about the so-called transmission line modeling method. This is a rather simple and intuitive method which is particularly useful when dealing with layered media. The reflection and transmission of plane waves from dielectric boundaries can be modeled as the reflection and transmission coefficients at the junction of two transmission lines with different characteristic impedances. Figure 4.9 shows the problem of reflection and transmission of a plane wave of arbitrary polarization on top and its equivalent transmission line model on the bottom. In the transmission line model, the propagation constant is denoted by β and the characteristic impedance is denoted by Z.

We start by giving some basic results for reflection and transmission of wave at the junction of two transmission lines. Assuming a source at $z < 0$, the voltage and current for the incident wave can be written as

$$V_i(z) = V_0 e^{-j\beta_i z} \tag{4.100}$$

$$I_i(z) = I_0 e^{-j\beta_i z} \tag{4.101}$$

where V_0 is the magnitude of the incident voltage and the characteristic impedance of the transmission line is the ratio of incident voltage to incident current:

$$Z_i = \frac{V_0}{I_0} \tag{4.102}$$

The total voltage and current for $z < 0$ due to incident and reflected waves are

$$V(z) = V_0\left[e^{-j\beta_i z} + Re^{j\beta_i z}\right], \quad z < 0 \tag{4.103}$$

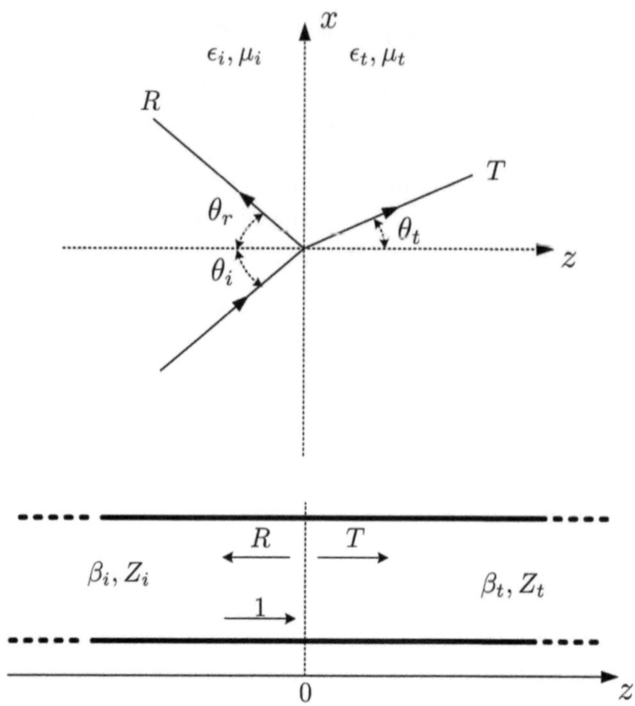

Fig. 4.9 Transmission line equivalent of reflection and refraction phenomena

$$I(z) = \frac{V_0}{Z_i}\left[e^{-j\beta_i z} - Re^{j\beta_i z}\right], \quad z < 0 \tag{4.104}$$

where R is the reflection coefficient at the junction as denoted in Fig. 4.9. The total voltage and current for $z > 0$ is

$$V(z) = TV_0 e^{-j\beta_t z}, \quad z > 0 \tag{4.105}$$

$$I(z) = \frac{TV_0}{Z_t} e^{-j\beta_t z}, \quad z > 0 \tag{4.106}$$

where T is the transmission coefficient at the junction. From basic transmission line theory, we know that the reflection and transmission coefficients at the junction are

$$R = \frac{Z_t - Z_i}{Z_t + Z_i} \tag{4.107}$$

$$T = \frac{2Z_t}{Z_t + Z_i} \tag{4.108}$$

The reflection and transmission of plane waves from dielectric boundaries can be expressed as an equivalent transmission line problem by treating the tangential electric field as the voltage and the tangential magnetic field as the current and defining the characteristic impedances accordingly. For example, in the case of perpendicular polarization the values of Z_i and Z_t are

$$Z_i = -\frac{E_y^{\text{inc}}}{H_x^{\text{inc}}} = \frac{\eta_i}{\cos\theta_i} \tag{4.109}$$

$$Z_t = -\frac{E_y^{\text{trn}}}{H_x^{\text{trn}}} = \frac{\eta_t}{\cos\theta_t} \tag{4.110}$$

Substituting (4.109) and (4.110) in (4.107) and (4.108) we will arrive at (4.93) and (4.94). Transmission line modeling can be easily generalized to multi-layered systems which will greatly simply the analysis of plane wave propagation in layered media which we will consider later in this chapter. From basic network analysis, we know that a segment of a transmission line can be modeled as a two-port network. Figure 4.10 shows a transmission line of length d with propagation constant β and characteristic impedance Z and its equivalent two-port network. The two-port network is characterized by its $ABCD$ matrix which is defined in terms of the total voltages and currents as shown in Fig. 4.10.

$$\begin{bmatrix} V_1 \\ I_1 \end{bmatrix} = \begin{bmatrix} A & B \\ C & D \end{bmatrix} \begin{bmatrix} V_2 \\ I_2 \end{bmatrix} \tag{4.111}$$

For the transmission line shown in Fig. 4.10 the $ABCD$ matrix is

$$A = \cos\beta d \tag{4.112}$$
$$B = jZ\sin\beta d \tag{4.113}$$
$$C = \frac{j\sin\beta d}{Z} \tag{4.114}$$
$$D = \cos\beta d \tag{4.115}$$

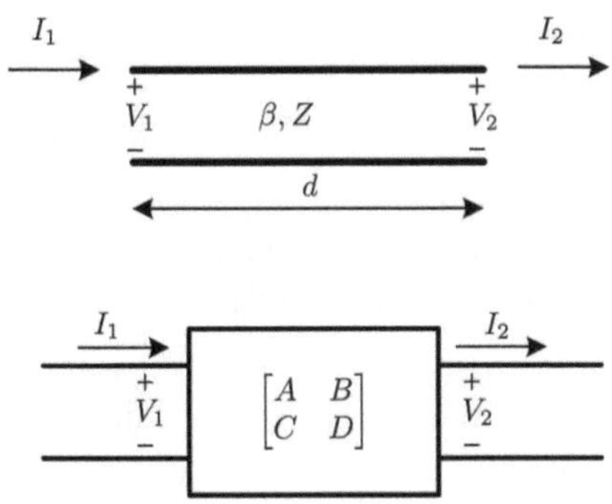

Fig. 4.10 Scattering matrix representation of a transmission line segment

An important property of *ABCD* matrix representation is that when multiple networks are cascaded, the *ABCD* matrix for the cascaded system is the product of *ABCD* matrices for individual networks. This property will be used later to analyze propagation in layered media.

4.1.6.4 Total Transmission and Brewster's Angle

Having derived the expressions for reflection and transmission coefficients, it is often of interest to study what conditions can lead to total transmission or total reflection of the incident wave. We start by examining the case of total transmission which requires vanishing reflection coefficient, i.e., $R = 0$. The incident angle at which the total transmission (zero reflection) condition is realized is referred to as the Brewster's angle and denoted by θ_B.

Considering the case of perpendicular polarization, the reflection coefficient is given in (4.93). Setting $R_\perp = 0$, we arrive at

$$\eta_2 \cos\theta_B = \eta_1 \cos\theta_t \tag{4.116}$$

Using Snell's law (4.91), we can solve for θ_B as

$$\sin^2\theta_B = \frac{1 - \frac{\epsilon_2\mu_1}{\epsilon_1\mu_2}}{1 - \left(\frac{\mu_1}{\mu_2}\right)^2} \tag{4.117}$$

For the incident angle to be physically realizable, the right-hand side of (4.117) must be between zero and one. Assuming non-magnetic materials where $\mu_1 = \mu_2 = \mu_0$ the expression in (4.117) becomes infinity and hence physically unrealizable. With the exception of ferromagnetic materials, the non-magnetic assumption is generally true for most dielectrics, hence *for non-magnetic materials total transmission cannot be achieved for perpendicular polarization.*

Further examination of (4.117) reveals that in the rare case of $\epsilon_1 = \epsilon_2$ and $\mu_1 \neq \mu_2$, Brewster's angle does exist for perpendicular polarization and its value is given by:

$$\theta_B = \tan^{-1}\left(\sqrt{\frac{\mu_2}{\mu_1}}\right) \tag{4.118}$$

Next, we consider parallel polarization. From (4.98) we set $R_\parallel = 0$ and use Snell's law (4.91) to solve for θ_B

$$\sin^2\theta_B = \frac{1 - \frac{\mu_2\epsilon_1}{\mu_1\epsilon_2}}{1 - \left(\frac{\epsilon_1}{\epsilon_2}\right)^2} \tag{4.119}$$

Considering the common case of non-magnetic materials where $\mu_1 = \mu_2 = \mu_0$ the expression in (4.119) is reduced to

$$\theta_B = \tan^{-1}\left(\sqrt{\frac{\epsilon_2}{\epsilon_1}}\right) \tag{4.120}$$

4.1.6.5 Total Reflection and Critical Angle

In the previous section, we looked derived the conditions for zero reflection at a planar interface. Here we consider the case of total reflection. We start by considering an incident beam obliquely incident at a dielectric interface with an angle if incident of θ_i. Using Snell's law (4.92) the refraction angle for the transmitted beam is

$$\sin\theta_t = \frac{n_1}{n_2}\sin\theta_i \tag{4.121}$$

We consider the case where light is traveling from a more dense medium to a less dense medium, i.e., $n_1 > n_2$. From (4.92) it is easy to observe that in such case $\theta_t > \theta_i$. A case of particular interest is when $\theta_t = \pi/2$ where the refracted wave flows along the boundary and no energy is transmitted into the second medium. The incident angle at which this phenomenon is observed is known at the critical angle θ_c and its value is obtained as

$$\theta_c = \sin^{-1}\left(\frac{n_2}{n_1}\right) \tag{4.122}$$

When the plane wave is incident with an angle larger than the critical angle, it is totally reflected and the refracted wave in the second medium becomes inhomogeneous.

4.1.7 Propagation in Layered Media

Consider a plane wave obliquely incident on a stack of N dielectric layers sandwiched between two semi-infinite media as shown in Fig. 4.11. The incident, reflected, and transmitted angles are denoted by θ_i, θ_r, and θ_t as shown in Fig. 4.11. Direct analysis of the geometry shown in Fig. 4.11 which requires separate treatment for parallel and perpendicular polarizations is quite cumbersome. It entails deriving boundary condition equations at all $N + 1$ interfaces and solving them simultaneously.However, the problem can be solved rather easily using transmission line modeling method which was introduced in the previous section. Figure 4.11 shows the equivalent transmission line problem which can be obtained by taking the tangential electric field to be the voltage and the tangential magnetic field to be the current.

Let us first consider the perpendicular polarization. The characteristic impedance and the propagation constant of the mth layer are

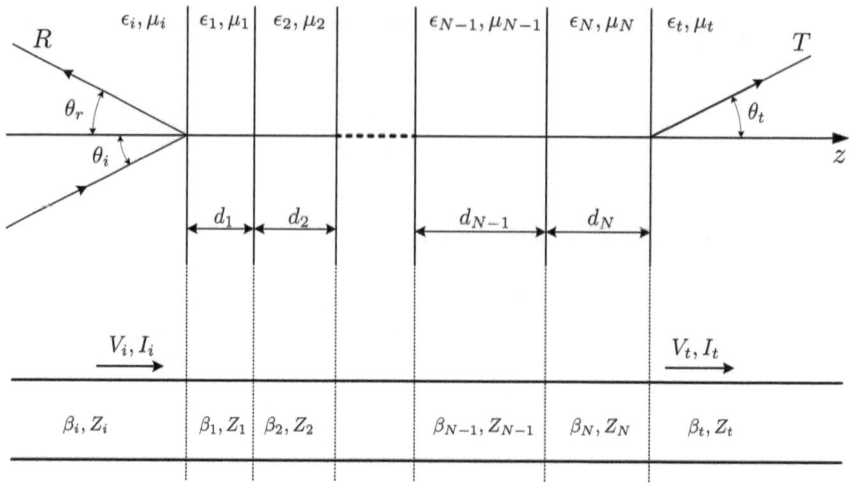

Fig. 4.11 Illustration of reflection and refraction from a general layered dielectric medium

$$\beta_m = \omega\sqrt{\epsilon_m\mu_m}\sqrt{1 - \frac{\epsilon_i\mu_i}{\epsilon_m\mu_m}\sin^2\theta_i} \tag{4.123}$$

$$Z_m = \frac{\omega\mu_m}{\beta_m} \tag{4.124}$$

Now using (4.115), the *ABCD* matrix for the *m*th layer is

$$A_m = \cos\beta_m d_m \tag{4.125}$$

$$B_m = jZ_m \sin\beta_m d_m \tag{4.126}$$

$$C_m = \frac{j\sin\beta_m d_m}{Z_m} \tag{4.127}$$

$$D_m = \cos\beta_m d_m \tag{4.128}$$

As it was noted, the *ABCD* matrix for a cascaded system is the product of *ABCD* matrices for individual networks, thus

$$\begin{bmatrix} V_i \\ I_i \end{bmatrix} = \begin{bmatrix} A & B \\ C & D \end{bmatrix}\begin{bmatrix} V_t \\ I_t \end{bmatrix} \tag{4.129}$$

where,

$$\begin{bmatrix} A & B \\ C & D \end{bmatrix} = \begin{bmatrix} A_1 & B_1 \\ C_1 & D_1 \end{bmatrix}\begin{bmatrix} A_2 & B_2 \\ C_2 & D_2 \end{bmatrix}\cdots\begin{bmatrix} A_N & B_N \\ C_N & D_N \end{bmatrix} \tag{4.130}$$

Reflection and transmission coefficients can be obtained by solving (4.129) leading to

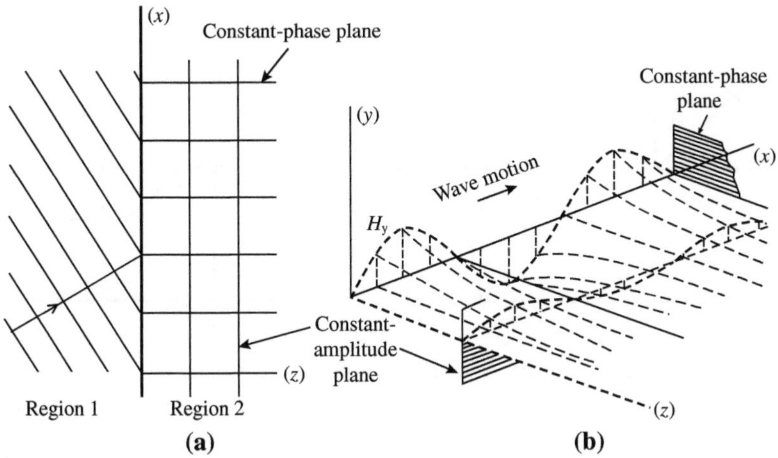

Fig. 4.12 A plane wave incident in the first medium with an angle larger than the critical angle. The plane wave in the second medium is inhomogeneous, **a** moving constant phase fronts in Regions 1 and 2, **b** amplitude of magnetic field component along propagation direction

$$R = \frac{A + \frac{B}{Z_t} - Z_i\left(C + \frac{D}{Z_t}\right)}{A + \frac{B}{Z_t} + Z_i\left(C + \frac{D}{Z_t}\right)} \tag{4.131}$$

$$T = \frac{2}{A + \frac{B}{Z_t} + Z_i\left(C + \frac{D}{Z_t}\right)} \tag{4.132}$$

For parallel polarization, the characteristic impedance is given by

$$Z_m = \frac{\beta_m}{\omega \epsilon_m} \tag{4.133}$$

However, all other results remain unchanged (Fig. 4.12).

4.1.8 Reflection from Inhomogeneous Layers

In this section, we will examine the reflection of plane waves from inhomogeneous dielectric layers characterized by a slowly varying permittivity profile. In particular, we will find the different equation satisfied by the reflection coefficient from such a medium.

Consider an inhomogeneous dielectric slab characterized by a relative permittivity profile which is a function of one space variable. For convenience, let us assume that the permittivity of the layer, ϵ is a function of z. The slab is infinite in x and y directions.

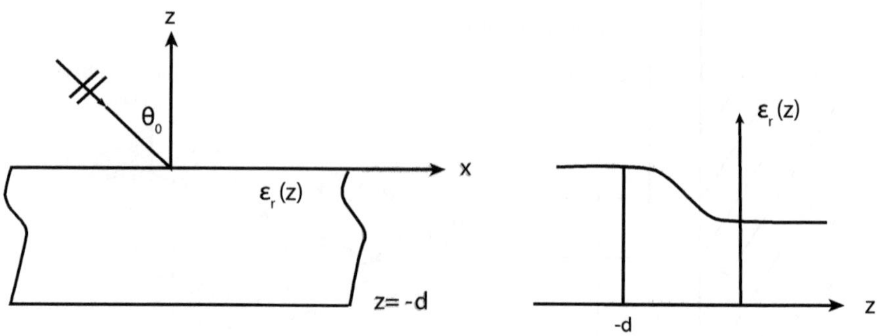

Fig. 4.13 An inhomogeneous dielectric layer with a permittivity profile $\epsilon(z)$

The layer is illuminated by a transverse magnetic (TM) plane wave (see Fig. 4.13)

$$\mathbf{E}^i = \widehat{y}e^{-jk_0 x \sin\theta_0 + jk_0 z \cos\theta_0} \tag{4.134}$$

where k_0 is the wavenumber and θ_0 is the angle of incidence. The electromagnetic fields satisfy the source-free Maxwell's equations everywhere

$$\begin{cases} \nabla \times \mathbf{E} = -j\omega\mu\mathbf{H} \\ \nabla \times \mathbf{H} = j\omega\epsilon\mathbf{E} \end{cases} \tag{4.135}$$

Since the layer is infinite in the y direction, the fields have translational symmetry in y and $\frac{\partial}{\partial y} = 0$. From the first of Maxwell's equations, we have

$$\begin{cases} \dfrac{\partial E_y}{\partial z} = j\omega\mu H_x \\[2ex] \dfrac{\partial E_y}{\partial x} = -j\omega\mu H_z \end{cases} \tag{4.136}$$

and from the second equation, we obtain

$$\frac{\partial H_x}{\partial z} - \frac{\partial H_z}{\partial x} = j\omega\epsilon E_y \tag{4.137}$$

Using the phase matching condition in the x direction, the phase constant in this direction is given by

$$k_x = k_0 \sin\theta_0 \tag{4.138}$$

and, therefore, $\frac{\partial}{\partial x} = -jk_x$. Thus, combining the second member of (4.136) with (4.137), we find that

$$\frac{\partial H_x}{\partial z} - j\left(\frac{k_z^2}{\omega\mu}\right)E_y = 0 \tag{4.139}$$

where

$$k_z = \sqrt{k_0^2\epsilon_r(z) - k_x^2} \tag{4.140}$$

is the phase constant in the z direction and ϵ_r is the relative permittivity of the medium.

In order to find an equation for the reflection coefficient, we define the incident (downward travelling) and reflected (upward travelling) fields at each point in the inhomogeneous medium as

$$E_y^i = A(z)e^{-jk_x x}, \quad H_x^i = \frac{k_z}{\omega\mu}A(z)e^{-jk_x x} \tag{4.141}$$

$$E_y^r = B(z)e^{-jk_x x}, \quad H_x^r = -\frac{k_z}{\omega\mu}B(z)e^{-jk_x x} \tag{4.142}$$

The local reflection coefficient in the medium is defined as

$$R \equiv B/A \tag{4.143}$$

By substituting the total fields (the sum of incident and reflected fields) from (4.141) and (4.142) in (4.137) and (4.139), we find

$$\frac{d}{dz}(A - B) = jk_z(A + B) - \frac{k_z'}{k_z}(A - B) \tag{4.144}$$

and

$$\frac{d}{dz}(A + B) = jk_z(A - B) \tag{4.145}$$

where the prime denotes differentiations with respect to the argument. Summing and subtracting the last two equations, we have

$$A' = jk_zA - \frac{k_z'}{2k_z}(A - B) \tag{4.146}$$

and

$$B' = -jk_zB + \frac{k_z'}{2k_z}(A - B) \tag{4.147}$$

Multiplying (4.146) by B and (4.147) by A, subtracting the resulting equations and dividing by A^2, we obtain

$$\frac{B'A - BA'}{A^2} = -jk_z \frac{2AB}{A^2} + \frac{k_z'}{2k_z} \frac{A^2 - B^2}{A^2} \tag{4.148}$$

The left hand side of this equation is identified as $(B/A)'$. Thus, in view of (4.143), the differential equation satisfied by R is given by

$$R' = -j2k_z R + \frac{k_z'}{2k_z}(1 - R^2) \tag{4.149}$$

This is a nonlinear differential equation of Riccati type valid for the TM polarization. In general, we may write

$$R' = -j2k_z R + \gamma(1 - R^2) \tag{4.150}$$

where γ depends on the polarization as follows

$$\gamma = \begin{cases} \dfrac{k_z'}{2k_z}, & \text{TE Polarization} \\[2ex] \dfrac{1}{2}\left(\dfrac{k_z}{\epsilon_r}\right)' \dfrac{\epsilon_r}{k_z}, & \text{TM Polarization} \end{cases} \tag{4.151}$$

In order to find a unique solution to the above equation, an initial (boundary) condition is required for R. For example, if the layer is terminated in a homogeneous half space, we may write

$$\lim_{z \to -\infty} R(z) = 0 \tag{4.152}$$

On the other hand, if the layer is terminated by a perfectly conducting plane at $z = -d$, we have

$$R(-d) = -1 \tag{4.153}$$

4.1.8.1 An Iterative Solution

When the reflection coefficient of the layer is weak, the Riccati differential equation can be solved iteratively. Assuming that the layer is terminated in a homogeneous half space, let us write (4.150) as

$$R' + j2k_z R = \gamma(z)(1 - R^2) \tag{4.154}$$

Multiplying both sides by the integrating factor $\exp\left(2j \int_{z0}^{z} k_z dz\right)$ and integrate from $-\infty$ to z to obtain

$$R(z) = \exp\left(-2j\int_{z_0}^{z} k_z dz\right)\left[\int_{-\infty}^{z} \gamma(z)[1 - R^2(z)]\exp\left(2j\int_{z_0}^{z} k_z dz\right)dz\right] \tag{4.155}$$

where z_0 is an arbitrary point in the medium where the reflection coefficient is desired. Using the initial (boundary) condition (4.152), we have

$$R(z_0) = \int_{-\infty}^{z_0} \gamma(z)(1 - R^2)\exp\left(2j\int_{z_0}^{z} k_z dz\right)dz \tag{4.156}$$

For small values of reflection coefficient, R, the first approximate solution is obtained by ignoring the quadratic term R^2 in comparison to unity. Thus, the first order approximation for the reflection coefficient is given by

$$R_1(z_0) = \int_{-\infty}^{z_0} \gamma(z)\exp\left(2j\int_{z_0}^{z} k_z dz\right)dz \tag{4.157}$$

Substituting R_1 into the right-hand side of (4.156), the second order approximation is obtained as

$$R_2(z_0) = \int_{-\infty}^{z_0} \gamma(z)(1 - R_1^2(z))\exp\left(2j\int_{z_0}^{z} k_z dz\right)dz \tag{4.158}$$

Higher order approximations are obtained by continuing the iteration process.

The above iteration method only works if the permittivity profile is continuous and there is no total reflection at any point along the profile.

4.1.8.2 An Impedance Formulation

Let us examine the same problem by considering an impedance formulation. We define the wave impedance at each point in the medium as

$$Z = E_y/H_x \tag{4.159}$$

To find an equation for the wave impedance, we differentiate the above equation with respect to z

$$Z' = \frac{E_y' H_x - E_y H_x'}{H_x^2} \tag{4.160}$$

Substituting from (4.136), we get

$$Z' = \frac{j\omega\mu H_x^2 - \frac{j}{\omega\mu} k_z^2 E_y^2}{H_x^2} \tag{4.161}$$

or, equivalently

$$Z' = jk_0 Z_0 [1 - (\frac{k_z}{k_0 Z_0})^2 Z^2] \tag{4.162}$$

Introducing the normalized impedance $\eta = Z/Z_0$, the last equation becomes

$$\eta' = jk_0 [1 - (\frac{k_z}{k_0})^2 \eta^2] \tag{4.163}$$

and using the change of variables $\xi = \eta / jk_0$, we obtain

$$\xi' = 1 + k_z^2 \xi^2 \tag{4.164}$$

which is again a Riccati differential equation for the wave impedance.

4.1.9 Wave Velocities

We saw before that in a plane wave

$$\psi(z, t) = A_0 e^{-j(kz - \omega t)}$$

the surfaces of constant phase move with a certain velocity known as the *phase velocity*. In a simple medium, the phase velocity is given by

$$v_p = \omega / k \tag{4.165}$$

where ω is the angular frequency of the wave, and k is the wavenumber. The phase velocity of a plane wave in vacuum is equal to the speed of light c. However, it differs from c in dielectric media. The ratio

$$n = c / v_p = \sqrt{\mu_r \epsilon_r} \tag{4.166}$$

is known as the *refractive index* of the medium (or index of refraction) where μ_r and ϵ_r are relative permeability and permittivity of the medium, respectively.

4.1.9.1 Group Velocity

For plane waves in a lossless medium, the phase constant $k = \omega\sqrt{\mu\epsilon}$ is a linear function of ω. As a consequence, the phase velocity v_p is a constant and independent

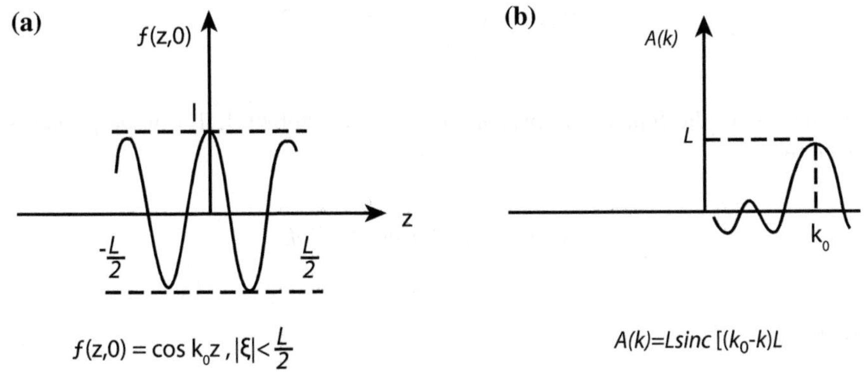

Fig. 4.14 A pulse of width L shown at $t = 0$ propagating in the positive z direction

of f. However, in some cases (such as wave propagation in a lossy dielectric, or in a waveguide) the phase constant is not a linear function of frequency. Waves of different frequencies will propagate at different phase velocities—a phenomenon known as *dispersion*.

A simple harmonic field contains no information. To communicate we have to impress information on, for example, a carrier, and whether we use amplitude, frequency or pulse modulation to do so, the resulting signal will have a spread of frequencies. Typically, this spread of frequencies is few percent. The concept of group velocity applies when the frequencies are confined to a narrow band.

In an infinitely expanded homogeneous medium, consider a time dependent signal propagating in the z direction. Using Fourier synthesis, we can write this signal as

$$\psi(z,t) = \int_{-\infty}^{\infty} A(k) e^{-j(kz-\omega t)} dk \tag{4.167}$$

where $A(k)$ is called the amplitude function. Assuming that $A(k)$ is negligible outside the region $k_c - \Delta k \leq k \leq k_c + \Delta k$, we write

$$\psi(z,t) = \int_{k_c-\Delta k}^{k_c+\Delta k} A(k) e^{-j(kz-\omega t)} dk \tag{4.168}$$

which is commonly called a *wave packet* or *group*. The wave packet can be regarded as the superposition of plane monochromatic waves with different wave vectors and frequencies.

Note that a concentration of the field in space actually implies a spread in the frequency or wavelength spectrum.

Example 4.4 Consider a pulse of length L propagating in the z-direction. The puls shape at $t = 0$ is given by (Fig. 4.14)

$$
\begin{aligned}
f(z,0) &= \cos k_0 z, \quad |z| < L/2 \\
&= 0, \quad |z| > L/2
\end{aligned}
$$

We may write the function in phasor form $f(z) = \Re\mathrm{e}\{e^{jk_0 z}\}$. The wave packet is given by

$$\psi(z,t) = \int_{-\infty}^{\infty} A(k) e^{j(\omega t - kz)} dk$$

where

$$A(k) = \int_{-\infty}^{\infty} e^{jk_0 z} e^{-jkz} dz = \int_{-L/2}^{L/2} e^{j(k_0 - k)z} dz$$

We find that

$$A(k) = 2\frac{\sin(k_0 - k)L/2}{k_0 - k}$$

Clearly, the requirement of narrow interval of significance in the spatial spectrum of $A(k)$ is equivalent to have a wide pulse L. The wave packet is given by

$$\psi(z,t) = L \int_{k_0 - \Delta k}^{k_0 + \Delta k} \mathrm{sinc}[(k_0 - k)L] e^{j(\omega t - kz)} dk \qquad \square$$

Assume that ω is a function of k, but deviates little from its value at k_c over the interval $2\Delta k$. Then

$$\omega(k) = \omega(k_c) + \frac{d\omega}{dk}|_{k_c}(k - k_c) + \cdots \tag{4.169}$$

so that

$$kz - \omega t = k_c z - \omega_c t + (k - k_c)\left(z - \frac{d\omega}{dk}|_{k_c} t\right) + \cdots \tag{4.170}$$

We can, therefore, write the wave packet as

$$\psi = \tilde{\psi} e^{-j(k_c z - \omega_c t)} \tag{4.171}$$

where the *spatial mean amplitude* $\tilde{\psi}$ is given by

$$\tilde{\psi}(z,t) = \int\limits_{k_c-\Delta k}^{k_c+\Delta k} A(k) e^{-j(k-k_c)\left(z-\frac{d\omega}{dk}|_{k_c} t\right)} dk \tag{4.172}$$

The mean amplitude is constant over the surfaces

$$z - \frac{d\omega}{dk}|_{k_c} t = \text{constant} \tag{4.173}$$

from which it is evident that the wave packet propagates with the velocity

$$v_g = \frac{d\omega}{dk}|_{k_c} \tag{4.174}$$

This is the *group velocity.*

In an infinite medium whose refractive index is frequency dependent, the group velocity differs from the phase velocity. The concept of group velocity is vital in connection with waveguides.

Example 4.5 In an unbounded region we have

$$\omega = k v_p = \frac{kc}{n}$$

where, n is the index of refraction. If n is independent of ω

$$v_g = \frac{c}{n} = v_p$$

which is also the phase velocity. But if $n = n(\omega)$, we have *material dispersion* and

$$v_g = \frac{c}{n} - \frac{kc}{n^2}\frac{dn}{d\omega} = v_p\left(1 - \frac{k}{n}\frac{dn}{d\omega}\right)$$ □

The group velocity is related to the phase velocity through the following relations

$$v_g = \frac{v_p}{1 - \dfrac{\omega}{v_p}\dfrac{dv_p}{d\omega}} = v_p\left(1 - \frac{k}{n}\frac{dn}{d\omega}\right) \tag{4.175}$$

We may classify various regions based on their dispersive behavior:

If $dv_p/d\omega = 0$, the phase velocity is independent of ω and the medium is non-dispersive.

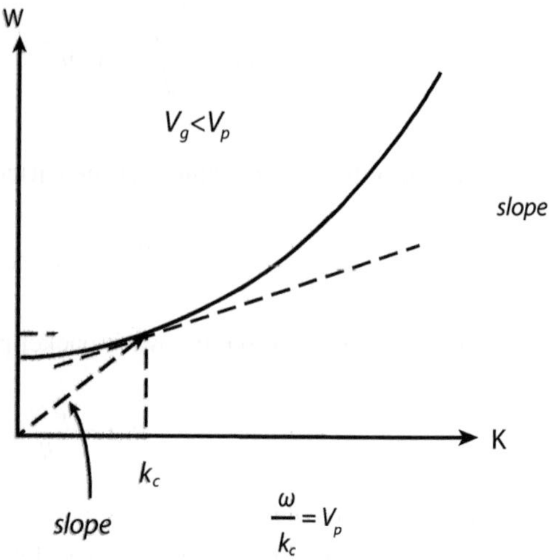

Fig. 4.15 The ω-k curve for a normally dispersive medium

If $dv_p/d\omega < 0$ or $dn/d\omega > 0$, we have normal dispersion and $v_g < v_p$.
If $dv_p/d\omega > 0$ or $dn/d\omega < 0$, we have anomalous dispersion and $v_g < v_p$.

An example of normal dispersion is shown in Fig. 4.15.

Example 4.6 In general, v_g is less than v_p, but it may exceed v_p in regions of anomalous dispersion where $dn/d\omega < 0$. For example, in a good conductor, we saw that

$$k \simeq \sqrt{\frac{\omega\mu\sigma}{2}}$$

Therefore, the effective index of refraction in this case is a function of frequency

$$n = c/v_p = kc/w = c\sqrt{\frac{\mu\sigma}{2\omega}}$$

and the medium has anomalous dispersion. □

In a region of normal dispersion, the group velocity is equal to the velocity of signal energy. This is not the case in anomalosly dispersive regions.

Dispersion produces pulse spreading, since different frequencies have different velocities of propagation.

4.2 Planar Waveguides

In this section, we will examine a number wave guiding structures whose modes are naturally expressible in terms of plane waves. The basic concepts and ideas are presented first in the context of a parallel plate waveguide.

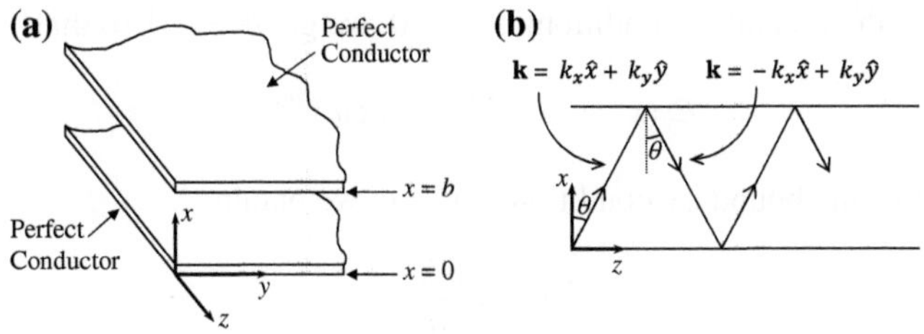

Fig. 4.16 A parallel plate waveguide. **a** Geometry of the waveguide. **b** The bouncing waves inside the waveguide

4.2.1 The Parallel Plate Waveguide

Consider a homogeneous dielectric slab with perfectly conducting boundaries at $x = 0, a$ as shown in Fig. 4.16. This constitutes a dielectric filled parallel plate waveguide. We seek the fields (modes) that propagate in the z direction down the guide.

Since the entire structure is independent of y, fields can exist which are independent of y. The simplest fields are those having either **E** in $\widehat{y}$-direction ($\mathbf{E} = \widehat{y}E_y$) or **H** in $\widehat{y}$-direction ($\mathbf{H} = \widehat{y}H_y$). The latter are referred to as transverse magnetic to $\widehat{z}$ (TM_Z), while the former are called transverse electric to $\widehat{z}$ (TE_Z). We will first consider the TE_Z case. The TM_Z modes can be determined by duality.

The electric field satisfies the homogeneous wave equation inside the guide

$$\left(\frac{\partial^2}{\partial x^2} + \frac{\partial^2}{\partial z^2} + k^2\right) E_y = 0 \tag{4.176}$$

subject to the boundary condition

$$E_y = 0 \quad x = 0, a \tag{4.177}$$

Consistent with the desire for propagation in the z-direction, assume that

$$E_y = e^{-j(k_x x + k_z z)} + Ae^{-j(-k_x x + k_z z)} \tag{4.178}$$

with the consistency relation

$$k_x^2 + k_z^2 = k^2 \tag{4.179}$$

By enforcing the boundary conditions at $x = 0$, we get $A = -1$ so that

$$E_y(x, z) = -2j \sin(k_x x) e^{-jk_z z} \tag{4.180}$$

Enforcing now the boundary condition at $x = a$, we obtain

$$k_x = \frac{m\pi}{a} \tag{4.181}$$

This is a condition on the transverse component of the propagation vector, and is called the transverse resonance condition.

The quantity of interest is k_z, the guide propagation constant. Knowing k_x, we have the characteristic equation

$$k_z = \sqrt{k^2 - \left(\frac{m\pi}{a}\right)^2} \tag{4.182}$$

A plot of normalized propagation constant k_z/k versus frequency is the mode diagram (Fig. 4.17). The characteristic equation shows that k_z can only take discrete values. It defines modes which

propagate if $k > m\pi/a \quad (a > m\lambda/2)$
attenuate if $k < m\pi/a \quad (a < m\lambda/2)$

This presents the cut-off phenomenon; each mode in general has a frequency below which it does not propagate. We order the modes according to their cut-off frequencies, with the fundamental mode having the lowest cut-off. There may or may not be a mode which propagates at all frequencies no matter how low. It would correspond to $m = 0$ above and is the TM_0 mode (the TE_0 mode is null).

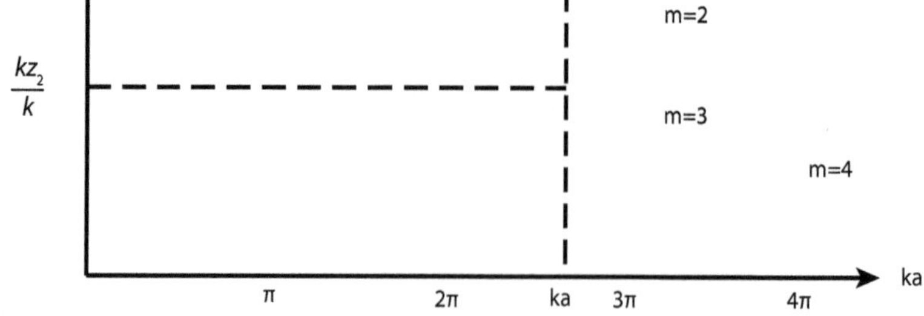

Fig. 4.17 Mode diagram for a parallel plate waveguide with wall separation a in the TE_Z case

For a given guide and given frequency there is a finite number of propagating modes (having $k_z^2 > 0$) and an infinite number of attenuated or evanescent modes (having $k_z^2 < 0$).

4.2.1.1 Phase Velocity

Phase velocity is the velocity of propagation down the guide and is given by

$$v_e = \frac{\omega}{k_z} = \frac{v_p}{\sqrt{1 - \left(\frac{m\pi}{ka}\right)^2}} \tag{4.183}$$

where $v_p = (\epsilon\mu)^{1/2} = c/n$ is the intrinsic velocity in the dielectric and n is the index of refraction. The phase velocity in the wave guide is mode dependent and frequency dependent. This causes *dispersion* in the guide. Also, since $v_e \geq v$, we refer to this type of guided wave as *fast wave*.

The phase velocity above can be used to define an equivalent refractive index. That is the refractive index of an infinite medium (without walls) having the same intrinsic velocity as the phase velocity of the guide. Thus

$$n_e = \frac{c}{v_e} = \frac{c}{v}\sqrt{1 - \left(\frac{m\pi}{ka}\right)^2} = n\sqrt{1 - \left(\frac{m\pi}{ka}\right)^2} \tag{4.184}$$

and we have $n_e \leq n$. This is the basis for the design of microwave lenses, and a parallel plate (TE mode) lens which has the same focusing properties as a standard convex lens in optics is shown in Fig. 4.18.

4.2.1.2 Group Velocity

The group velocity is important in the study of guided modes. In a parallel plate waveguide, we have

$$\frac{1}{v_g} = \frac{dk_z}{d\omega} = \frac{d}{d\omega}\sqrt{\left(\frac{\omega n}{c}\right)^2 - \left(\frac{m\pi}{a}\right)^2} \tag{4.185}$$

Fig. 4.18 A concave TE mode microwave lens acting as a focusing lens

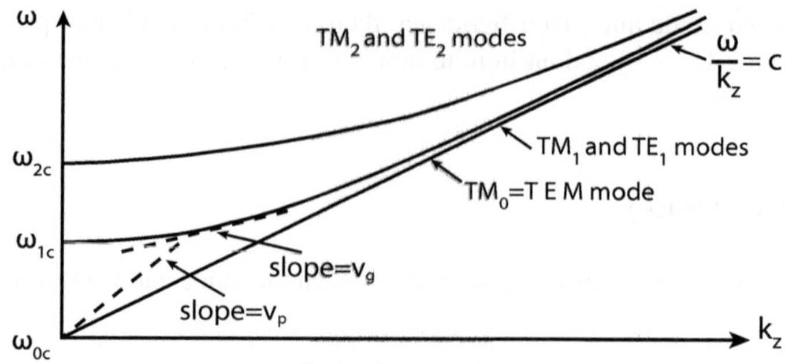

Fig. 4.19 The $\omega - k_z$ plot for the parallel plate waveguide

If n is independent of ω, we get modal dispersion only. Thus

$$\frac{1}{v_g} = \frac{\omega(\frac{n}{c})^2}{\sqrt{(\frac{\omega n}{c})^2 - (\frac{m\pi}{a})^2}} = \frac{v_e}{v_p^2} \tag{4.186}$$

which is always less than the speed of light. Comparing the expressions for v_e and v_g, we find that

$$v_e v_g = v_p^2 \tag{4.187}$$

A plot of ω versus k_z is as shown in Fig. 4.19. When $k_z \rightarrow 0$, phase velocity approaches infinity while the group velocity becomes zero. On the ohter hand, when $k_z \rightarrow \infty$, or $\omega \rightarrow \infty$, the phase and group velocities both approach the velocity of light in the medium. This is the TEM wave velocity.

If the index of refraction is a function of ω, we get an additional contribution to v_g from material dispersion coming from $dn/d\omega$.

Modal dispersion in a waveguide can be a limiting factor as regards the usable bandwidth of the channel capacity for communication purposes. Though we can minimize material dispersion by the appropriate choice of material and frequency, we cannot escape the modal dispersion.

The sketch of the instantaneous fields for the dominant TE_1 and TM_1 modes are shown in Figs. 4.20 and 4.21, respectively. The tangential electric field is zero on the plates while the tangential magnetic field is not zero on the plates and gives the surface current distribution on the waveguide walls ($\mathbf{K} = \widehat{n} \times \mathbf{H}$). The direction of propagation is given by $\mathbf{E} \times \mathbf{H}$.

4.2.2 Grounded Dielectric Slab

A geometry of much interest in waveguides is a homogeneous non-metallic dielectric slab of thickness a and refractive index n_1 with a metal backing on the lower

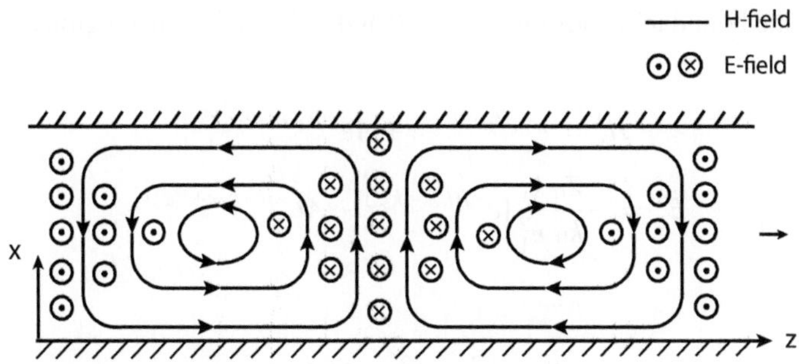

Fig. 4.20 The sketch of the field distribution for the dominant TE_1 mode

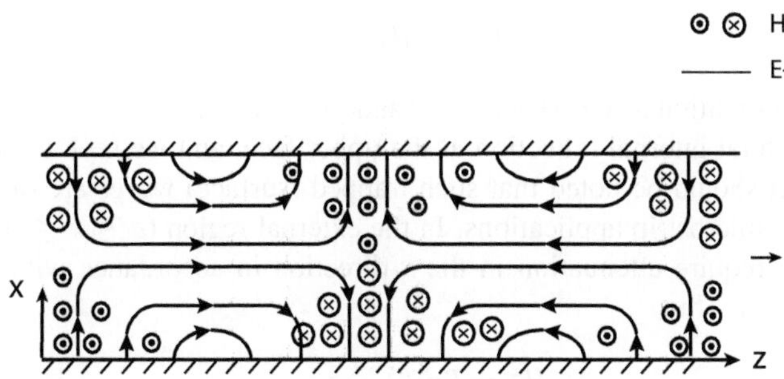

Fig. 4.21 The sketch of the field distribution for the dominant TM_1 mode

Fig. 4.22 A metal-backed dielectric slab waveguide

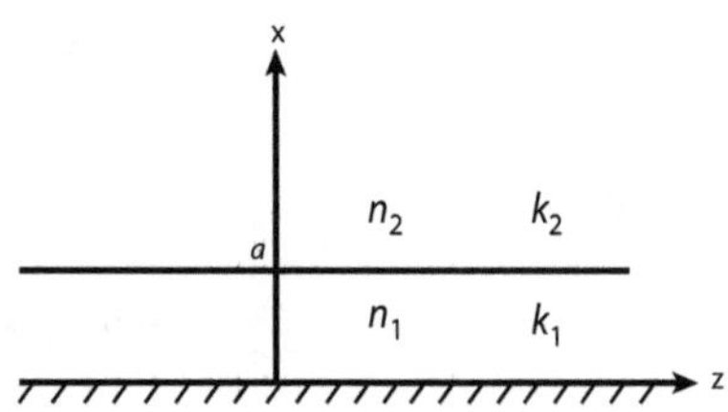

surface (Fig. 4.22). There is a wide variety of modes that can exist, but only a subset correspond to lossless propagation within the slab.

The simplest modes propagating in the z-direction are those having **E** in the y-direction (TE_z) and those having **H** in the y-direction (TM_z).

We will consider the TM_z modes first. We have

$$\mathbf{H} = \hat{y} H_y \tag{4.188}$$

$$E_x = -\frac{Z_0}{jk_0 n_1^2}\frac{\partial H_y}{\partial z}, \qquad E_z = \frac{Z_0}{jk_0 n_1^2}\frac{\partial H_y}{\partial x}, \quad i = 1, 2 \tag{4.189}$$

subject to the boundary conditions at $x = 0$ and a. In the internal region ($0 \leq x \leq a$) we have

$$H_y^1 = e^{-j(k_x x + k_z z)} + Ae^{-j(-k_x x + k_z z)} \tag{4.190}$$

$$E_z^1 = -\frac{Z_0}{k_0}\frac{k_x}{n_1^2}\{e^{-j(k_x x + k_z z)} - Ae^{-j(-k_x x + k_z z)}\} \tag{4.191}$$

where

$$k_x^2 + k_z^2 = k_1^2 = (n_1 k_0)^2 \tag{4.192}$$

Since we require a real propagation constant ($k_z^2 > 0$), we should have

$$k_x < n_1 k_0 \tag{4.193}$$

The boundary condition at $x = 0$ is $E_z = 0$ and gives $A = 1$.

We require total internal reflection at the upper face, and we seek to determine these modes. It should be noted that such trapped (surface) waves are disadvantageous for most microstrip applications. In the external region ($a \leq x \leq \infty$) on the other hand we require attenuation in the x-direction in accordance with the total internal reflection. Thus

$$H_y^2 = Be^{-px - jk_z z} \tag{4.194}$$

Therefore

$$E_z^2 = -\frac{Z_0}{jk_0}\frac{p}{n_2^2}Be^{-px - jk_z z} \tag{4.195}$$

where

$$-p^2 + k_z^2 = k_2^2 = (n_2 k_0)^2 \tag{4.196}$$

In order to find B and p, we impose continuity conditions on tangential components of electromagnetic fields at the dielectric interface. The continuity of tangential electric fields yields

$$E_z^1 = E_z^2, \quad x = a \tag{4.197}$$

$$2j\frac{Z_0 k_x}{k_0 n_1^2}\sin k_x a = -\frac{Z_0}{jk_0}\frac{p}{n_2^2}Be^{-p} \tag{4.198}$$

while the continuity of tangential magnetic field gives

$$H_y^1 = H_y^2, \quad x = a \tag{4.199}$$

$$2\cos k_x a = Be^{-pa} \tag{4.200}$$

Dividing the above two equations, we obtain

$$k_x \tan k_x a = pa(\frac{n_1}{n_2})^2 \tag{4.201}$$

or equivalently,

$$k_x a \tan k_x a = (\frac{n_1}{n_2})^2 \sqrt{(k_0 a)^2 (n_1^2 - n_2^2) - (k_x a)^2} \tag{4.202}$$

The above is a transcendental equation in $k_x a$. For a given guide configuration, $k_0 a$, n_1 and n_2, this equation has solutions for a finite number of values for k_x. A simple graphical solution can be obtained if we write

$$X \tan X = (\frac{n_1}{n_2})^2 \sqrt{V^2 - X^2} \tag{4.203}$$

where $X = k_x a$ and $V = k_0 a \sqrt{n_1^2 - n_2^2}$. Intersections of $X \tan X$ with the circles of radius V gives the solution to (4.203) as shown in Fig. 4.23.

From this graphical presentation, it is clear that m modes will be excited if

$$(m-1)\pi < V < m\pi \tag{4.204}$$

with cutoff frequencies given by

$$V_{nc} = n\pi, \quad 0 \le n < m-1 \tag{4.205}$$

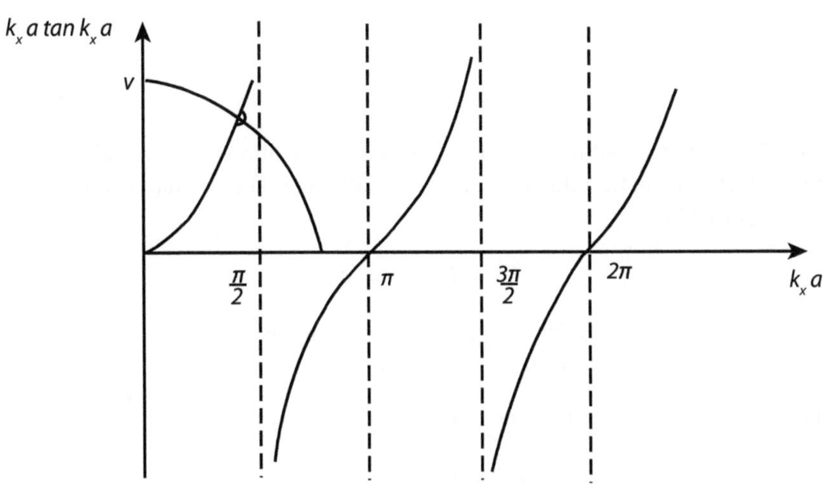

Fig. 4.23 Graphical solution of the transcendental dispersion equation (4.203)

The parameter V is known as the numerical aperture in optics and is directly related to the frequency of operation, slab thickness and the dielectric contrast.

Consider now the TE_z modes. In this case, we have

$$\mathbf{E} = \hat{y}E_y \tag{4.206}$$

and

$$E_x = -\frac{Z_0}{jk_0 n_1^2}\frac{\partial H_y}{\partial z}, \quad E_z = \frac{Z_0}{jk_0 n_1^2}\frac{\partial H_y}{\partial x}, \quad i = 1, 2 \tag{4.207}$$

subject to the boundary conditions at $x = 0$ and a. In the internal region ($0 \leq x \leq a$) we have

$$E_y^1 = e^{-j(k_x x + k_z z)} + Ce^{-j(-k_x x + k_z z)} \tag{4.208}$$

$$E_z^1 = -\frac{Z_0}{k_0}\frac{k_x}{n_1^2}\{e^{-j(k_x x + k_z z)} - Ae^{-j(-k_x x + K - zz)}\} \tag{4.209}$$

where the consistency relation holds between k_x, k_z and k_1. Enforcing the boundary condition at $x = 0$ gives $A = 1$.

In the external region, we write

$$H_y^2 = Be^{-px - jk_z z} \tag{4.210}$$

Therefore

$$E_z^2 = -\frac{Z_0}{jk_0}\frac{p}{n_2^2}Be^{-px - jk_z z} \tag{4.211}$$

where

$$-q^2 + k_z^2 = k_2^2 = (n_2 k_0)^2 \tag{4.212}$$

In order to find D and q, we impose continuity conditions on tangential components of electromagnetic fields at the dielectric interface. The continuity of tangential electric field yields

$$E_z^1 = E_z^2, \quad x = a \tag{4.213}$$

$$2j\frac{Z_0 k_x}{k_0 n_1^2}\sin k_x a = -\frac{Z_0}{jk_0}\frac{p}{n_2^2}Be^{-p} \tag{4.214}$$

while the continuity of the tangential magnetic field gives

$$H_y^1 = H_y^2, \quad x = a \tag{4.215}$$

$$2\cos k_x a = Be^{-pa} \tag{4.216}$$

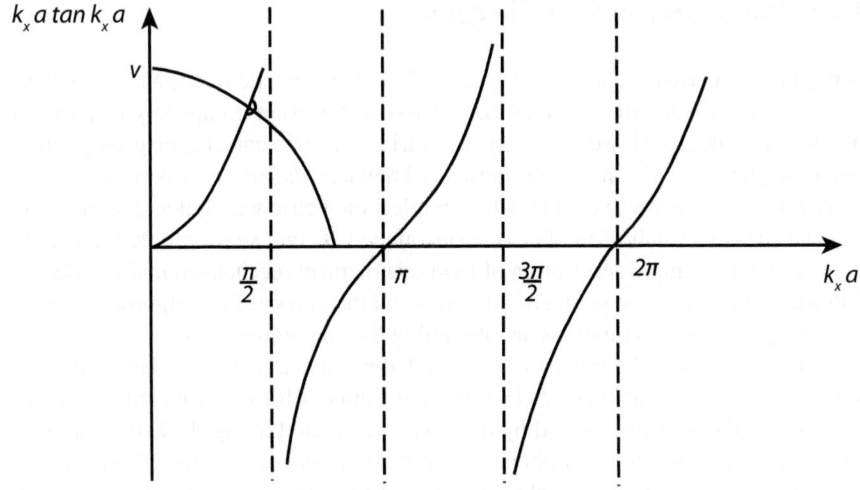

Fig. 4.24 Graphical solution of the transcendental dispersion equation (4.219)

Dividing the above two equations, we obtain

$$k_x \tan k_x a = pa(\frac{n_1}{n^2})^2 \tag{4.217}$$

or equivalently,

$$k_x a \tan k_x a = (\frac{n_1}{n_2})^2 \sqrt{(k_0 a)^2 (n_1^2 - n_2^2) - (k_x a)^2} \tag{4.218}$$

$$X \tan X = (\frac{n_1}{n_2})^2 \sqrt{V^2 - X^2} \tag{4.219}$$

A graphical solution of (4.219) is shown in Fig. 4.24.

From this graphical presentation, it is clear that m modes will be excited if

$$(m-1)\pi < V < m\pi \tag{4.220}$$

with cutoff frequencies given by

$$V_{nc} = n\pi, \quad 0 \leq n < m - 1 \tag{4.221}$$

The parameter V is known as the numerical aperture in optics and is directly related to the frequency of operation, slab thickness and the dielectric contrast.

4.2.3 The Dielectric Slab Waveguide

Waveguides are used to contain the energy of electromagnetic waves and guide them in a given direction. All examples considered so far, were completely or partially composed of metals. However purely dielectric structures can also be used to guide electromagnetic waves. In general these are known as *dielectric waveguides*.

The dielectric slab waveguide is the simplest dielectric waveguide. It consists of a planar dielectric slab of thickness d surrounded by two semi-infinite media. We start by giving a simple description of field confinement mechanism in the dielectric slab waveguide based on geometrical optics and then proceed to a rigorous analysis by solving Maxwell's equations and obtaining the confined modes.

Figure 4.27 shows the trajectory of a guided wave in a dielectric slab with index of refraction n_1 and thickness d. The refractive indices for two semi-infinite media surrounding the slab are denoted by n_2 and n_3 as denoted in Fig. 4.27. In general n_2 and n_3 are not equal and the problem is referred to as the *asymmetric dielectric slab waveguide*. If $n_2 = n_3$ the problem is referred to as the *symmetric dielectric slab waveguide*.

The guided wave is incident at the two boundaries with an angle of θ_1 and refraction angles for two semi-infinite media are denoted by θ_2 and θ_3 as denoted in Fig. 4.27. Without loss of generality, we assume $n_2 \geq n_3$. Now let us assume that the dielectric slab is denser than both media surrounding it

$$n_1 > n_2 \geq n_3 \tag{4.222}$$

Assuming (4.222) is true, total reflection can occur if the incident angle θ_1 is larger than critical angles at two boundaries. Denoting the critical angles at the top and bottom boundaries θ_c^1 and θ_c^2, respectively, we have

$$\theta_c^1 = \sin^{-1}\left(\frac{n_3}{n_1}\right) \tag{4.223}$$

$$\theta_c^2 = \sin^{-1}\left(\frac{n_2}{n_1}\right) \tag{4.224}$$

Since it was assumed $n_2 \geq n_3$, we can also conclude that $\theta_c^2 \geq \theta_c^1$. Thus, if θ_1 is larger than θ_c^2, total reflection occurs at both boundaries and wave is completely confined in the dielectric slab. As it will be shown, the condition in (4.222) which requires the dielectric slab be denser than both media surrounding it, is the necessary condition to achieve mode confinement. Otherwise total reflection will not be possible and power will leak out from one or both boundaries. We now proceed to a more rigorous treatment of the problem starting from Maxwell's equations and using the boundary conditions at the interfaces. For simplicity, we consider the problem of *symmetric dielectric slab waveguide*. The extension of results to the asymmetric case is straight forward.

Figure 4.25 shows the physical setup of a symmetric dielectric slab waveguide of thickness d. It is assumed that the slab is in the y-z plane and extends from $x = -d/2$

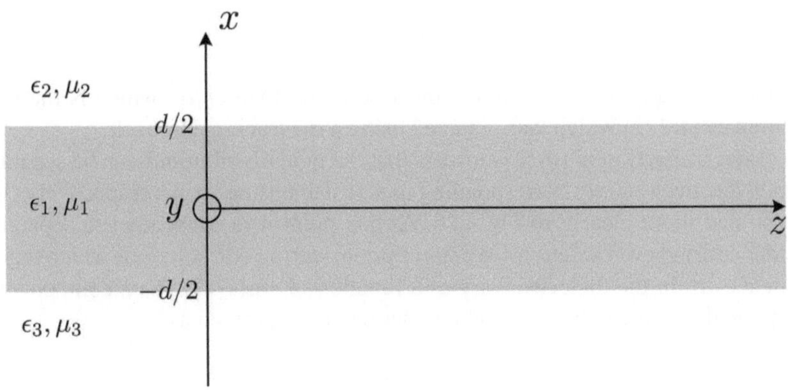

Fig. 4.25 Dielectric Slab Waveguide; propagation is along z-axis

to $x = d/2$. The permittivity and permeability of the slab and the surrounding half-spaces are denoted according to Fig. 4.25. Without of generality we assume there is no variation along y, hence

$$\frac{\partial}{\partial y} = 0 \tag{4.225}$$

The modes of the slab can be classified as TE and TM modes. TE modes have no electric field component in the direction of wave propagation and TM modes have no magnetic field component in the direction of wave propagation.

4.2.3.1 TM Modes

The only non-zero field components for TM modes are H_y, E_x, and E_z. For waves to be guided inside the slab, fields must be evanescent in $\pm\hat{x}$ directions. The z component of electric field in three regions can be expressed as

$$E_z = E^+ e^{-\alpha(x-d/2)} e^{-jk_z z} \quad x \geq d/2 \tag{4.226}$$
$$E_z = (A\sin(k_x x) + B\cos(k_x x))\, e^{-jk_z z} \quad -d/2 \leq x \leq d/2 \tag{4.227}$$
$$E_z = E^- e^{\alpha(x+d/2)} e^{-jk_z z} \quad x \leq -d/2 \tag{4.228}$$

where, A, B, E^+, and E^- are constants to be determined. Wave vector components k_x, k_z, and α must obey the dispersion relations

$$k_z^2 + k_x^2 = \omega^2 \epsilon \mu \tag{4.229}$$
$$k_z^2 - \alpha^2 = \omega^2 \epsilon_0 \mu_0 \tag{4.230}$$

We can eliminate k_z from (4.229) and (4.230) arriving at

$$k_x^2 + \alpha^2 = \omega^2 (\epsilon\mu - \epsilon_0\mu_0) \tag{4.231}$$

In order for k_x and α to be real, $\epsilon\mu$ must be larger than $\epsilon_0\mu_0$, which is the same condition as (4.222) which was obtained using a ray tracing approach.

A closer inspection of (4.227) reveals that the field distributions can be separated into odd and even modes corresponding to sine and cosine terms, respectively. This is illustrated more clearly in Fig. 4.26. At this point it is convenient to separately consider odd and even solutions. We start by considering odd solutions which require that fields within the dielectric vary as a function of $\sin(k_x x)$. The z component of electric field in three regions for odd modes can be expressed as

$$E_z = E_o^+ e^{-\alpha(x-d/2)} e^{-jk_z z} \quad x \geq d/2 \tag{4.232}$$

$$E_z = A \sin(k_x x) e^{-jk_z z} \quad -d/2 \leq x \leq d/2 \tag{4.233}$$

$$E_z = E_o^- e^{\alpha(x+d/2)} e^{-jk_z z} \quad x \leq -d/2 \tag{4.234}$$

Using Maxwell's equations, we can obtain E_x and H_y as

$$E_x = -\frac{jk_z}{\alpha} E_o^+ e^{-\alpha(x-d/2)} e^{-jk_z z} \quad x \geq d/2 \tag{4.235}$$

$$E_x = -\frac{jk_z}{k_x} A \cos(k_x x) e^{-jk_z z} \quad -d/2 \leq x \leq d/2 \tag{4.236}$$

$$E_x = \frac{jk_z}{\alpha} E_o^- e^{\alpha(x+d/2)} e^{-jk_z z} \quad x \leq -d/2 \tag{4.237}$$

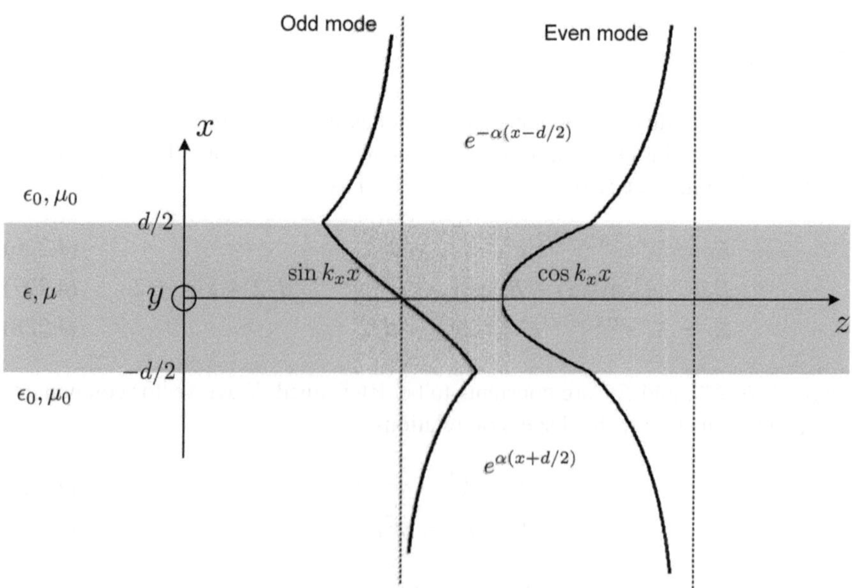

Fig. 4.26 Two lowest order modes of a symmetric dielectric slab waveguide

and

$$H_y = -\frac{j\omega\epsilon_0 E_o^+}{\alpha} e^{-\alpha(x-d/2)} e^{-jk_z z} \quad x \geq d/2 \tag{4.238}$$

$$H_y = -\frac{j\omega\epsilon A}{\alpha} \cos(k_x x) e^{-jk_z z} \quad -d/2 \leq x \leq d/2 \tag{4.239}$$

$$H_y = \frac{j\omega\epsilon_0 E_o^+}{\alpha} e^{\alpha(x+d/2)} e^{-jk_z z} \quad x \leq -d/2 \tag{4.240}$$

Applying boundary conditions at $x = \pm d/2$ and solving the resulting equations will lead to

$$\alpha = \frac{\epsilon_0}{\epsilon} k_x \tan\left(\frac{k_x d}{2}\right) \tag{4.241}$$

Next, we consider the even modes. The z component of electric field in three regions for even modes can be expressed as (Fig. 4.27)

$$E_z = E_e^+ e^{-\alpha(x-d/2)} e^{-jk_z z} \quad x \geq d/2 \tag{4.242}$$

$$E_z = B \cos(k_x x) e^{-jk_z z} \quad -d/2 \leq x \leq d/2 \tag{4.243}$$

$$E_z = E_e^- e^{\alpha(x+d/2)} e^{-jk_z z} \quad x \leq -d/2 \tag{4.244}$$

We follow a similar procedure which was performed for odd modes, eventually arriving at

$$\alpha = -\frac{\epsilon_0}{\epsilon} k_x \cot\left(\frac{k_x d}{2}\right) \tag{4.245}$$

In order to obtain k_x and α for odd and even modes, (4.241) and (4.245) must be simultaneously solved with (4.231) which requires solving non-linear systems. Figure 4.28 shows a graphical solution for (4.241) and (4.245) where V denotes the radius of the circle and its value is given by

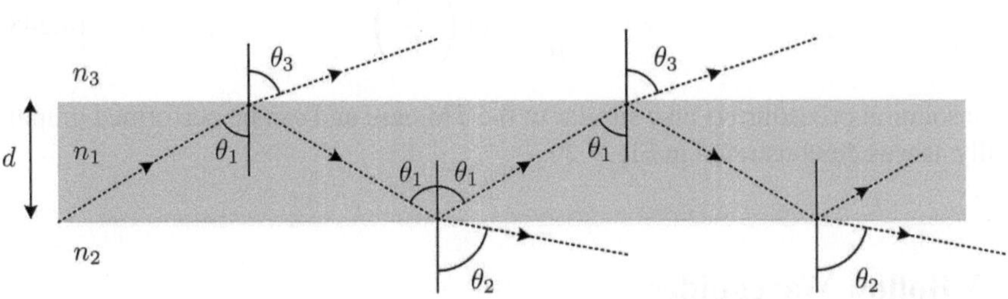

Fig. 4.27 A dielectric slab waveguide of thickness d

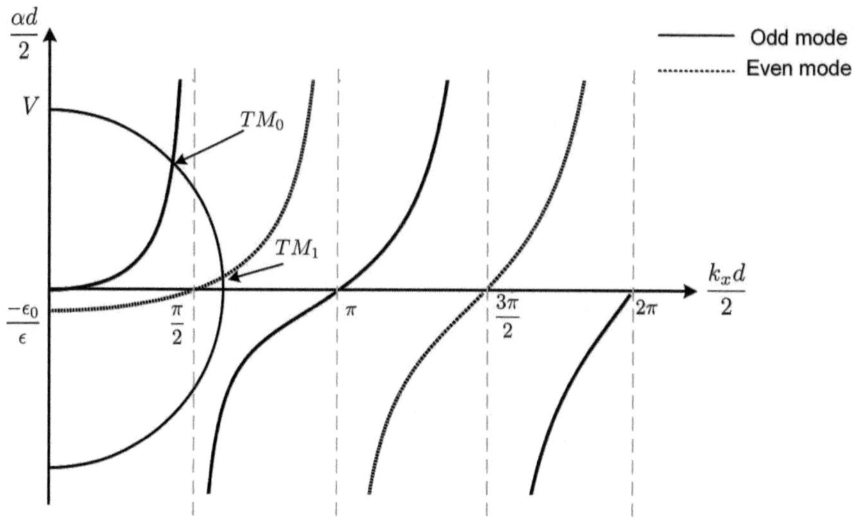

Fig. 4.28 Graphical solution of the transcendental dispersion equation (4.241) and (4.245)

$$V = \frac{\omega d}{2}\sqrt{\epsilon\mu - \epsilon_0\mu_0} \tag{4.246}$$

4.2.3.2 TE Modes

The same procedure is performed for TE solutions. For the sake of brevity we forgo all the derivations and just include the final results. For odd TE modes, the dispersion equation is

$$\alpha = \frac{\mu_0}{\mu}k_x \tan\left(\frac{k_x d}{2}\right) \tag{4.247}$$

and for even TE modes, the resulting equation is

$$\alpha = -\frac{\mu_0}{\mu}k_x \cot\left(\frac{k_x d}{2}\right) \tag{4.248}$$

The solution procedure is also similar to the TM case and can be performed graphically as was demonstrated in Fig. 4.29.

4.3 Hollow Waveguides

In an infinitely long cylindrical hollow waveguide, the spatial variations of the fields in the z-direction may be assumed to take the form

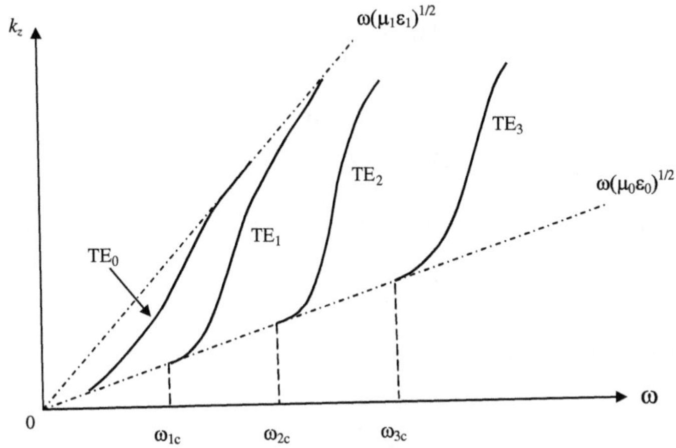

Fig. 4.29 Schematic dispersion diagram of different modes for a dielectric slab waveguide

$$\mathbf{E}(x, y, z) = \mathbf{E}(x, y)e^{-jk_z z}$$
$$\mathbf{H}(x, y, z) = \mathbf{H}(x, y)e^{-jk_z z} \tag{4.249}$$

With $k_z > 0$, the above represents waves propagating in the $+z$-direction. With the z-dependence separated out this way, all field quantities can be expressed in terms of the longitudinal components, E_z and H_z.

4.3.1 Waveguide Modes

The fields in a hollow waveguide, as depicted in (Fig. 4.30) generally consist of a superposition of transverse electric and transverse magnetic (to z) types. These two field types can be written as

4.3.1.1 TE$_z$ Modes

$$H_z = \psi_h \tag{4.250}$$

Fig. 4.30 A cylindrical hollow waveguide

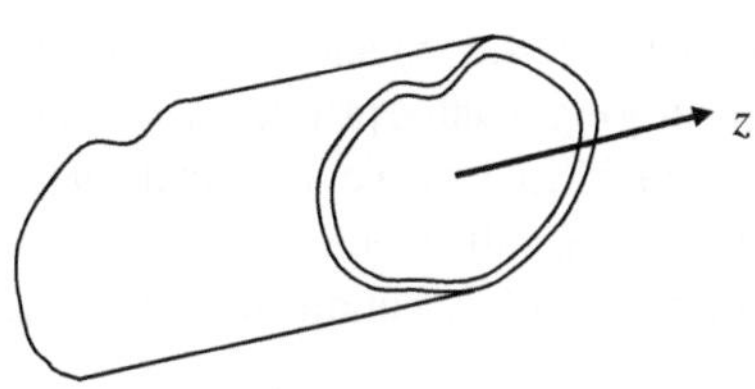

$$\mathbf{H}_t = -\frac{j\beta}{\Gamma^2}\nabla_t\psi_h \tag{4.251}$$

$$E_z = 0 \tag{4.252}$$

$$\mathbf{E}_t = Z_{TE}(\mathbf{H}_t \times \widehat{z}) \tag{4.253}$$

$$Z_{TE} = kZ/\beta \tag{4.254}$$

where ψ_h satisfies the Helmholtz equation

$$(\nabla_t^2 + \Gamma^2)\psi_h = 0 \tag{4.255}$$

subject to the boundary condition

$$\frac{\partial\psi_h}{\partial n}|_S = 0 \tag{4.256}$$

and Γ satisfies the compatibility condition

$$\Gamma^2 = k^2 - \beta^2 \tag{4.257}$$

4.3.1.2 TM$_Z$ Modes

$$E_z = \psi_e \tag{4.258}$$

$$\mathbf{E}_t = -\frac{j\beta}{\Gamma^2}\nabla_t\psi_e \tag{4.259}$$

$$H_z = 0 \tag{4.260}$$

$$\mathbf{H}_t = \frac{\widehat{z} \times \mathbf{E}_t}{Z_{TM}} \tag{4.261}$$

$$Z_{TM} = \beta Z/k \tag{4.262}$$

where ψ_e satisfies the Helmholtz equation

$$(\nabla_t^2 + \Gamma^2)\psi_e = 0 \tag{4.263}$$

subject to the boundary condition

$$\psi_e|_S = 0 \tag{4.264}$$

Equations (4.255) and (4.263) subject the boundary conditions (4.256) and (4.264) specify eigenvalue problems. Because ψ must be oscillatory in order to satisfy boundary conditions on a cylindrical surface, Γ^2 must be non-negative. For a certain value of $\Gamma = \Gamma_i$, called the *eigenvalue*, there corresponds a solution ψ_i satisfying the appropriate boundary condition, called the *eigenfunction*. In general, there are an infinite number of discrete eigenvalues and corresponding number of eigenfunctions. In waveguide problems, it is customary to call the eigenfunctions as *modes*.

4.3.2 Cutoff Frequency

For a given frequency ω, the propagation constant in the guide is determined for each mode as

$$\beta_i^2 = k^2 - \Gamma_i^2 \tag{4.265}$$

By definition, the cutoff or no propagation for the mode i occurs when $\beta_i = 0$ at the *cutoff frequency* ω_{ci} such that

$$k_{ci} = \Gamma_i \tag{4.266}$$

and

$$\omega_{ci} = c\Gamma_i \tag{4.267}$$

where c is the speed of light. Thus

$$f_{ci} = \frac{c}{2\pi}\Gamma_i \tag{4.268}$$

Using the above, the ith mode propagation constant can be expressed as

$$\begin{aligned} \beta_i &= k(1 - \frac{\omega_{ci}^2}{\omega^2})^{1/2} \\ &= k(1 - \frac{f_{ci}^2}{f^2})^{1/2} \end{aligned} \tag{4.269}$$

Note that for $f > f_{ci}$, the propagation constant β_i is real and the waves of the ith mode can propagate without attenuation along the guide; while for $f < f_{ci}$, β_i is imaginary and such modes cannot propagate and are called cutoff or *evanescent* modes.

4.3.3 Guide Wavelength

The guide wavelength for the ith mode is defined as

$$\lambda_{gi} = 2\pi/\beta_i \tag{4.270}$$

and we obtain from (4.269)

$$\lambda_{gi} = \lambda(1 - \frac{f_{ci}^2}{f^2})^{1/2} \tag{4.271}$$

which indicates the wavelength in the guide is always greater than the free space wavelength.

The phase velocity in the guide for the ith mode is given by

$$\begin{aligned} v_{pi} &= \frac{\omega}{\beta_i} = \frac{\omega}{k} \frac{1}{(1 - f_{ci}^2/f^2)^{1/2}} \\ &= \frac{c}{(1 - f_{ci}^2/f^2)^{1/2}} \end{aligned} \tag{4.272}$$

where c is the phase velocity (speed of light) in the free medium. It is noted that the phase velocity in the guide v_{pi} is greater than c above cutoff and is infinite at cutoff.

The group velocity of the ith mode is given by

$$v_{gi} = \frac{\partial \omega}{\partial \beta_i} = \frac{\partial (ck)}{\partial \beta_i} = c \frac{\partial k}{\partial \beta_i} \tag{4.273}$$

Therefore, we find that

$$v_{gi} = \frac{\beta_i}{k} = c(1 - \frac{f_{ci}^2}{f^2})^{1/2} \tag{4.274}$$

indicating that the group velocity is zero at cutoff and

$$v_{pi} v_{gi} = c^2 \tag{4.275}$$

4.3.4 Orthogonality of Modes

The mode functions possess some orthogonality properties over the cross section of the guide. Consider the cross sectional area of the guide as S bounded by the contour C. Take two different mode functions ψ_i and ψ_j where $i \neq j$ and ψ_i and ψ_j may be either E or H type.

Since ψ_i and ψ_j are two distinct eifenfunctions, it can be shown by applying the Green's second theorem that their inner product is zero. Thus

$$\int_S \psi_i \psi_j \mathbf{ds} = 0, \quad i \neq j \tag{4.276}$$

Recalling that ψ represent the longitudinal components E_z and/or H_z, we find that the axial components of the fields for two different modes are orthogonal and there is no coupling between the two.

Also by applying the Green's first identity, we find that

$$\int_S \nabla_t\psi_i \cdot \nabla_t\psi_j \mathbf{ds} = 0, \quad i \neq j \tag{4.277}$$

Recalling that $\mathbf{E}_t$ is related to $\nabla_t\psi$ for E-modes and $\mathbf{H}_t$ is related to $\nabla_t\psi$ for H-modes, we find that the transverse electric fields for two different E-modes, and the transverse magnetic fields for two different H-modes are orthogonal.

We also have

$$\int_S (\nabla_t\psi_i \times \widehat{z}) \cdot (\nabla_t\psi_j \times \widehat{z})\mathbf{ds} = 0, \quad i \neq j \tag{4.278}$$

Since $\mathbf{H}_t$ is related to $\nabla_t\psi \times \widehat{z}$ for E-modes, and $\mathbf{E}_t$ is related to $\nabla_t\psi \times \widehat{z}$ for H-modes, the above implies that the transverse magnetic fields for two different E-modes, and the transverse electric fields of two different H-modes are orthogonal.

Finally, we note that

$$\int_S (\nabla_t\psi_{hi} \times \widehat{z}) \cdot \nabla_t\psi_{ej}\mathbf{ds} = 0, \quad i \neq j \tag{4.279}$$

showing that the transverse electric fields of one H-mode and one E-mode are orthogonal. Similarly, for the transverse magnetic fields.

In summary, the transverse electric fields for two different E-modes or for one E-mode and H-mode, are mutually orthogonal. The same is true for the transverse magnetic fields for any two (E, H, E and H) modes.

The above orthogonality properties break down for guides with finitely conducting walls as well as for those filled inhomogeneously. In the former case, modes are coupled through imperfect walls, and in the latter, the normal modes are neither E nor H—rather a combination of both (hybrid modes).

In a lossless guide, the power flow is the sum of the power carried by each mode separately.

4.3.5 The Rectangular Hollow Waveguide

Rectangular waveguides were some of the earliest types of transmission lines used in many practical radio-frequency systems. Figure 4.31 shows the geometry of a rectangular waveguide filled with a material with permittivity ϵ and permeability μ. The walls of the waveguide are assumed to be PEC. The dimensions of the waveguide along the x and y axis are denoted by a and b respectively, and it is assumed that the waveguide extends infinitely along the z axis. By convention it is assumed that $a > b$, however this assumption does not affect the generality of our results.

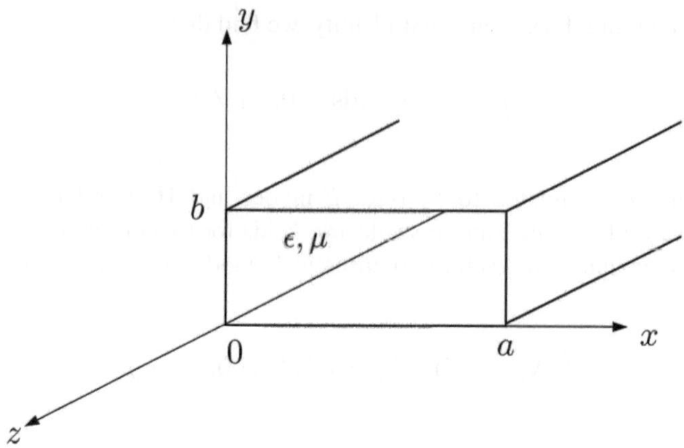

Fig. 4.31 A rectangular hollow waveguide

It can be shown that the boundary conditions of the rectangular hollow waveguide do not support TEM modes. However TE_z and TM_Z modes can propagate along the waveguide under certain conditions and we study them separately.

4.3.5.1 TE_z Modes

The TE_z modes are characterized by $E_z = 0$ and $H_z = \psi_h(x, y)e^{-j\beta z}$ where $\psi_h(x, y)$ must satisfy (4.255). Writing (4.255) in the Cartesian system we arrive at the reduced wave equation

$$\left(\frac{\partial^2}{\partial x^2} + \frac{\partial^2}{\partial y^2} + \Gamma^2\right) \psi_h(x, y) = 0 \tag{4.280}$$

where Γ satisfies the compatibility condition

$$\Gamma^2 = k^2 - \beta^2 \tag{4.281}$$

Using the method of separation of variables, we let

$$\psi_h(x, y) = X(x)Y(y) \tag{4.282}$$

Substituting (4.282) into (4.280) we arrive at

$$\frac{1}{X}\frac{d^2X}{dx^2} + \frac{1}{Y}\frac{d^2Y}{dy^2} + \Gamma^2 = 0 \tag{4.283}$$

Equation (4.283) can be decoupled into two ordinary linear second order differential equations by introducing separation constants k_x and k_y

$$\frac{d^2X}{dx^2} + k_x^2 X = 0 \tag{4.284}$$

$$\frac{d^2Y}{dy^2} + k_y^2 Y = 0 \tag{4.285}$$

subject to constraint

$$k_x^2 + k_y^2 = \Gamma^2 \tag{4.286}$$

Hence the general solution for $\psi_h(x, y)$ has the following form

$$\psi_h(x, y) = (A \cos k_x x + B \sin k_x x) \left(C \cos k_y y + D \sin k_y y\right) \tag{4.287}$$

To evaluate (4.287), we must apply boundary conditions which requires the tangential components of electric field to vanish at waveguide walls. That is

$$E_x(x, 0, z) = 0 \tag{4.288}$$

$$E_x(x, b, z) = 0 \tag{4.289}$$

$$E_y(0, y, z) = 0 \tag{4.290}$$

$$E_y(a, y, z) = 0 \tag{4.291}$$

Solving the boundary conditions, it can be shown that $D = B = 0$ and

$$k_x = \frac{m\pi}{a} \qquad m = 0, 1, 2, \ldots \tag{4.292}$$

$$k_y = \frac{n\pi}{b} \qquad n = 0, 1, 2, \ldots \tag{4.293}$$

To further simplify the solution, we can combine constants $AC \equiv A_{mn}$. Thus, the general solution for H_z of the TE_{mn} mode is

$$H_z = A_{mn} \cos \frac{m\pi x}{a} \cos \frac{n\pi y}{b} e^{-j\beta z} \tag{4.294}$$

The remaining field components can be obtained using Maxwell's equations and are

$$E_x = \frac{j\omega\mu n\pi}{\Gamma^2 b} A_{mn} \cos \frac{m\pi x}{a} \sin \frac{n\pi y}{b} e^{-j\beta z} \tag{4.295}$$

$$E_y = \frac{-j\omega\mu m\pi}{\Gamma^2 a} A_{mn} \sin \frac{m\pi x}{a} \cos \frac{n\pi y}{b} e^{-j\beta z} \tag{4.296}$$

$$E_z = 0 \tag{4.297}$$

$$H_x = \frac{j\beta m\pi}{\Gamma^2 a} A_{mn} \sin\frac{m\pi x}{a} \cos\frac{n\pi y}{b} e^{-j\beta z} \tag{4.298}$$

$$H_y = \frac{j\beta n\pi}{\Gamma^2 b} A_{mn} \cos\frac{m\pi x}{a} \sin\frac{n\pi y}{b} e^{-j\beta z} \tag{4.299}$$

Here it is important to note that the special case of $m = n = 0$ is excluded since it will lead to a trivial solution. The propagation constant β for the TE_{mn} mode is

$$\beta = \sqrt{k^2 - \left(\frac{m\pi}{a}\right)^2 - \left(\frac{n\pi}{b}\right)^2} \tag{4.300}$$

The cutoff frequency for the TE_{mn} mode is denoted by $(f_c)_{mn}$ and given by

$$(f_c)_{mn} = \frac{1}{2\pi\sqrt{\epsilon\mu}}\sqrt{\left(\frac{m\pi}{a}\right)^2 + \left(\frac{n\pi}{b}\right)^2} \tag{4.301}$$

Since it was assumed that $a > b$, it can easily be shown that the mode with the lowest cutoff frequency is the TE_{10} mode with

$$(f_c)_{10} = \frac{1}{2a\sqrt{\epsilon\mu}} \tag{4.302}$$

The mode with the lowest cutoff frequency is also called the dominant mode. The cutoff frequency of the dominant mode is particularly important since only waves with frequencies above it can propagate along the waveguide.

4.3.5.2 TM_Z Modes

For TM_Z modes we follow a similar procedure that was performed for TE_z modes. For the sake of brevity, we forgo the derivations and just show the results. The general solution for the field components of the TM_{mn} mode are

$$E_x = \frac{-j\beta m\pi}{a\Gamma^2} B_{mn} \cos\frac{m\pi x}{a} \sin\frac{n\pi y}{b} e^{-j\beta z} \tag{4.303}$$

$$E_y = \frac{-j\beta n\pi}{b\Gamma^2} B_{mn} \sin\frac{m\pi x}{a} \cos\frac{n\pi y}{b} e^{-j\beta z} \tag{4.304}$$

$$E_z = B_{mn} \sin\frac{m\pi x}{a} \sin\frac{n\pi y}{b} e^{-j\beta z} \tag{4.305}$$

$$H_x = \frac{j\omega\epsilon n\pi}{b\Gamma^2} B_{mn} \sin\frac{m\pi x}{a} \cos\frac{n\pi y}{b} e^{-j\beta z} \tag{4.306}$$

$$H_y = \frac{-j\omega\epsilon m\pi}{a\Gamma^2} B_{mn} \cos\frac{m\pi x}{a} \sin\frac{n\pi y}{b} e^{-j\beta z} \tag{4.307}$$

$$H_z = 0 \tag{4.308}$$

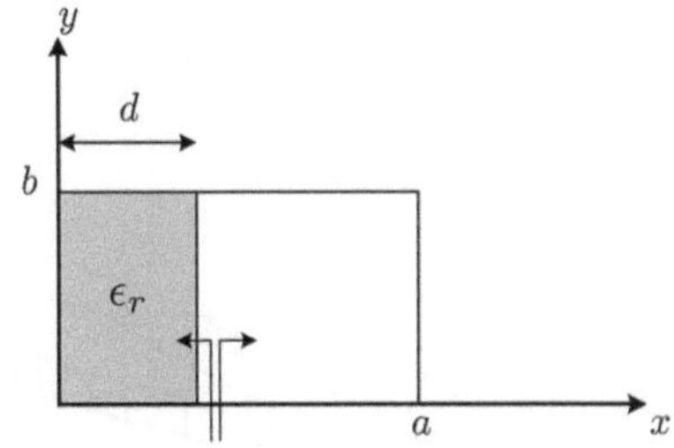

Fig. 4.32 A partially filled rectangular waveguide

The propagation constant β is given by the same expression as the TE_z case in (4.300).

A close inspection of field expressions for TM_Z modes reveals that TM_{00}, TM_{10}, and TM_{01} all lead to trivial solutions. Hence the dominant mode is the TM_{11} mode with a cutoff frequency of (Fig. 4.32)

$$(f_c)_{11} = \frac{1}{2\sqrt{\epsilon\mu}}\sqrt{\left(\frac{1}{a}\right)^2 + \left(\frac{1}{b}\right)^2} \tag{4.309}$$

4.3.6 The Corrugated Rectangular Waveguide

Corrugated horns are widely known to be excellent antennas for a wide range of microwave applications. They serve as primary feeds in microwave reflector antennas. A typical corrugated rectangular waveguide is shown in Fig. 4.33.

Corrugator horns taper the electric field so that the diffraction from the edges of the horn and side-lobe levels are reduced. The overall main beam becomes smooth and almost rotationally symmetric. This is crusial when the horn is used as a feed to a reflector antenna, because it maximazies the cross-polarization discrimination of the antenna.

The key to the analysis of this type of waveguides is the ability to model the scattering matrix at a given junction between two corrugations shown in Fig. 4.10. The derivation of the scattering matrix at a junction involves the matching of the total power in all modes in both sides of the junction. The number of models on the left-hand side of the junction and the number of modes on the right-hand side of the junction can in general be arbitrary. It can be shown that the number of modes on both sides of the junction must obey the conditions of relative convergence for good convergence of the results. However, the analysis and the computational procedure are simplified if the number of modes is the same on both sides of the junction (Fig. 4.33).

Each uniform waveguide section in the junction contains travelling waves in which the transverse electric fields can be represented as a spectrum of modes. The transverse electric and magnetic modal functions on the left-hand side of the junction are represented by the subscript L (that is e_{nL} and h_{nL}), and those on the right-hand side of the junction by the subscript R.

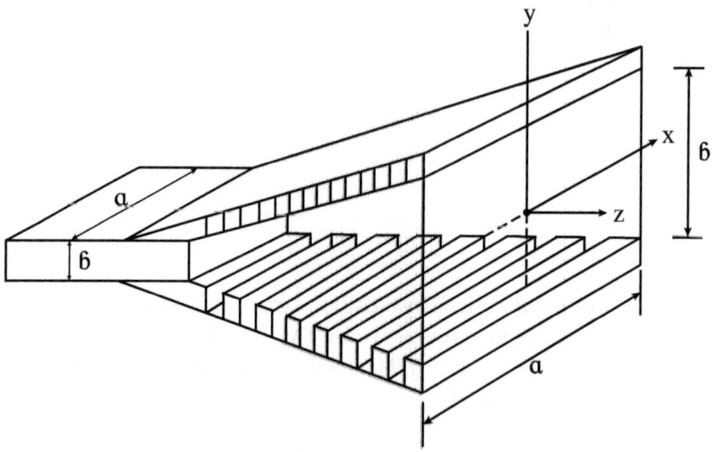

Fig. 4.33 A corrugated rectangular waveguide

On the left-hand side of the junction, the total transverse electric and magnetic fields are given by

$$\mathbf{E}_L = \sum_{n=1}^{N_L} [A_n e^{-\gamma_n z} + B_n e^{\gamma_n z}] e_{nL} \tag{4.310}$$

and

$$\mathbf{H}_L = \sum_{n=1}^{N_L} [A_n e^{-\gamma_n z} - B_n e^{\gamma_n z}] h_{nL} \tag{4.311}$$

where A_n and B_n are the forward and reflected amplitude coeeficients of mode n on the left-hand side of the junction and γ_n is its corresponding propagation constant. Also, N_L is the number of modes retained for the left-hand side of the junction. On the right-hand side of the junction, the total transverse electric and magnetic fields have the form

$$\mathbf{E}_R = \sum_{n=1}^{N_R} [D_n e^{-\gamma_n z} + C_n e^{\gamma_n z}] e_{nR} \tag{4.312}$$

and

$$\mathbf{H}_R = \sum_{n=1}^{N_R} [D_n e^{-\gamma_n z} - C_n e^{\gamma_n z}] h_{nR} \tag{4.313}$$

where C_n and D_n are the forward and reflected amplitude coefficients of mode n on the right-hand side of the junction, looking into the junction and γ_n is the corresponding propagation constant. The number of modes retained for the right-hand side of the junction is denoted by N_R.

The total transverse fields must match across the junction. If the cross-section area of the waveguide on the left-hand side of the junction is S_L and that on the right-hand side of the junction is S_R, the boundary conditions give that the transverse electric fields over the area $S_R - S_L$ will be zero. The tangential fields over the area S_L will be continuous. The expansions (4.198)–(4.201) are used for enforcing the continuity of the tangential fields across the junction which is described by

$$\mathbf{E}_R = \begin{cases} \mathbf{E}_L & on\ S_L \\ 0 & on\ S_R - S_L \end{cases} \tag{4.314}$$

and

$$\mathbf{H}_R = \mathbf{H}_L \text{on } S_L \tag{4.315}$$

Imposing the above boundary conditions at $z = 0$, we find

$$\sum_{n=1}^{N_R}(C_n + D_n)e_{nR} = \begin{cases} \sum\limits_{n=1}^{N_L}(A_n + B_n)e_{nL} & on\ S_L \\ \\ 0 & on\ S_R - S_L \end{cases} \tag{4.316}$$

$$\sum_{n=1}^{N_R}(C_n + D_n)\mathbf{h}_{nR} = \sum_{n=1}^{N_L}(A_n - B_n)\mathbf{h}_{nL} \quad on\ S_L \tag{4.317}$$

The vector product of the terms in the electric field continuity Eq. (4.204) is taken with a magnetic mode function from the right-hand side of the junction $\widehat{h}_{nR}$, and integrated over the surface S_R. Also, the vector product of the terms in the magnetic field continuity Eq. (4.205) is taken with an electric mode function from the left-hand side of the junction $\widehat{e}_{nL}$ and integrated over the surface S_L. Using the orthogonality of modes, this leads to a pair of simultaneous matrix equations

$$\begin{aligned} P^T(A + B) &= Q(C + D) \\ R(A - B) &= P(D - C) \end{aligned} \tag{4.318}$$

where $[A]$ and $[B]$ are N_L-element column vectors in the section on the left-hand side of the junction containing the unknown modal coefficients A_{nL} and B_{nL}. Similarly, $[C]$ and $[D]$ are N_R-element column vectors in the section for the right-hand side of the junction containing the unknown modal coefficients C_{nR} and D_{nR}. The matrix $[P]$ in an $N_L \times N_R$ square matrix whose elements are integrals representing the mutual coupled power between mode i on the left-hand side and mode j on the right-hand side

$$P_{ij} = \int_{S_L} (e_{iL} \times \mathbf{h}_{jR}).\widehat{z}ds \tag{4.319}$$

The matrix $[P]^T$ is the transpose of $[P]$. The matrix $[Q]$ is an $N_R \times N_R$ diagonal matrix describing the self-coupled power between modes on the right-hand side of the junction. The elements of $[Q]$ are given by

$$Q_{ii} = \int_{S_R} (e_{iR} \times \mathbf{h}_{iR}).\widehat{z}ds \tag{4.320}$$

Similarly the matrix $[R]$ is an $N_L \times N_L$ diagonal matrix describing the self-coupled power between modes on the left-hand side of the junction. The elements are expressed as

$$R_{ii} = \int_{S_L} (e_{iL} \times \mathbf{h}_{iL}).\widehat{z}ds \tag{4.321}$$

The power-coupling integrals in (4.207)–(4.209) contain the information about the type of waveguide on either side of the junction. They must be evaluated for the appropriate homogeneous or inhomogeneous cross-section. This can be done either analytically or numerically. Numerical evaluation reduces the amount of mathematical effort but increases the computational time since the integrals need to be evaluated for all modal combinations at each junction. Analytical evaluation is only possible in some cases and may involve considerable mathematics, but where it is possible the saving in computer time can be considerable.

Equation (4.206) can be rearranged into the scattering matrix form as

$$\begin{bmatrix} [B] \\ D \end{bmatrix} = \begin{bmatrix} S_{11} & S_{12} \\ S_{21} & S_{22} \end{bmatrix} \begin{bmatrix} [A] \\ C \end{bmatrix} \tag{4.322}$$

where the elements of $[S]$ are given by

$$\begin{aligned}
[S_{11}] &= \{[R] + [P][Q]^{-1}[P]^T\}^{-1}\{[R] - [P][Q]^{-1}[P]^T\} \\
[S_{12}] &= 2\{[R] + [P][Q]^{-1}[P]^T\}^{-1}[P] \\
[S_{21}] &= 2\{[Q] + [P]^T[R]^{-1}[P]\}^{-1}[P]^T \\
[S_{22}] &= -\{[Q] + [P]^T[R]^{-1}[P]\}^{-1}\{[Q] - [P]^T[R]^{-1}[P]\}
\end{aligned} \tag{4.323}$$

The scattering matrix $[S]$ described above is called the *generalized scattering matrix*. This matrix should not be confused with the conventional scattering matrix $[S_c]$, which relates only wave amplitudes of propagating modes. If it is desirable to compute the $[S_c]$ matrix for a waveguide junction, then the corresponding generalized $[S]$ matrix should be computed using (ideally) an infinite number of modes.

Subsequently, only the scattering matrix elements of $[S]$, which relate propagating modes, should be selected to construct the conventional scattering matrix $[S_c]$.

The above analysis assumes that the area S_R is greater than the area S_L. If this is not the case (the area S_L is greater than the area S_R), the elements of $[S]$ become

$$\begin{aligned}
[S_{11}] &= -\{[R] + [P]^T[Q]^{-1}[P]\}^{-1}\{[R] - [P]^T[Q]^{-1}[P]\} \\
[S_{12}] &= 2\{[R] + [P]^T[Q]^{-1}[P]\}^{-1}[P]^T \\
[S_{21}] &= 2\{[Q] + [P][R]^{-1}[P]^T\}^{-1}[P] \\
[S_{22}] &= \{[Q] + [P][R]^{-1}[P]^T\}^{-1}\{[Q] - [P][R]^{-1}[P]^T\}
\end{aligned} \tag{4.324}$$

The elements of the matrices $[P]$, $[Q]$, and $[R]$ are given by

$$P_{ij} = \int_{S_R} (\mathbf{e}_{iR} \times \mathbf{h}_{jL}) \cdot \mathbf{ds}_R \tag{4.325}$$

$$Q_{ii} = \int_{S_R} (\mathbf{e}_{iR} \times \mathbf{h}_{iR}) \cdot \mathbf{ds}_R \tag{4.326}$$

$$R_{ii} = \int_{S_L} (\mathbf{e}_{iL} \times \mathbf{h}_{iL}) \cdot \mathbf{ds}_L \tag{4.327}$$

This completes the description of the general modal-matching technique for analyzing any combination of rectangular junctions and waveguide sections.

There are usually some uniform sections of guides between the junctions. The scattering matrix elements for a uniform section of guide are

$$[S_{11}] = [S_{22}] = 0, \quad [S_{12}] = [S_{21}] = [V] \tag{4.328}$$

where $[V]$ is an $N \times N$ diagonal matrix with elements

$$V_{nn} = e^{-\gamma_n \ell} \tag{4.329}$$

where ℓ is the length of the section and γ_n is the propagation constant for the nth mode in the waveguide $(1 < n < N)$. In principle, the waveguide sections could contain lossy material so that γ_n is complex. However, this would lead to extensive computations and is generally unnecessary since the influence of lossy materials can usually be adequately accounted for by a perturbation approach. The propagation coefficient is normally either purely imaginary ($\gamma_n = j\beta_n$) for travelling modes or is purely real ($\gamma_n = \alpha_n$) for evanescent modes. A substantial number of evanescent modes must be included in the analysis. This is because the uniform sections will often be relatively short in length so that the amplidtude of a decaying wave may still be significant by the time the wave reaches the next junction.

Suppose now that there are a number of waveguide junctions and some uniform sections (between them) whose generalized scattering matrices are known. The calculation of the generalized scattering matrix of the overall system may be found as follows. Let the two scattering matrices $[S^a]$ and $[S^b]$ shown in the Fig. 4.28, have elements

$$[S^a] = \begin{bmatrix} [S^a_{11}] & [S^a_{12}] \\ [S^a_{21}] & [S^a_{22}] \end{bmatrix} \tag{4.330}$$

$$\left[S^b\right] = \begin{bmatrix} [S^b_{11}] & [S^b_{12}] \\ [S^b_{21}] & [S^b_{22}] \end{bmatrix} \tag{4.331}$$

Denoting the cascaded matirx by $[S^c]$, we have

$$[S^c] = \begin{bmatrix} [S^c_{11}] & [S^c_{12}] \\ [S^c_{21}] & [S^c_{22}] \end{bmatrix} \tag{4.332}$$

where

$$\begin{aligned}
\left[S^c_{11}\right] &= [S^a_{11}] + [S^a_{12}]\{[I] - [S^b_{11}][S^a_{11}]\}^{-1}[S^b_{11}][S^a_{11}] \\
\left[S^c_{12}\right] &= [S^a_{12}]\{[I] - [S^b_{11}][S^a_{22}]\}^{-1}[S^b_{12}] \\
\left[S^c_{21}\right] &= [S^b_{21}]\{[I] - [S^a_{22}][S^b_{11}]\}^{-1}[S^a_{21}] \\
\left[S^c_{22}\right] &= [S^b_{22}] + [S^b_{21}]\{[I] - [S^a_{22}][S^b_{11}]\}^{-1}[S^a_{22}][S^b_{12}]
\end{aligned} \tag{4.333}$$

where $[I]$ is the unit matrix. The cascading process has the advantage that the exact number of junctions and sections does not have to be known at the start of the analysis as the process proceeds from the input to the output in a logical fashion.

The transverse components of the electric and magnetic fields are given by

$$\mathbf{E}_t = \sum_i V_i e_i = \sum_i V_i^{(h)} e_i^{(h)} + \sum_i V_i^{(e)} e_i^{(h)} \tag{4.334}$$

$$\mathbf{H}_t = \sum_i I_i h_i = \sum_i I_i^{(h)} h_i^{(h)} + \sum_i I_i^{(e)} h_i^{(h)} \tag{4.335}$$

The magnetic field eigenvectors $\widehat{h_i}$ are related to the $\widehat{e_i}$ by

$$\mathbf{h_i} = \widehat{z} \times \mathbf{e_i} \tag{4.336}$$

The functions $\widehat{e_i}^{(h)}$ and $\widehat{e_i}^{(e)}$ possess the vector orthogonality properties

$$\int e_i^{(e)} \cdot e_j^{(e)} ds = \begin{cases} 1 & \text{for } i = j \\ 0 & \text{for } i \neq j \end{cases} \tag{4.337}$$

and

$$\int e_i^{(e)} \cdot e_j^{(h)} ds = 0 \tag{4.338}$$

The electric field eigenvectors $e_i^{(h)}$ and $e_i^{(e)}$ are obtained from the H-mode and E-mode potentials

$$e_i^{(h)} = \hat{z} \times \nabla_t \psi_i^{(h)} \tag{4.339}$$

$$e_i^{(e)} = -\nabla_t \psi_i^{(e)} \tag{4.340}$$

The E-mode potential for a uniform waveguide of rectangular cross section $(a \times b)$ is given by

$$\psi_i^{(e)} = \sqrt{\frac{\epsilon_{0m}\epsilon_{0n}/ab}{(m\pi/a)^2 + (n\pi/b)^2}} \sin\frac{m\pi x}{a} \sin\frac{n\pi y}{b} \tag{4.341}$$

where $m, n = 1, 2, 3, \ldots$ and ϵ_{0n} is the Neumann's number

$$\epsilon_{0n} = \begin{cases} 1 & n = 0 \\ 2 & n > 0 \end{cases} \tag{4.342}$$

The H-mode potential for the waveguide is given by

$$\psi_i^{(e)} = \sqrt{\frac{\epsilon_{0m}\epsilon_{0n}/ab}{(m\pi/a)^2 + (n\pi/b)^2}} \cos\frac{m\pi x}{a} \cos\frac{n\pi y}{b} \tag{4.343}$$

where $m, n = 1, 2, 3, \ldots$ and the mode $m = n = 0$ is excluded. The transverse electric fields in the rectangular waveguide with dimensions $a \times b$ are thus given as

$$e_i^{(h)} = \frac{\sqrt{\epsilon_{0m}\epsilon_{0n}/ab}}{(m\pi/a)^2 + (n\pi/b)^2} \cdot \left[\hat{x}\frac{n\pi}{b}\cos\frac{m\pi x}{a}\sin\frac{n\pi y}{b} - \hat{y}\frac{m\pi}{a}\sin\frac{m\pi x}{a}\cos\frac{n\pi y}{b}\right] \tag{4.344}$$

$$e_i^{(e)} = \frac{\sqrt{\epsilon_{0m}\epsilon_{0n}/ab}}{(m\pi/a)^2 + (n\pi/b)^2} \cdot \left[\hat{x}\frac{m\pi}{a}\cos\frac{m\pi x}{a}\sin\frac{n\pi y}{b} + \hat{y}\frac{n\pi}{b}\sin\frac{m\pi x}{a}\cos\frac{n\pi y}{b}\right] \tag{4.345}$$

The mode propagation constant is given by

$$\gamma_i = \begin{cases} j\sqrt{k_0^2 - K_{ci}^2} & k_0 > k_{ci} \\ \sqrt{k_{ci}^2 - K_0^2} & k_0 < k_{ci} \end{cases}$$

where

$$k_{ci} = \sqrt{(m\pi/a)^2 + (n\pi/b)^2} \tag{4.346}$$

The mode characteristic impedances are expressed as

$$Z_i^{(e)} = \frac{V_i^{(e)}}{I_i^{(e)}} = \frac{\gamma_i}{j\omega\epsilon_0} \tag{4.347}$$

$$Z_i^{(h)} = \frac{V_i^{(h)}}{I_i^{(h)}} = \frac{j\omega\mu_0}{\gamma_i} \tag{4.348}$$

4.4 Radiation from Sources in a Plane

Let $\psi(x, y, x)$ be a scalar function satisfying Helmholtz equation. We assume that ψ is known in the entire plane $z = 0$ and seek ψ in the half-space $z \geq 0$ (Fig. 4.34).

In the region $z \geq 0$, the most general expression for ψ is given by (4.21). Thus,

$$\psi(x, y, z) = \iint_C f(k_x, k_y)e^{-j(k_x x+k_y y+k_z z)}dk_x dk_y \tag{4.349}$$

with

$$k_z = \pm\sqrt{k^2 - k_x^2 - k_y^2} \tag{4.350}$$

for arbitrary weighting function f and for suitably chosen contour of integration C. The radiation condition requires that

$$\Re\text{e}k_z \geq 0, \quad \Im\text{m}k_z \leq 0 \tag{4.351}$$

Hence, we have

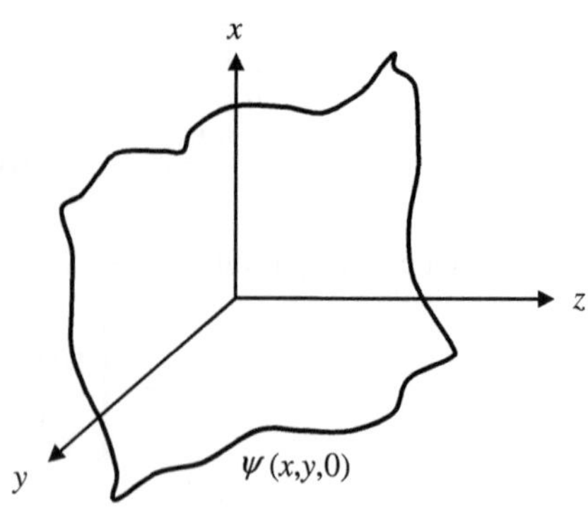

Fig. 4.34 Radiation From sources in the plane $z = 0$

$$\psi(x, y, z) = \iint\limits_{-\infty}^{\infty} f(k_x, k_y) e^{-jz\sqrt{k^2-k_x^2-k_y^2}} e^{-j(k_x x + k_y y)} dk_x dk_y \tag{4.352}$$

everywhere in $z \geq 0$. In particular, when $z = 0$

$$\psi(x, y, 0) = \iint\limits_{-\infty}^{\infty} f(k_x, k_y) e^{-j(k_x x + k_y y)} dk_x dk_y \tag{4.353}$$

which is the Fourier transform of the function f. Therefore, we may write

$$f(k_x, k_y) = \frac{1}{4\pi} \iint\limits_{-\infty}^{\infty} \psi(x, y, 0) e^{j(k_x x + k_y y)} dx dy \tag{4.354}$$

Since $\psi(x, y, 0)$ is known, the solution is now complete.

The exponential factor in the representation of $\psi(x, y, z)$ is a plane wave whose direction of propagation is defined by the direction cosines k_x/k, k_y/y and k_z/k, and f is the amplitude or weighting factor of each such wave. For this reason, the representation is often referred to as *angular spectrum* of plane waves, and the spectrum f is simply the Fourier transform of the field distribution in the aperture $z = 0$. This method is widely used in electromagnetics and optics.

From (4.352), it is observed that each spectral frequency pair (k_x, k_y) in the spectrum of the aperture distribution corresponds (and gives rise) to a plane wave in a particular direction in space. The wave propagates in the z direction only if $k_x^2 + k_y^2 < k^2$. In this case, the plane wave is a uniform one radiating energy away from the plane $z = 0$. But if $k_x^2 + k_y^2 > k^2$, the wave is attenuated in the z direction, and represents a non-uniform one, associated with reactive (or stored) power only. In optics, the latter is referred to as the invisible region, while the former is called the visible region.

Example 4.7 Consider the two-dimensional problem of parallel plate waveguide terminated with an open flange procluding a narrow slit aperture as shown Fig. 4.35. Assume a TEM mode in parallel plate waveguide having

$$\mathbf{H} = \widehat{y} H_y$$

Because of symmetry, there is no y dependence and $\mathbf{H}$ is in y direction everywhere $z \geq 0$. From Helmholtz'z equation

$$(\nabla^2 + k^2) H_y = 0$$

we find that

$$k_x^2 + k_z^2 = k^2$$

The plane wave angular spectral representation of the magnetic field is written as

$$H_y(x,z)=\int_{-\infty}^{\infty} f(k_x)e^{-jz\sqrt{k^2-k_x^2}}e^{-jk_x x}dk_x$$

But, from Maxwell's equation, the electric field components are given by

$$E_x(x,z)=-\frac{Z}{jk}\frac{\partial H_y}{\partial z}\quad E_z(x,z)=\frac{Z}{jk}\frac{\partial H_y}{\partial x}$$

Therefore, we have

$$E_x(x,z)=Z\int_{-\infty}^{\infty}\frac{\sqrt{k^2-k_x^2}}{k}f(k_x)e^{-jz\sqrt{k^2-k_x^2}}e^{-jk_x x}dk_x$$

$$E_z(x,z)=-Z\int_{-\infty}^{\infty}\frac{k_x}{k}f(k_x)e^{-jz\sqrt{k^2-k_x^2}}e^{-jk_x x}dk_x$$

In order to find an expression for $f(k_x)$, we impose the boundary condition on the tangential component of the electric field over the conducting surface.

$$E_x=0,\quad z=0\quad \text{(on the perfect conductor)}$$

Thus, taking inverse Fourier transform, we have

$$\frac{\sqrt{k^2-k_x^2}}{k}f(k_x)=\frac{Y}{2\pi}\int_{-\infty}^{\infty}E_x(x,0)e^{jk_x x}dx=\frac{Y}{2\pi}\int_{-a/2}^{a/2}E_x(x,0)e^{jk_x x}dx$$

Approximating the aperture field with a TEM wave

$$E_z=0,\qquad E_x(x,0)=1,\quad |x|<a/2$$

we find

$$\frac{\sqrt{k^2-k_x^2}}{k}f(k_x)=\frac{Y}{2\pi}\int_{-a/2}^{a/2}e^{jk_x x}dx=\frac{Ya}{2\pi}\frac{\sin k_x a/2}{k_x a/2}$$

or

$$f(k_x)=\frac{Ya}{2\pi}\frac{k}{\sqrt{k^2-k_x^2}}\mathrm{sinc}(k_x a/2)$$

Substituting this result in the expression for the magnetic field, we finally find

$$H_y(x,z) = \frac{Ya}{2\pi}\int_{-\infty}^{\infty}\frac{k}{\sqrt{k^2-k_x^2}}\text{sinc}(k_xa/2)e^{-jz\sqrt{k^2-k_x^2}}e^{-jk_xx}dk_x, \quad z \geq 0$$

This integral cannot be evaluated exactly in closed form. At large distances from the aperture, it can be evaluated asymptotically using the stationary phase (or steepest descent) technique. The presence of the $\text{sinc}(k_xa/2)$ term signifies the importance of small values of k_x. □

4.5 Cylindrical Waves

Consider the wave equation in cylindrical coordinate system

$$\frac{1}{\rho}\frac{\partial}{\partial\rho}(\rho\frac{\partial\Psi}{\partial\rho}) + \frac{1}{\rho^2}\frac{\partial^2\Psi}{\partial\phi^2} + \frac{\partial^2\Psi}{\partial z^2} + k^2\Psi = 0 \tag{4.355}$$

We solve this partial differential equation by separation of variables. The most general solution can be synthesized using the separated solutions

$$\Psi(\rho,\phi,z) = R(\rho)\Phi(\phi)Z(z) \tag{4.356}$$

Substituting in the equation, we get

$$\frac{1}{R}\frac{1}{\rho}\frac{d}{d\rho}(\rho\frac{dR}{d\rho}) + \frac{1}{\Phi}\frac{1}{\rho^2}\frac{d^2\Phi}{d\phi^2} + \frac{1}{Z}\frac{d^2Z}{dz^2} + k^2 = 0 \tag{4.357}$$

The third term in the right hand side is a function of z alone, but from the rest of the equation it is also independent of z. It must therefore be a constant. We call this constant $-k_z^2$ and write

$$\frac{d^2Z}{dz^2} + k_z^2Z = 0 \tag{4.358}$$

with the solution

$$Z(z) = e^{\pm jk_zz} \tag{4.359}$$

Let $k_\rho^2 = k^2 - k_z^2$. On multiplying the original equation by ρ^2, we get

$$\frac{1}{R}\rho\frac{d}{d\rho}(\rho\frac{dR}{d\rho}) + k_\rho^2\rho^2 + \frac{1}{\Phi}\frac{d^2\Phi}{d\phi^2} = 0 \tag{4.360}$$

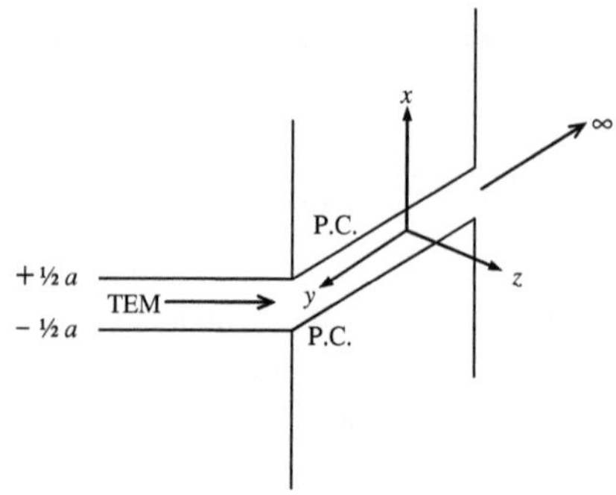

Fig. 4.35 A narrow slit aperture in a perfectly conducting screen excited by the TEM mode of a parallel plate waveguide

By the same line of reasoning, the third term is a constant $-\nu^2$, say

$$\frac{d^2\Phi}{d\phi^2} + \nu^2\Phi = 0 \tag{4.361}$$

with the solution

$$\Phi(\phi) = e^{\pm j\nu\phi} \tag{4.362}$$

The one remaining equation is for $R(\rho)$ and it is

$$\rho\frac{d}{d\rho}(\rho\frac{dR}{d\rho}) + [(k_\rho\rho)^2 - \nu^2]R = 0 \tag{4.363}$$

or, equivalently

$$\frac{d^2R}{d\rho^2} + \frac{1}{\rho}\frac{dR}{d\rho} + [k_\rho^2 - \frac{\nu^2}{\rho^2}]R = 0 \tag{4.364}$$

This is the Bessel's equation of order ν. Like all other separated equations, it is a second order differential equation, and it requires two linearly independent solutions. The equation is singular at $\rho = 0$, so one of the solutions must be singular at $\rho = 0$ and the other finite there. Finally, it is noted that the solutions are functions of $k_\rho\rho$. A typical solution has the form

$$\Psi(\rho, \phi, a) = B_\nu(k_\rho\rho)e^{\pm j\nu\phi}e^{\pm jk_z z} \tag{4.365}$$

with

$$k_\rho^2 + k_z^2 = k^2 \tag{4.366}$$

and this is the general form of a wave function in cylindrical coordinates (Fig. 4.35).

Bessel Functions

The Bessel's equation is given by

$$\frac{d^2B_\nu}{d\rho^2}+\frac{1}{\rho}\frac{dB_\nu}{d\rho}+[k_\rho^2-\frac{\nu^2}{\rho^2}]B_\nu=0 \tag{4.367}$$

In all problems of practical interest, ν is real ($\nu \in \mathcal{R}$), and since the equation involves ν^2, it is sufficient to take $\nu \geq 0$. We now seek two independent solutions.

The solution which is finite at $\rho = 0$ is the *Bessel function of the first kind.* Its series representation is expressed as

$$J_\nu(k_\rho\rho)=\sum_{m=0}^{\infty}\frac{(-1)^m(k_\rho\rho/2)^{(\nu+2m)}}{m!(m+\nu)!} \tag{4.368}$$

The small and large argument approximations for the Bessel function of the first kind are given resepectively as

$$\begin{aligned} k_\rho\rho \ll 1 J_\nu(k_\rho\rho) &\simeq \frac{1}{\nu!}(\frac{k_\rho\rho}{2})^\nu \\ k_\rho\rho \gg 1 J_\nu(k_\rho\rho) &\sim \sqrt{\frac{2}{\pi k_\rho\rho}}\cos(k_\rho\rho-\frac{\nu\pi}{2}-\frac{\pi}{6}) \end{aligned} \tag{4.369}$$

The asymptotic behavior of the Bessel function looks like a damped standing wave with zeros π apart in $k_\rho\rho$, and decaying to zero at infinity.

If ν is non-integer, a second solution which is clearly independent of the first is $J_{-\nu}(k_\rho\rho)$, and this is infinite at $\rho = 0$. However, if ν is an integer, $\nu = n$, then

$$J_{-n}(k_\rho\rho)=(-1)^nJ_n(k_\rho\rho) \tag{4.370}$$

and this solution is no longer independent of the first. The second solution must be constructed separately, and the one usually chosen is the *Neumann function*

$$N_n(k_\rho\rho)=\lim_{\nu\to n}\frac{J_\nu(k\rho)\cos(\nu\pi)-J_{-\nu}(k\rho)}{\sin(\nu\pi)} \tag{4.371}$$

The small argument approximation for the Neumann function is

$$\begin{aligned} k_\rho\rho \ll 1\ N_0(k_\rho\rho) &\simeq \frac{2}{\pi}\ln(\frac{\gamma k_\rho\rho}{2}) \\ N_n(k_\rho\rho) &\simeq -\frac{(n-1)!}{\pi}(\frac{2}{k_\rho\rho})^n, \quad n>0 \end{aligned} \tag{4.372}$$

where $\gamma = 1.78107\ldots$ is Euler's constant and $\ln\gamma = 0.57722$. Thus, N_n is infinite at $\rho = 0$ for all n. The asymptotic behavior of the Neumann function is

$$k_\rho \rho \gg 1 \quad N_n \simeq \sqrt{\frac{2}{\pi k_\rho \rho}} \sin(k_\rho \rho - n\pi/2 - \pi/4) \tag{4.373}$$

Any combination of J_n and N_n can serve as the second solution, and based on the behavior for large $k_\rho \rho$, there are two combinations of particular importance. These are the *Hankel functions of the first kind*

$$H_n^{(1)}(k_\rho \rho) = J_n(k_\rho \rho) + jN_n(k_\rho \rho) \tag{4.374}$$

and the *Hankel function of the second kind*

$$H_n^{(2)}(k_\rho \rho) = J_n(k_\rho \rho) - jN_n(k_\rho \rho) \tag{4.375}$$

The Hankel function of the first kind behaves asymptotically as

$$H_n^{(1)}(k_\rho \rho) \simeq \sqrt{\frac{2}{\pi k_\rho \rho}} e^{j(k_\rho \rho - n\pi/2 - \pi/4)} \tag{4.376}$$

For k_ρ real and positive, and with our time convention $e^{j\omega t}$, this behaves like an inward-travelling wave. We also have

$$H_n^{(2)}(k_\rho \rho) \simeq \sqrt{\frac{2}{\pi k_\rho \rho}} e^{-j(k_\rho \rho - n\pi/2 - \pi/4)} \tag{4.377}$$

which is particularly important for our purposes since, for k_ρ real and with our time convention, it looks like an outward-travelling wave. It therefore satisfies the radiation condition.

4.5.1 Line Sources

An electric line source of oscillating current radiates cylindrical waves. Consider a line source of strength I_0 located along the z-axis as shown in Fig. 4.36. The current density associated with the line source is given by

$$\mathbf{J}(\rho) = \hat{z} I_0 \delta(x) \delta(y) \tag{4.378}$$

Considering the problem in cylindrical coordinate system, we have

Fig. 4.36 An infinite line source

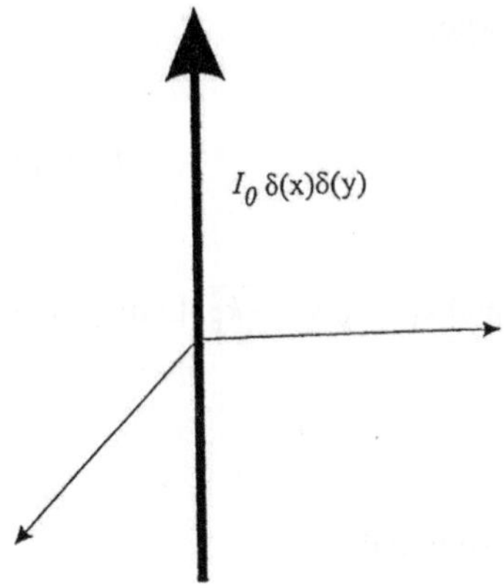

$$\mathbf{J}(\rho) = \hat{z} I_0 \frac{\delta(\rho)}{2\pi\rho} \tag{4.379}$$

Because of axial symmetry, the radiated electric field is independent of ϕ and z. Thus, from Maxwell's equations, we have

$$\begin{aligned} \frac{dH_z}{d\rho} &= -j\omega\epsilon E_\phi \\ \frac{1}{\rho}\frac{d(\rho E_\phi)}{d\rho} &= -j\omega\mu H_z \end{aligned} \tag{4.380}$$

and

$$\begin{aligned} \frac{dE_z}{d\rho} &= -j\omega\mu H_\phi \\ \frac{1}{\rho}\frac{d(\rho H_\phi)}{d\rho} &= J_z + j\omega\epsilon E_z \end{aligned} \tag{4.381}$$

while $E_\rho = H_\rho = 0$. The first system of equations is independent of the second system and it is source-free. We conclude that $H_z = E_{phi} = 0$. Combining equations of the second system, we obtain

$$\frac{1}{\rho}[\frac{d}{d\rho}\frac{\rho dE_z}{d\rho}] + k^2 E_z = j\omega\mu I_0 \frac{\delta(\rho)}{2\pi\rho} \tag{4.382}$$

subject to the radiation condition for large ρ.

In order to solve this equation, we apply the Fourier transform to both sides of (4.382) to find

$$\frac{d^2\widetilde{E}_z}{dy^2} = k_y^2 \widetilde{E}_z = j\omega\mu I_0 \delta(y) \tag{4.383}$$

where

$$\widetilde{E}_z(k_x, y) = \int_{-\infty}^{\infty} E_z(x, y) e^{-jk_x x} dx \tag{4.384}$$

and $k_y = \sqrt{k^2 - k_x^2}$. Letting now

$$\widetilde{G} = -\frac{\widetilde{E}_z}{j\omega\mu I_0} \tag{4.385}$$

we find that

$$\frac{d^2\widetilde{G}}{dy^2} + k_y^2 \widetilde{G} = -\delta(y) \tag{4.386}$$

subject to the radiation condition. But, the solution to this Green's function problem is given by (4.40). That is

$$\widetilde{G} = e^{\frac{-jk_y|y|}{2jk_y}}, \quad \Im k_y < 0 \tag{4.387}$$

Taking the inverse transform

$$G = \frac{1}{2\pi} \int_{-\infty}^{\infty} e^{\frac{-jk_y|y|}{2jk_y}} e^{jk_x x} dk_x \tag{4.388}$$

from the integral presentation of the Hankel function, we know that

$$\frac{1}{\pi} \int_{-\infty}^{\infty} e^{\frac{-j\sqrt{k^2-k_x^2}|y|}{\sqrt{k^2-k_x^2}}} e^{jk_x x} dk_x = H_0^2[k\sqrt{x^2 + y^2}] \tag{4.389}$$

Thus, the Green's function is given by

$$G(\rho) = \frac{1}{4j} H_0^{(2)}(k\rho) \tag{4.390}$$

This is known as the *two-dimensional free space Green's function*. The electric field due to an infinite line source located on the z-axis is thus given by

$$E_z = -\frac{\omega\mu I_0}{4} H_0^{(2)}(k\rho) \tag{4.391}$$

From (4.381), we obtain the magnetic field

$$H_\phi = -\frac{jkI_0}{4} H_1^{(2)}(k\rho) \tag{4.392}$$

If the line source is located at ρ', the electric field is given by

$$E_z = \frac{-\omega\mu I_0}{4} H_0^{(2)}(k|\rho - \rho'|) \tag{4.393}$$

4.5.2 Cylindrical Wave Transformation

Consider a plane wave propagating in positive x-direction and polarized in the z-direction

$$\mathbf{E} = \widehat{z}E_0 e^{-jkx} = \widehat{z}E_0 e^{-jk\rho\cos\phi} \tag{4.394}$$

This wave can be represented as an infinite sum of cylindrical waves. Since the expansion should remain finite at the origin, we choose

$$\mathbf{E} = \widehat{z}E_0 \sum_{n=-\infty}^{\infty} a_n J_n(k_\rho \rho) e^{jn\phi} \tag{4.395}$$

In order to find the coefficients a_n, we invoke the orthogonality of exponential functions $e^{jn\phi}$. Thus, multiplying both sides by $e^{-jm\phi}$ and integrating from 0 to 2π

$$\begin{aligned}\int_0^{2\pi} e^{-j(k\rho\cos\phi + m\phi)} d\phi &= \int_0^{2\pi} [\sum_n a_n J_n(k\rho) e^{jn\phi}] e^{-jm\phi} d\phi \\ &= \sum_n a_n J_n(k\rho) \int_0^{2\pi} e^{j(n-m)\phi} d\phi \end{aligned} \tag{4.396}$$

Using the identities

$$\int_0^{2\pi} e^{j(x\cos\phi + m\phi)} d\phi = 2\pi j^m J_m(x) \tag{4.397}$$

$$\int_0^{2\pi} e^{j(n-m)\phi} d\phi = 2\pi\,\delta_{mn} \tag{4.398}$$

where δ_{mn} is the Kronecker delta function, we have

$$\int_0^{2\pi} e^{-j(k\rho\cos\phi + m\phi)} d\phi = 2\pi j^{-m} J_{-m}(-k\rho) \tag{4.399}$$

Therefore, using (4.370), we obtain

$$2\pi j^{-m} J_m(k\rho) = 2\pi a_m J_m(k\rho) \tag{4.400}$$

resulting in the following expression for the coefficients

$$a_m = j^{-m} \tag{4.401}$$

Finally, the plane wave can be expressed in the following form

$$E_z = E_0 e^{-jkx} = E_0 e^{-jk\rho\cos\phi} = \sum_{n=-\infty}^{\infty} j^{-n} J_n(k\rho) e^{jn\phi} \tag{4.402}$$

4.5.3 Addition Theorem

Consider the wave function

$$\begin{aligned} \psi &= H_0^{(2)}(k|\boldsymbol{\rho} - \boldsymbol{\rho}'|) \\ &= H_0^{(2)}(k\sqrt{\rho^2 + \rho'^2 - 2\rho\rho'\cos(\phi - \phi')}) \end{aligned} \tag{4.403}$$

We may consider ψ as being due to a line source located at P' with a position vector $\boldsymbol{\rho}'$. We would like to express the above in terms of wave functions referred to the origin O due to a line source along the z-axis.

We note that for $\rho < \rho'$, ψ is finite at $\rho = 0$ and periodic in 2π in ϕ. Therefore, permissible wave functions are of the form $J_n(\rho)e^{jn\phi}$. On the other hand, for $\rho > \rho'$, ψ must represent an outgoing wave and permissible wave functions are of the form $H_n^{(2)}(\rho)e^{jn\phi}$. We also note that ψ must be symmetric in primed and unprimed coordinates.

We now let

$$\psi = \begin{cases} \sum\limits_{n=-\infty}^{\infty} b_n H_n^{(2)}(k\rho') J_n(k\rho) e^{jn(\phi-\phi')} & \rho < \rho' \\ \sum\limits_{n=-\infty}^{\infty} b_n J_n(k\rho') H_n^{(2)}(k\rho) e^{jn(\phi-\phi')} & \rho > \rho' \end{cases} \tag{4.404}$$

where b_n are expansion coefficients. To evaluate $\{b_n\}$, we shall make use of the asymptotic expansion for the Hankel function (4.377) for $\rho' \to \infty$, $\phi' = 0$. That is

$$H_n^{(2)}(k|\boldsymbol{\rho} - \boldsymbol{\rho}'|) \simeq \sqrt{\frac{2}{\pi k\rho'}} e^{-j[k(\rho' - \rho\cos\phi) - \frac{2n+1}{4}\pi]} \tag{4.405}$$

Using the above, we have

$$\psi \simeq \sqrt{\frac{2}{\pi k\rho'}} e^{-j(k\rho' - \pi/4)} e^{jk\rho\cos\phi} \tag{4.406}$$

On the other hand, we have for $\rho' > \rho$

$$\begin{aligned}\psi &= \sum_{n=-\infty}^{\infty} b_n H_n^{(2)}(k\rho') J_n(k\rho) e^{jn(\phi-\phi')} \\ &\simeq \sqrt{\frac{2}{\pi k\rho'}} e^{-j(k\rho' - \pi/4)} \sum_{n=-\infty}^{\infty} b_n j^n J_n(k\rho) e^{jn\phi}\end{aligned} \tag{4.407}$$

Knowing that

$$e^{jk\rho\cos\phi} = \sum_{n=-\infty}^{\infty} j^n J_n(k\rho) e^{jn\phi} \tag{4.408}$$

we find by comparing the last two equations that

$$b_n = 1 \tag{4.409}$$

Thus

$$H_0^{(2)}(k|\boldsymbol{\rho} - \boldsymbol{\rho}'|) = \begin{cases} \sum\limits_{n=-\infty}^{\infty} H_n^{(2)}(k\rho') J_n(k\rho) e^{jn(\phi-\phi')} & \rho < \rho' \\ \sum\limits_{n=-\infty}^{\infty} J_n(k\rho') H_n^{(2)}(k\rho) e^{jn(\phi-\phi')} & \rho > \rho' \end{cases} \tag{4.410}$$

This is known as the addition theorem for the Hankel functions. The theorem is valid for $H_0^{(1)}$ provided that the Hankel functions of the first kind is used on the right hand side.

The addition theorem for the Bessel function can be written as

$$J_0(k|\boldsymbol{\rho} - \boldsymbol{\rho}'|) = \sum_{n=-\infty}^{\infty} J_n(k\rho') J_n(k\rho) e^{jn(\phi-\phi')} \quad \forall \rho \tag{4.411}$$

4.5.4 The Circular Metallic Waveguide

A very simple waveguide is a circular metallic pipe of radius a filled with a homogeneous lossless dielectric or air. We seek the modes corresponding to lossless propagation down the guide in the $+z$-direction.

The modes can be divided to two different types, namely, those having the magnetic or electric field transverse to z.

4.5.4.1 Transverse Magnetic Modes

In this case, the magnetic field is transverse to the axis of the guide and there is no longitudinal magnetic field component. Thus

$$H_z = 0 \tag{4.412}$$

The electric Hertz vector

$$\pi_e = \widehat{z}\psi(\rho, \phi, z) \tag{4.413}$$

satisfies the wave equation

$$(\nabla^2 + k^2)\psi = 0 \tag{4.414}$$

and the electromagnetic fields are expressed in terms of the Hertz potential as

$$\begin{aligned} \mathbf{E} &= \nabla\nabla \cdot \pi_e + k^2 \pi_e \\ &= \frac{\partial^2 \psi}{\partial \rho \partial z}\widehat{\rho} + \frac{1}{\rho}\frac{\partial^2 \psi}{\partial \phi \partial z}\widehat{\phi} + \left(k^2 - k_z^2\right)\psi \widehat{z} \end{aligned} \tag{4.415}$$

and

$$\begin{aligned} \mathbf{H} &= jkY\nabla \times \pi_e \\ &= jkY\left(\frac{1}{\rho}\frac{\partial \psi}{\partial \phi}\widehat{\rho} - \frac{\partial \psi}{\partial \rho}\widehat{\phi}\right) \end{aligned} \tag{4.416}$$

As required, **H** has no component in the z direction.

The task is to construct the solution ψ of scalar wave equation to satisfy the constraints, which are

a. lossless propagation in the $+z$ direction,
b. finite fields everywhere inside the guide including $\rho = 0$, and
c. vanishing tangential electric fields at the perfectly conducting walls.

To satisfy (a), we choose

$$Z(z) = e^{-jk_z z} \tag{4.417}$$

Since the field must be single-valued as a function of ϕ, it is necessary that

$$\Phi(2\nu\pi + \phi) = \Phi(\phi) \quad \forall\phi \tag{4.418}$$

requiring that ν be an integer $\nu = n$. It is sufficient that $n \geq 0$. Thus

$$\Phi(\phi) = a_n \cos n\phi + b_n \sin n\phi \tag{4.419}$$

Finally, because of (b), it is necessary that

$$R(\rho) = J_n(k_\rho \rho) \tag{4.420}$$

giving

$$\psi(\rho, \phi, z) = J_n(k_\rho \rho)(a_n \cos n\phi + b_n \sin n\phi)e^{-jk_z z} \tag{4.421}$$

with

$$k_\rho^2 + k_z^2 = k^2 \tag{4.422}$$

for any non-negative integer n.

The only constraints still to be satisfied are the boundary conditions at $\rho = a$, and from the examination of the field components, we observe that these are satisfied if

$$J_n(k_\rho a) = 0 \tag{4.423}$$

The Bessel function $J_n(x)$ has infinite number of discrete zeros spaced approximately π apart. We may sort these zeros in ascending order and denote them by

$$X_{np} \quad p = 1, 2, \ldots$$

where n is the order of the Bessel function and p is the order of the zero. The resulting values of the propagation constant k_z for the TM_{np} modes are

$$k_z = \sqrt{k^2 - (\frac{X_{np}}{a})^2} \tag{4.424}$$

The cutoff frequencies are therefore

$$f_{cnp} = \frac{cX_{np}}{2\pi a} \tag{4.425}$$

and the cutoff wavelengths

$$\lambda_{cnp} = \frac{2\pi a}{X_{np}} \tag{4.426}$$

where c is the speed of light in the medium.

The dominant TM mode is TM_{01} mode and this has the advantage that it is symmetric in ϕ.

4.5.4.2 Transverse Electric Modes

The analysis, in this case, is very similar to that of the TM_Z modes and the modes are derived from the magnetic Herta potential

$$\pi_m = \widehat{z}\psi(\rho, \phi, z) \tag{4.427}$$

implying by duality

$$\mathbf{E} = -jkZ\left(\frac{1}{\rho}\frac{\partial\psi}{\partial\phi}\widehat{\rho} - \frac{\partial\psi}{\partial\rho}\widehat{\phi}\right) \tag{4.428}$$

$$\mathbf{H} = \frac{\partial^2\psi}{\partial\rho\partial z}\widehat{\rho} + \frac{1}{\rho}\frac{\partial^2\psi}{\partial\phi\partial z}\widehat{\phi} + \left(k^2 + \frac{\partial^2}{\partial z^2}\right)\psi\widehat{z} \tag{4.429}$$

With the expression for ψ as before, the boundary condition to be satisfied at $\rho = a$ is $E_\phi = 0$, requiring that

$$J_n'(k_\rho a) = 0 \tag{4.430}$$

where the prime denotes the derivative with respect to the argument. If the zeros of $J_n'(x)$ are denoted by

$$X_{np}' \quad p = 1, 2, \ldots$$

we then have

$$k_z = \sqrt{k^2 - (\frac{X_{np}'}{a})^2} \tag{4.431}$$

and the cutoff frequencies and wavelenghts of the TE_{np} modes are

$$f_{cnp} = \frac{cX_{np}'}{2\pi a} \tag{4.432}$$

and

$$\lambda_{cnp} = \frac{2\pi a}{X'_{np}} \tag{4.433}$$

4.5.4.3 Fundamental Mode

The smallest X'_{np} is $X'_{11} = 1.841$, and since this is smaller than any X_{np}, the dominant or *fundamental* mode in the circular cylindrical waveguide is the TE_{11} mode.

The only drawback with this mode is that in an actual guide with some wall losses, it does not have the lowest attenuation. In fact, several TE_{0p} modes have lower attenuation but higher cutoff frequencies. However, to excite (and sustain) one of these without also exciting the TE_{mn} mode is (in practice) very difficult.

4.5.5 Circular Corrugated Horns

Figure 4.37 depicts a typical conical corrugated horn in microwave frequencies. The major advantages of such horns are low side lobe lebels and high cross-polarization discriminations. The corrugated horn satisfies the need for enabling orthogonal polarizations to be separated at levels down to −40 dB over bandwidths of 10–15 %.

The buliding block for a corrugated horn is the corrugated waveguide shown in cross sectional view in Fig. 4.38. There is a simple relationship between the depth and the free space wavelength

$$d = \lambda_0/4 \tag{4.434}$$

The field distributions in the direction of propagation for TM- and TE-modes in a circular waveguide are respectively expressed as

$$E_z^{TM} = \sum_{n=1}^{\infty} J_n(k_c^{TM} R)[C_1 \sin(n\phi) + C_2 \cos(n\phi)]e^{-j\beta_{TM} z} \tag{4.435}$$

$$H_z^{TE} = \sum_{n=1}^{\infty} J_n(k_c^{TE} R)[C_3 \sin(n\phi) + C_4 \cos(n\phi)]e^{-j\beta_{TE} z} \tag{4.436}$$

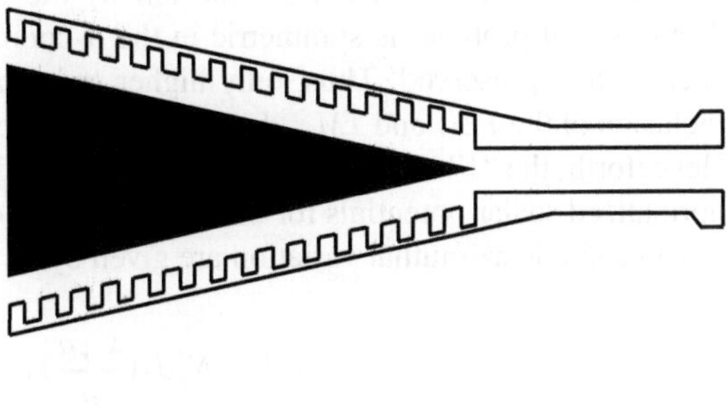

Fig. 4.37 A typical cylindrical corrugated horn

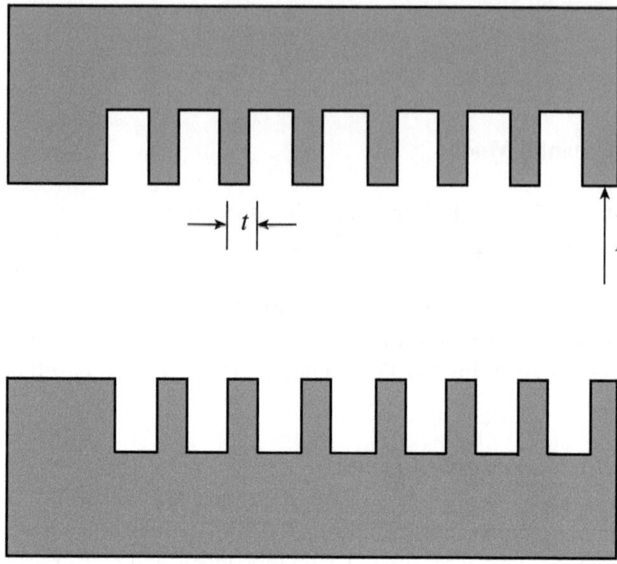

Fig. 4.38 A section of a corrugated circular horn

The radial and circumferential distributions for these modes are obtained from the Maxwell's equations.

It can be shown that the synthesis of modes travelling inside an ordinary circular waveguide of radius R, by combining different modes is not possible in such waveguides, owing to the different propagation constants k_z and cutoff wavenumbers k_c of these modes. The reason for this difference is the different kind of boundary condition imposed on E- and H-fields by a metallic surface. However, substituting these boundary conditions with the surface impedance boundary condition which is realized by a correctly corrugated surface, allowing E-fields in the direction of propagation but no circumferential component, allows TE- and TM-modes to travel at equal speeds, forming the so called hybrid modes. Under these conditions it is possible to combine the modes inside a circular corrugated waveguide in a form to yield a linearly polarized distribution of the E-field, tapering to zero towards the waveguide wall.

In order to analyze this waveguide, we consider the cocentric step discontinuity shown in Fig. 4.38 excited from the left by the TE_{11} mode with unit amplitude. The physical problem is symmetric in the ϕ coordinate and the ϕ variation of the excitation is preserved. Thus, only higher order radial modes are needed. That is inclusion of the TE_{1r} and TM_{1r}, for $r = 1, 2, \ldots$ modes in the waveguide is required. Henceforth, the "1" in the modal indices is dropped for notational convenience. The normalized scalar potentials for the natural modes of a circular waveguide of radius a with a single azimuthal variation are given by

$$\psi_r^h = N_r^h J_1(\frac{x'_{1r}\rho}{a}) \cos\phi \tag{4.437}$$

$$\psi_r^e = N_r^e J_1(\frac{x_{1r}\rho}{a}) \sin\phi \tag{4.438}$$

where the normalization factors N_r^h and N_r^e are given by

$$N_r^h = \sqrt{\frac{2}{\pi}} \frac{1}{\sqrt{x_{r1}'^2 - 1} J_1(x_{1r}')} \tag{4.439}$$

and

$$N_r^e = \sqrt{\frac{2}{\pi}} \frac{1}{\sqrt{x_{r1}^2 - 1} J_2(x_{1r})} \tag{4.440}$$

with x_{1r} and x_{1r}' the rth zeros of J_1 and J_1', respectively. The transverse electric field modal vectors are expressed as

$$e_r^{(h)} = \widehat{\rho} \frac{N_r^h}{\rho} J_1(\frac{x_{1r}'}{a}) \sin\phi + \widehat{\phi} \frac{N_r^h x_{1r}'}{a} J_1'(\frac{x_{1r}'\rho}{a}) \cos\phi \tag{4.441}$$

$$e_r^{(e)} = -\widehat{\rho} \frac{N_r^e x_{1r}}{a} J_1'(\frac{x_{1r}\rho}{a}) \sin\phi - \widehat{\phi} \frac{N_r^e}{\rho} J_1(\frac{x_{1r}\rho}{a}) \cos\phi \tag{4.442}$$

and the cutoff wave-numbers for the H- and E-modes are respectively given by

$$k_{cr}^h = \frac{x_{1r}'}{a} \tag{4.443}$$

and

$$k_{cr}^e = \frac{x_{1r}}{a} \tag{4.444}$$

The inner product over the circular aperture between the TE_k mode of the left (smaller) waveguide and the TE_m mode of the right (larger) waveguide in Fig. 4.39 is given by

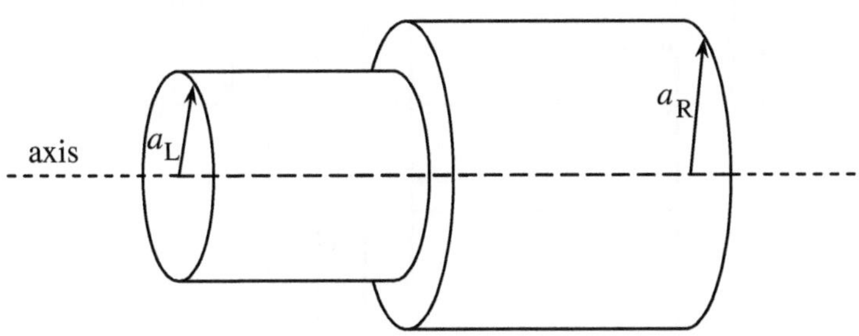

Fig. 4.39 A circular cocentric step discontinuity

$$\langle e_k^{(h)}, e_m^{(h)} \rangle = \frac{2}{a_L a_R} \frac{J_1'(x_{1m}' \frac{a_L}{a_R})}{J_1(x_{1m}')} \frac{x_{1k}'^2 x_{1m}'}{\sqrt{(x_{1k}'^2)(x_{1m}'^2 - 1)}} \frac{1}{(\frac{x_{1k}'}{a_L})^2 - (\frac{x_{1m}'}{a_R})^2} \tag{4.445}$$

Also

$$\langle e_k^{(e)}, e_m^{(e)} \rangle = \frac{2}{a_R^2} \frac{J_1(x_{1m} \frac{a_L}{a_R})}{J_2(x_{1m})} \frac{1}{(\frac{x_{1k}'}{a_L})^2 - (\frac{x_{1m}'}{a_R})^2} \tag{4.446}$$

$$\langle e_k^{(h)}, e_m^{(e)} \rangle = \frac{-2J_1(x_{1m} \frac{a_L}{a_R})}{x_{1m}\sqrt{x_{1k}'^2 - 1} J_2(x_{1m})} \tag{4.447}$$

$$\langle e_k^{(e)}, e_m^{(h)} \rangle = 0 \tag{4.448}$$

The characteristic equation of the corrugated waveguide is expressed as

$$\frac{1}{u^3} \frac{J_n(u)}{J_n'(u)} \left[(\frac{uJ_n'(u)}{J_n(u)})^2 - n^2 + (\frac{nu}{kR})^2 \right]^2 = -\frac{1}{\tan(kd) \left[1 - \frac{t}{p} \right] kR} \tag{4.449}$$

where

$$u = k_c R \tag{4.450}$$

and k is the wavenumber, p is the period, d is the slot depth, and t is the ridge width of the corrugations (see Fig. 4.37). Solving (4.449) for u yields the cutoff wavenumber k_c.

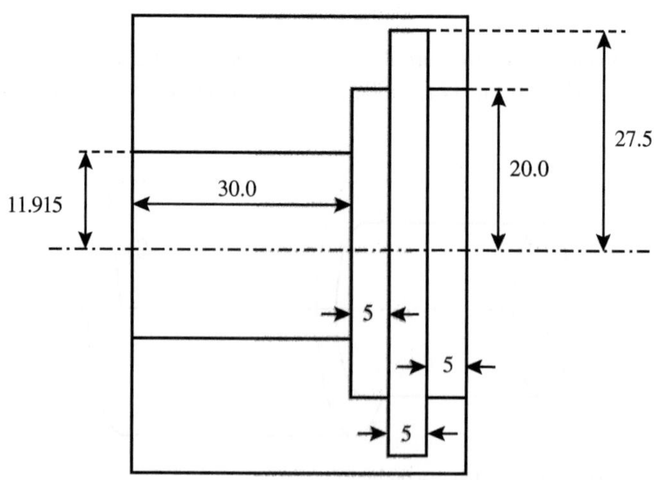

Fig. 4.40 A simple corrugated circular horn. All dimensions are in millimeters

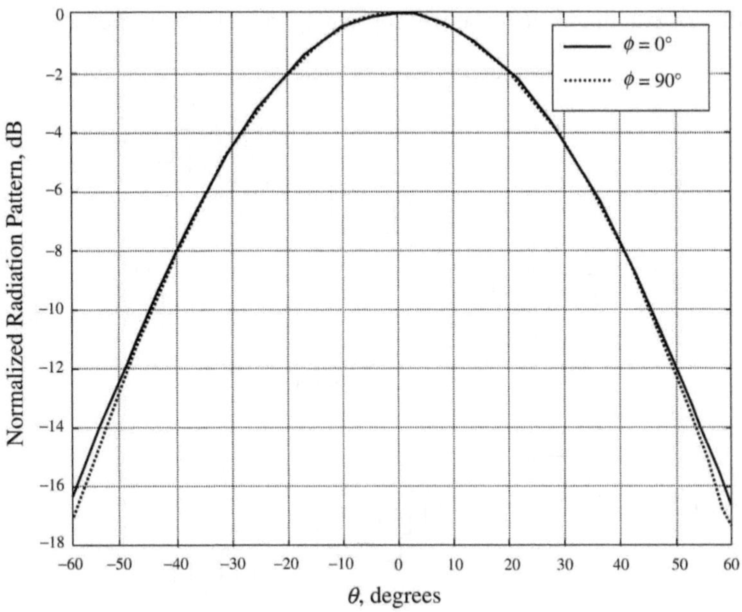

Fig. 4.41 The radiation pattern of the circular horn shown in Fig. 4.39 at 10 GHz

We will present sample numerical results for a simple corrugated cylindrical horn at X-band. The shape and the geometrical parameters of the horn are depicted in Fig. 4.40. The normalized radiation pattern calculated by the cylindrical mode matching method is shown in Fig. 4.41 at 10 GHz.

As a second example, a conical corrugated horn shown in Fig. 4.42 is analyzed by the cylindrical mode-matching method. This method is useful for small flare angle horns. The normalized radiation pattern of the antenna is also shown.

4.5.6 The Coaxial Waveguide

As in the case of a rectangular waveguide, all of the modes in a circular waveguide have non-zero cut-off frequency. On the other hand, for a parallel plate waveguide we found that there was one mode (TE_0) which is, in fact, a TEM mode, and has zero cut-off. This suggests that we need a second wall to achieve zero cut-off in a round guide, leading to the idea of a coaxial waveguide.

Consider a circular metallic waveguide of radius a with a concentric inner conductor of radius b (Fig. 4.43). The space between the two conductors is filled with a lossless homogeneous dielectric and we seek the modes corresponding to lossless propagation in this region. The required boundary conditions are

a. $E_z = 0$ at $\rho = a, b$ requiring $k_\rho^2\psi = 0$ at $\rho = a, b$
b. $E_\phi = 0$ at $\rho = a, b$ requiring $\partial\psi/\partial\phi = 0$ at $\rho = a, b$

We will now examine various modes propagating in such a waveguide.

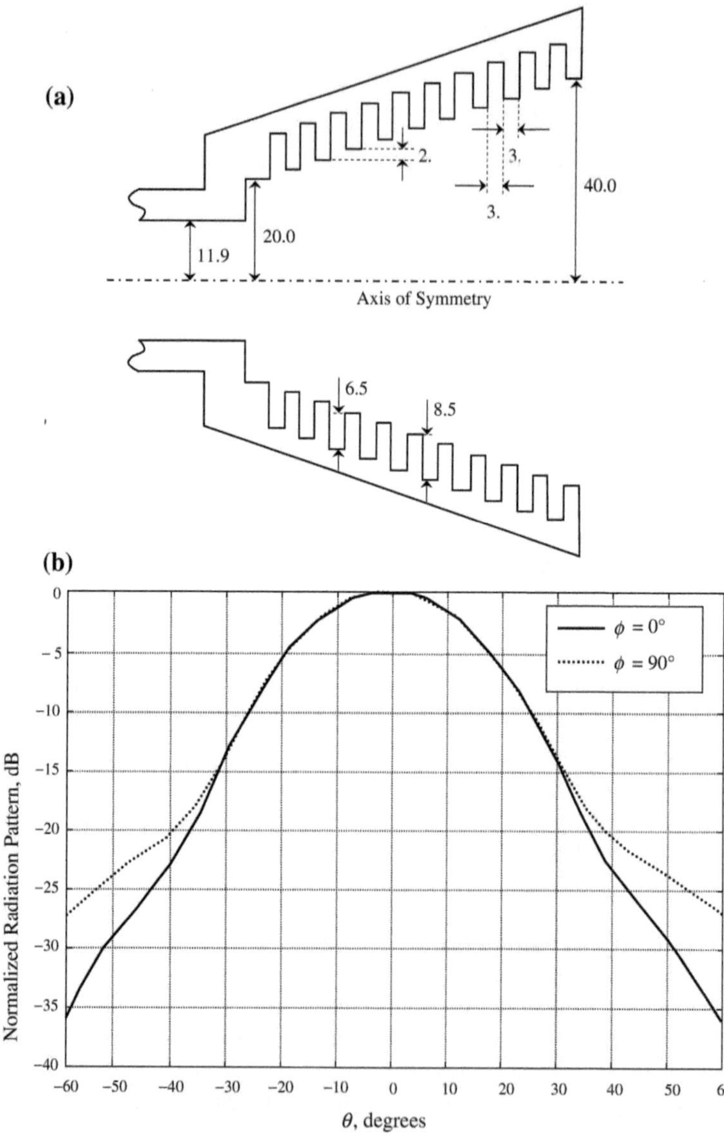

Fig. 4.42 **a** A small flare angle conical horn. All dimensions are in millimeters. **b** The radiation pattern of the horn at 10 GHz

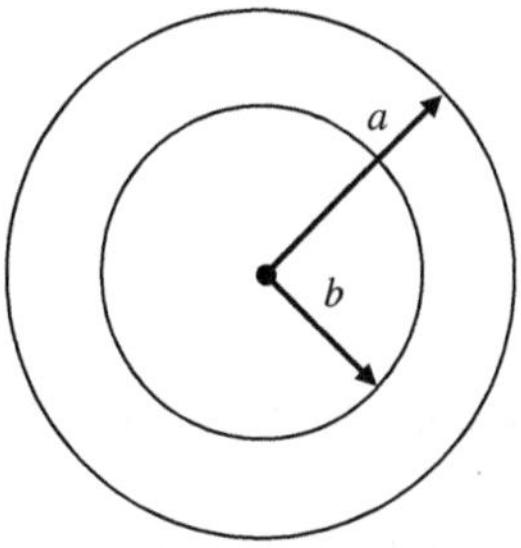

Fig. 4.43 A coaxial circular waveguide

4.5.6.1 Fundamental TEM Mode

Examining the boundary conditions, we note that if we let $k_\rho = 0$, corresponding to $k_z = k$, the boundary condition for E_z is automatically satisfied. Moreover, if we choose $n = 0$ with

$$\Phi(\phi) = \cos n\phi = 1 \quad n = 0 \tag{4.451}$$

the boundary condition on E_ϕ is also satisfied. In this case, a transverse electromagnetic mode (TEM) can exist with $E_z = H_z \equiv 0$ and

$$\mathbf{E} = \widehat{\rho} E_0 \frac{e^{-jkz}}{k\rho} \tag{4.452}$$

$$\mathbf{H} = \widehat{\phi} \frac{E_0}{Z} \frac{e^{-jkz}}{k\rho} \tag{4.453}$$

where Z is the intrinsic impedance of the medium filling the guide and k is the corresponding wavenumber. This particular mode has no cutoff and all frequencies propagate. Thus, it is the fundamental mode.

We may compute the characteristic impedance of the TEM mode from the induced voltage

$$V_b - V_a = -\int_a^b E_\rho d\rho = \frac{E_0}{k} \ln(a/b) \tag{4.454}$$

and current

$$I = \oint_c \mathbf{H} \cdot d\ell = \int_0^{2\pi} \frac{E_0}{Zka} a d\phi = \frac{2\pi E_0}{Zk} \tag{4.455}$$

as

$$Z_c = \frac{V_b - V_a}{I} = \frac{Z}{2\pi} \ln(a/b) \tag{4.456}$$

For a dielectric filled waveguide, the characteristic impedance is given by

$$Z_c = \frac{60}{\sqrt{\epsilon_r}} \ln(a/b) \tag{4.457}$$

where ϵ_r is the relative permittivity of the filling material.

4.5.6.2 Higher Order Modes

The higher order modes can be divided to transverse magnetic and transverse electric modes. The transverse magnetic modes are derived from $\pi_e = \widehat{z}\psi(\rho, \phi, z)$ satisfying

$$(\nabla_t^2 + k_\rho^2)\psi = 0 \tag{4.458}$$

subject to the boundary conditions. Since $\rho = 0$ is not in the region considered, we may choose

$$\psi(\rho, \phi, z) = [J_n(k_\rho \rho) + A_n N_n(\mathbf{k}_\rho \rho)] \begin{matrix}\sin \\ \cos\end{matrix} n\phi \; e^{-jk_z z} \tag{4.459}$$

for some constant A_n where

$$k_\rho^2 + k_z^2 = k^2 \tag{4.460}$$

as before. Thus

$$J_n(k_\rho a) + A_n N_n(k_\rho a) = 0 \tag{4.461}$$
$$J_n(k_\rho b) + A_n N_n(k_\rho b) = 0 \tag{4.462}$$

and these have a solution for A_n if and only if

$$\frac{N_n(k_\rho a)}{N_n(k_\rho b)} = \frac{J_n(k_\rho a)}{J_n(k_\rho b)} \tag{4.463}$$

which is the characteristic equation for the guide. For any integer $n \geq 0$, this equation has an infinity of discrete solutions $k_\rho = k_\rho^p$, $p = 1, 2, \ldots$, leading to the corresponding propagation constants k_z and, hence, the modes. For the $n = 0$ mode, the characteristic equation is given by

$$\frac{N_0(k_\rho a)}{N_0(k_\rho b)} = \frac{J_0(k_\rho a)}{J_0(k_\rho b)} \tag{4.464}$$

Note that if $k_\rho \to 0$, that is, $\mathbf{k}_z \to k$, we have $N_0(k_\rho a) = N_0(k_\rho b)$ whose only solution is $k_\rho = 0$ (for $a \neq b$) which would be the TEM mode. In the other extreme, $k_z \ll k$, and k_ρ is large. Using the asymptotic expansions for the Bessel functions

$$J_0(k_\rho a) \simeq \sqrt{\frac{2}{\pi k_\rho a}} \cos(k_\rho a - \pi/4) \tag{4.465}$$

$$N_0(k_\rho a) \simeq \sqrt{\frac{2}{\pi k_\rho a}} \sin(k_\rho a - \pi/4) \tag{4.466}$$

we obtain

$$\sin[k_\rho(a-b)] = 0 = \sin m\pi, \quad m = 1, 2, \ldots \quad (m \neq 0) \tag{4.467}$$

Therefore

$$k_{\rho 0m} = \frac{m\pi}{a-b} \tag{4.468}$$

and the propagation constant of the TM_{0m} mode is given by

$$k_{z0m} = \sqrt{k^2 - (\frac{m\pi}{a-b})^2} \tag{4.469}$$

with the cutoff wavelength

$$\lambda_{c0m} = \frac{2\pi}{k_{\rho 0m}} = \frac{2(a-b)}{m} \tag{4.470}$$

For example for the TM_{01} mode, we have

$$\lambda_{c01} = 2(a-b) \tag{4.471}$$

and for the next mode

$$\lambda_{c11} = \pi(a+b) \tag{4.472}$$

which is the mean circumference of the coaxial waveguide. This implies that the cutoff frequency of the TM_{11} mode is lower than that of TM_{01}.

The TE modes can be derived in a similar fashion. The characteristic equation is

$$\frac{N_n'(k_\rho a)}{N_n'(k_\rho b)} = \frac{J_n'(k_\rho a)}{J_n'(k_\rho b)} \quad n \neq 0 \tag{4.473}$$

which gives the mode propagation constants. These mode have non-zero frequency cutoff as well. It can be shown that for $n = 1$, the solution to the above equation is approximately

$$k_{\rho 1m} \simeq \frac{2\pi m}{a+b} \tag{4.474}$$

with the cutoff wavelength

$$\lambda_{c1m} = \frac{2\pi}{k_{\rho 1m}} \simeq \frac{\pi(a+b)}{m} \tag{4.475}$$

For the TM_{11} mode,

$$\lambda_{c11} \simeq \pi(a+b) \tag{4.476}$$

The coaxial line may operate in TEM mode upto a wavelength $\pi(a+b)$ without exciting higher order modes

4.5.7 The Dielectric Rod

A dielectric cylinder whose refractive index exceeds that of the surrounding medium could be used to guide modes which radiate through the outer medium (Fig. 4.44). These modes are, therefore, referred to as *leaky modes*.

In the dielectric rod, the TE- and TM-modes do not exist except when the fields have no azimuthal (ϕ) variation. In other words, the modes in general have six nonzero components. The modes can be constructed by using both the electric and magnetic Hertz vectors

$$\pi_e = \widehat{z}\psi_e(\rho, \phi, z), \quad \pi_m = \widehat{z}\psi_m(\rho, \phi, z)$$

By choosing the appropriate form of ψ_e and ψ_m in the two regions, and by imposing the boundary conditions at the dielectric interface $\rho = a$, the characteristic equation may be shown to be

$$\left\{\frac{1}{u}\frac{J_n'(u)}{J_n(u)} - \frac{1}{v}\frac{H_n^{(2)'}(v)}{H_n^{(2)}(v)}\right\}\left\{\frac{n_1^2}{u}\frac{J_n'(u)}{J_n(u)} - \frac{n_2^2}{v}\frac{H_n^{(2)'}(v)}{H_n^{(2)}(v)}\right\} = \left\{\frac{nk_z}{k_0}\left(\frac{1}{u^2} - \frac{1}{v^2}\right)\right\}^2 \tag{4.477}$$

where

$$u = k_{\rho 1}a, \quad v = k_{\rho 2}a \tag{4.478}$$

and

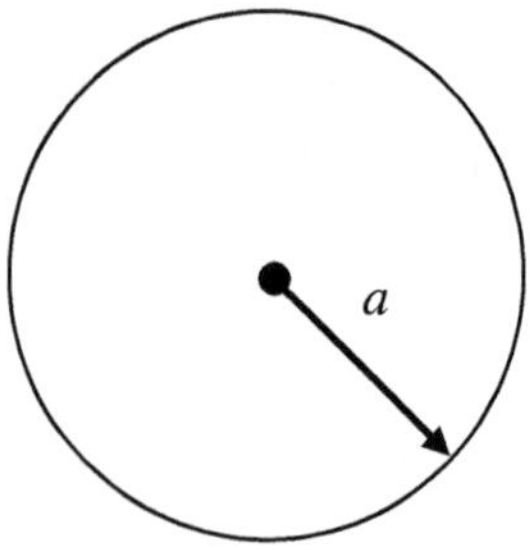

Fig. 4.44 A circular dielectric rod of radius a

$$k_{\rho 1}^2 + k_z^2 = (n_1 k_0)^2, \quad k_{\rho 2}^2 + k_z^2 = (n_2 k_0)^2 \tag{4.479}$$

For azimuthal symmetric case, $n = 0$, and the characteristic equation decouples. The vanishing of the first factor corresponds to TE-modes, while the vanishing of the second factor corresponds to TM-modes.

In general, as in the dielectric slab waveguide, a subset of the modes propagates unattenuated down the rod, and for these, it is necessary that the field is attenuated in the outer medium.

4.6 Spherical Waves

Consider the Helmholtz wave equation in spherical coordinates,

$$\frac{1}{r^2}\frac{\partial}{\partial r}(r^2\frac{\partial \Psi}{\partial r}) + \frac{1}{r^2 \sin\theta}\frac{\partial}{\partial \theta}(\sin\theta\frac{\partial \Psi}{\partial \theta}) + \frac{1}{r^2\sin^2\theta}\frac{\partial^2\Psi}{\partial \phi^2} + k^2\Psi = 0 \tag{4.480}$$

Employing separation of variables, we write

$$\Psi(r, \theta, \phi) = R(r)H(\theta)\Phi(\phi) \tag{4.481}$$

Substituting in the equation for Ψ,

$$\frac{H\Phi}{r^2}\frac{d}{dr}(r^2\frac{dR}{dr}) + \frac{R\Phi}{r^2\sin\theta}\frac{d}{d\theta}(\sin\theta\frac{dH}{d\theta}) + \frac{RH}{r^2\sin^2\theta}\frac{d^2\Phi}{d\phi^2} + k^2\Psi = 0 \tag{4.482}$$

Multiplying by $r^2\sin^2\theta$ and dividing by Ψ, we have

$$\frac{\sin^2\theta}{R}\frac{d}{dr}(r^2\frac{dR}{dr}) + \frac{\sin\theta}{H}\frac{d}{d\theta}(\sin\theta\frac{dH}{d\theta}) + \frac{1}{\Phi}\frac{d^2\Phi}{d\phi^2} + k^2r^2\sin^2\theta = 0 \tag{4.483}$$

The third term is only a function of ϕ while the other terms are independent of ϕ. It should therefore be a constant. Let

$$\frac{\Phi''}{\Phi} = -\nu^2 \tag{4.484}$$

where ν is the separation constant. Substituting in the previous equation and dividing by $\sin^2\theta$, we have

$$\frac{1}{R}\frac{d}{dr}(r^2\frac{dR}{dr}) + \frac{1}{H\sin\theta}\frac{d}{d\theta}(\sin\theta\frac{dH}{d\theta}) - \frac{\nu^2}{\sin^2\theta} + (kr)^2 = 0 \tag{4.485}$$

Introducing a new separation constant μ, we write

$$\frac{1}{H\sin\theta}\frac{d}{d\theta}(\sin\theta\frac{dH}{d\theta}) - \frac{\nu^2}{\sin^2\theta} = -\mu^2 \tag{4.486}$$

and

$$\frac{1}{R}\frac{\partial}{\partial r}(r^2\frac{dR}{\partial r}) + k^2r^2 = \mu^2 \tag{4.487}$$

The choice of separation constants ν and μ is governed by the physical requirement that at any fixed point in space the solution must be single valued.

In the following discussion, we assume that ϕ varies in the range $0 \le \phi \le 2\pi$. This implies that the potential must be periodic in ϕ with a period of 2π in order to be single valued. Thus, ν must be an integer m.

$$\frac{d^2\Phi}{d\phi^2} + m^2\Phi = 0 \tag{4.488}$$

This is the familiar harmonic equation.

Consider now the equation satisfied by $H(\theta)$

$$\frac{1}{\sin\theta}\frac{d}{d\theta}(\sin\theta\frac{dH}{d\theta}) + [\mu^2 - \frac{m^2}{\sin^2\theta}]H = 0 \tag{4.489}$$

which is a Sturm-Liouville differential equation. We consider the following cases for the solution of this equation.

Azimuthal Symmetry

In this case, there is no ϕ variation and we set $m = 0$.

$$\frac{1}{\sin\theta}\frac{d}{d\theta}(\sin\theta\frac{dH}{d\theta}) + \mu^2 H = 0 \tag{4.490}$$

We must select functions $H_n(\theta)$ which together constitute a complete set. Fortunately, the solutions of the above equation form a complete set when subjected to the condition that $H(\theta)$ be finite for $\theta = 0$ and $\theta = \pi$. Under these circumstances this equation has solutions if and only if μ is an integar such that

$$\mu^2 = n(n+1), \quad n = 0, 1, 2, \ldots \tag{4.491}$$

This choice for μ also causes the series solution to the differential equation to terminate in a finite number of terms (A polynomial of degree n). We therefore have the *Legendre equation*

$$\frac{1}{\sin\theta}\frac{d}{d\theta}(\sin\theta\frac{dH}{d\theta}) + n(n+1)H = 0 \tag{4.492}$$

and the corresponding functions are the *Legendre polynomials*

$$H_n(\theta) = P_n(\theta) = \frac{1}{2^n n!}\left[\frac{d}{d(\cos\theta)}\right]^n (\cos^2\theta - 1)^n \tag{4.493}$$

They form a complete set in the interval $-1 \le \cos\theta \le 1$. Therefore, any arbitrary wave function can be represented by a series of Legendre polynomials.

The first few polynomials are

$$\begin{aligned} P_0(\cos\theta) &= 1 \\ P_1(\cos\theta) &= \cos\theta \\ P_2(\cos\theta) &= \frac{1}{2}(3\cos^2\theta - 1) \\ P_3(\cos\theta) &= \frac{1}{2}(5\cos^3\theta - 3\cos\theta) \end{aligned} \tag{4.494}$$

We may also write $z = \cos\theta$ so that

$$P_n(z) = \frac{1}{2^n n!}(\frac{d}{dz})^n (z^2 - 1)^n \tag{4.495}$$

which is the *Rodrigues' formula*. The orthogonality relation for the Legendre polynomials is

$$\int_0^\pi P_n(\cos\theta)P_\ell(\cos\theta)\sin\theta d\theta = \frac{2}{2n+1}\delta_{n\ell} \tag{4.496}$$

and the corresponding normalized functions are $(\frac{2n+1}{2})^{1/2}P_n(\cos\theta)$.

Fourier-Legendre Series

Since $P_n(\cos\theta)$ form a complete orthogonal set, any function can be expanded in terms of them using the generalized Fouries series representation. Thus, if we have a function

$$f(\theta), \quad 0 \le \theta \le \pi$$

we write

$$f(\theta) = \sum_{n=0}^{\infty} a_n P_n(\cos\theta), \quad 0 \le \theta \le \pi \tag{4.497}$$

where

$$a_n = \frac{2n+1}{2} \int_0^{\pi} f(\theta) P_n(\cos\theta) \sin\theta d\theta \tag{4.498}$$

No Azimuthal Symmetry

In this case, we have the *associated Legendre equation*

$$\frac{1}{\sin\theta} \frac{d}{d\theta} (\sin\theta \frac{dH}{d\theta}) + [n(n+1) - \frac{m^2}{\sin^2\theta}]H = 0 \tag{4.499}$$

The solution to this equation are $P_n^m(\cos\theta)$ and $Q_n^m(\cos\theta)$, the *associated Legendre functions* of the first and second kind.

Except for $P_n^m(\cos\theta)$ with m and n integers, all other forms of the solutions are singular at $\theta = 0$ and $\theta = \pi$. Notice that

$$P_n^m(\cos\theta) = \sin^m\theta P_n^{(m)}(\cos\theta) \quad m, n > 0 \quad m \leq n \tag{4.500}$$

where $P_n^{(m)}$ is the mth derivative of the Legendre function of order n and

$$P_n^m(\cos\theta) = 0 \quad m > n \tag{4.501}$$

The product functions,

$$P_n^m(\cos\theta)e^{jm\phi} = T_{mn}^e(\theta, \phi) + jT_{mn}^o(\theta, \phi) \tag{4.502}$$

are also of interest as they apear in the solution of the wave equation. Here

$$T_{mn}^e(\theta, \phi) = P_n^m(\cos\theta) \cos m\phi \tag{4.503}$$

$$T_{mn}^o(\theta, \phi) = P_n^m(\cos\theta) \sin m\phi \tag{4.504}$$

are called tesseral harmonics and they form a complete orthogonal set on the surface of a sphere. Therefore, any wave function that is defined over spherical surface can be expressed by a series of tesseral harmonics. The orthogonality relation for T^e and T^o are given by

$$\int_0^{2\pi} \int_0^{\pi} T_{mn}^e(\theta, \phi) T_{pq}^o(\theta, \phi) \sin\theta d\theta d\phi = 0 \tag{4.505}$$

$$\int_0^{2\pi} \int_0^{\pi} T_{mn}^i(\theta, \phi) T_{pq}^i(\theta, \phi) \sin\theta d\theta d\phi = 0 \quad mn \neq pq, i = e, o \tag{4.506}$$

$$\int_0^{2\pi}\int_0^{\pi} |T^i_{mn}(\theta,\phi)|^2 \sin\theta d\theta d\phi = \frac{2\pi}{2n+1}\frac{(n+m)}{(n-1)}(1+\delta_{0m}), i = e, o \quad (4.507)$$

where δ_{0m} is the Kronecker delta function.

The last of the separated equations is the *spherical Bessel's equation*

$$\frac{d}{dr}(r^2\frac{dR}{dr}) + [(kr)^2 - n(n+1)]R = 0 \quad (4.508)$$

Let $R = \sqrt{\frac{\pi}{2kr}}B$. Then the above equation yields

$$\frac{d^2B}{dr^2} + \frac{1}{r}\frac{dB}{dr} + [k^2 - \frac{(n+1/2)}{r^2}]B = 0 \quad (4.509)$$

which is the Bessel's equation of order $n+1/2$ and B is the corresponding cylindrical Bessel function. The solutions are referred to as the *spherical Bessel functions* $b_n(kr)$ which are given by

$$R_n(r) = b_n(kr) = \sqrt{\frac{\pi}{2kr}}B_{n+1/2}(kr) \quad (4.510)$$

The general behavior of the spherical Bessel functions are similar to their cylindrical counterparts.

Also, it is noted that for $n = 0$, the spherical Bessel functions take particularly simple forms

$$j_0(kr) = \frac{\sin kr}{kr} n_0(kr) = -\frac{\cos kr}{kr} \quad (4.511)$$

$$h_0^{(1)}(kr) = \frac{e^{jkr}}{jkr} h_0^{(2)}(kr) = -\frac{e^{-jkr}}{jkr} \quad (4.512)$$

The only spherical Bessel functions which are finite at the origin $r = 0$ are the $j_n(kr)$.

Going back to the wave equation, the general solution is given by

$$\Psi = \sum_m \sum_n C_{mn} b_n(kr) L_n^m(\cos\theta) h(m\phi) \quad m, n \in \mathcal{I} \quad (4.513)$$

Thus, the finite field inside a spherical region (including $r = 0$) is given by

$$\Psi_{in} = \sum_{mn} C_{mn} j_n(kr) P_n^m(\cos\theta) e^{jm\phi} \quad (4.514)$$

and the finite field outside a sphere is given by ($r \to \infty$ included)

$$\Psi_{out} = \sum_{mn} C_{mn} h_n^{(2)}(kr) P_n^m(\cos\theta) e^{jm\phi} \tag{4.515}$$

4.6.1 Spherical Wave Transformation

Consider an incident electric field polarized in the $\widehat{x}$ direction and propagating in the $\widehat{z}$ direction

$$\mathbf{E} = \widehat{x} e^{-jkz} = \widehat{x} e^{-jkr\cos\theta} \tag{4.516}$$

we would like to represent the field (a plane wave) in terms of spherical waves.

Since there is no ϕ dependence, $m = 0$ and since the field is finite at the origin, we write

$$e^{-jkr\cos\theta} = \sum_{n=0}^{\infty} a_n j_n(kr) P_n(\cos\theta) \tag{4.517}$$

Recognizing this as a Fouries-Legendre expansion, we find

$$a_n = j^{-n}(2n+1) \tag{4.518}$$

4.6.2 Point Sources

Almost all sources used in practice are considered to excite spherical waves. Intuitively, it is clear that any finite source looks like a point from a great distance and a point source has spherical symmetry and emits spherical waves. However, the most prominent source of spherical waves is the infinitesimal dipole current.

Consider a small electric current element

$$\mathbf{J}dv = I_0 d\ell e^{j\omega t} \widehat{z} \tag{4.519}$$

located at the origin. Thus, the electric Hertz vector is given by

$$\begin{aligned} \pi_e(\mathbf{r}) &= -j\frac{Z}{k} I_0 d\ell \frac{e^{-jkr}}{4\pi r} \widehat{z} \\ &= -\widehat{z} \frac{Z}{4\pi} I_0 d\ell h_0^{(2)}(kr) \end{aligned} \tag{4.520}$$

where $h_0^{(2)}(kr)$ is the spherical Hankel function of the second kind.

Here, π_e is a "true" spherical wave, having its amplitude constant on the spherical surface $r =$ constant which is also the phase front, propagating in the $\widehat{r}$ direction. The electromagnetic fields due to this current element in the far zone is given by

$$\mathbf{E} = \nabla\nabla \cdot \pi + k^2 \pi$$

$$\simeq -\widehat{\theta} j \frac{kZ_0 I d\ell}{4\pi r} \sin\theta e^{-jkr} \tag{4.521}$$

$$\mathbf{H} \simeq -\widehat{\phi} j \frac{kId\ell}{4\pi r} \sin\theta e^{-jkr} \tag{4.522}$$

Energy propagation is still in the direction of $\mathbf{E} \times \mathbf{H}$, that is $\widehat{r}$. However, $|\mathbf{E}|$ and $|\mathbf{H}|$ are not constant on the spherical surface (they vary with θ). Uniform and isotropic spherical wave where the field vectors are constant on the spherical phase front is not possible in the electromagnetic case. (It is possible in scalar case).

4.6.2.1 Current and Dipole Moments

Consider a small linear current element of length ℓ carrying a constant current I_0 (actually $I_0 e^{j\omega t}$), and oriented along the $\widehat{z}$ direction. The (total) *current moment* is given by

$$\mathbf{p}_i = \int_V \mathbf{J}(\mathbf{r}')dv' = \widehat{z} \int_{-\ell/2}^{\ell/2} I(z')dz' = I_0 \ell \widehat{z} \tag{4.523}$$

Now imagine two changes (time varying) $+q$ and $-q$ separated by a distance ℓ and placed along the z axis, as shown. For small ℓ, we now have a time-varying electric dipole. The *dipole moment* is defined by

$$\mathbf{P} = q\ell\widehat{z} \tag{4.524}$$

Since $I_0 = \frac{dq}{dt} = j\omega q$, we have

$$\mathbf{P} = \frac{I_0 \ell}{j\omega}\widehat{z} \tag{4.525}$$

Thus, we can write the following relation between the current and dipole moments

$$\mathbf{p}_i = j\omega\mathbf{P} \tag{4.526}$$

4.6.3 Addition Theorem

The electric Hertz vector for a z-directed Hertzian dipole located away from the origin at $\mathbf{r}'$ is given by

$$\pi_e = -\widehat{z}\frac{Z}{4\pi} I_0 d\ell h_0^{(2)}(k|\mathbf{r} - \mathbf{r}'|) \tag{4.527}$$

It is often necessary to express π_e (and the corresponding fields) in terms of spherical wave functions. This can be accomplished using the addition theorem

$$h_0^{(2)}(k|\mathbf{r}-\mathbf{r}'|) = \begin{cases} \sum\limits_{n=0}^{\infty}(2n+1)h_n^{(2)}(kr')j_n(kr)P_n(\cos\xi) & r<r' \\ \sum\limits_{n=0}^{\infty}(2n+1)h_n^{(2)}(kr)j_n(kr')P_n(\cos\xi) & r>r' \end{cases} \tag{4.528}$$

where

$$\cos\xi = \cos\theta\cos\theta' + \sin\theta\sin\theta'\cos(\phi-\phi') \tag{4.529}$$

Exercises

4.1: Find the direction of propagation of the wave

$$\mathbf{E}(\mathbf{r};t) = \cos(100t + 4x + 2y + 4z)$$

4.2: Find the far field radiated field of a small Hertzian dipole The instantaneous expression of an elliptically polarized wave is given by

$$\mathbf{E}(z,t) = 3\cos(\omega t - k_0 z)\widehat{x} + 4\sin(\omega t - k_0 z)\widehat{y} \quad (V/m)$$

(a) Find the complex phasor $\mathbf{E}(z)$ of this wave.
(b) What is the instantaneous value and the average value of the Poynting vector pertaining to this wave?

4.3: An elliptically polarized wave is described by the superposition of two circularly polarized waves

$$\mathbf{E} = (\widehat{x} + j\widehat{y})e^{jkz} + 2(\widehat{x} - j\widehat{y})e^{j(kz+\pi/4)} \quad (V/m)$$

(a) Find the expression for the instantaneous value of the electric field.
(b) What is the average Poynting vector associated with each circularly polarized wave and the one with the entire wave?

4.4: The electric field in a source free region has only one component $E_y(x,t)$ which is independent of y and z. The region under consideration is a lossy dielectric of conductivity σ. If $E_y(x,t)$ is a harmonically oscillating function of time so that

$$E_y(x,t) = \Re e E_y(x)e^{j\omega t}$$

what is the differential equation for $E_y(x)$?

4.5: The electrical parameters of sea water at 400 MHz are

$$\mu_r \simeq 1, \quad \epsilon_r \simeq 81, \quad \sigma \simeq 1$$

(a) Can this medium be considered a good dielectric or a good conductor at this frequency?
(b) Find the phase constant, attenuation constant and skin depth.

4.6: A 10 kHz plane wave of unit amplitude is incident upon sea water from air at 30°. The relative permittivity and conductivity of sea water are assumed to be 81 and 4 S/m, respectively.

(a) Find the expression for the transmitted wave and the real angle of transmission.
(b) What is the wavelength of the wave in sea water?
(c) What is the velocity of propagation in sea water?

4.7: For a lossy dielectric with

$$\epsilon_r'' \gg \epsilon_r', \quad \mu_r = 1, \quad \sigma = 0$$

(a) show that the attenuation coefficient is given by

$$\alpha = \frac{2\pi}{\lambda_0}\sqrt{\frac{\epsilon_r''}{2}}$$

(b) find the distance in wavelength over which the amplitude of a plane wave is attenuated to half of its initial value entering the medium.

4.8: For a lossy dielectric with

$$\epsilon_r' \gg \epsilon_r'', \quad \mu_r = 1, \quad \sigma = 0$$

(a) show that the attenuation coefficient is given by

$$\alpha = \frac{\pi}{\lambda_0}\sqrt{\frac{\epsilon_r''}{\epsilon_r'}}$$

(b) Assuming that the loss tangent is 0.001, find the distance in wavelength over which the amplitude of a plane wave is attenuated to half of its initial value entering the medium.

4.9: A plane wave of the form

$$\mathbf{E} = 2\hat{y}e^{-\alpha z}\cos(10^8 t - \beta z) \quad (V/M)$$

is propagating in a medium charactrized by

$$\epsilon_r = 2, \quad \mu_r = 20, \quad \sigma = 3\,\text{S/m}$$

(a) Classify this medium and claculate the attenuaion and phase constants.
(b) Find the magnetic field intensity.
(c) Find the instantaneous net power flux density entering the rectangular box of dimensions $a \times b \times d$.

4.10: Verify (4.70).

4.11: A plane wave is incident on a dielectric half space of refractive index n.

(a) Find the voltage standing wave ratio (VSWR).
(b) Power reflection coefficient.

4.12: To reduce the probability of detection of a fighter plane at 100 MHz, the body is coated by a 10 mm thick layer of magnetic titanate with relative permittivity and permeablity $\epsilon_r = \mu_r = 120 - j60$.

(a) Find the refractive index of the layer.
(b) Assuming normal incidence, how much is the reflected wave attenuated in dB?

4.13: Fresnel's rhomb is a device for using total internal reflection at an interface to produce 100 % conversion of a linearly polarized plane wave to a circularly polarized one.

(a) Show how this can be done.
(b) For the interface between the two media having $n_1 = 3.0$ and $n_2 = 1$ (air), find the angle(s) of incidence θ_1 required to achieve this effect.

4.14: Consider a two-media problem with incidence from denser medium ($x < 0$) to the other medium ($x > 0$). Above the critical angle θ_{1c}, $|R| = 1$. But this does not imply that there is no field in $x > 0$ when $\theta_1 > \theta_{1c}$. Using the complex Poynting vector, examine the power flows for all angles of incidence (above and below θ_{1c})

(a) In $X < 0$,
(b) In $X > 0$.

4.15: Show that the resistive part of the surface impedance Z_s in (4.74) is the equivalent dc resistance of a square sheet of metal of thickness equal to the skin depth δ and with conductivity σ.

4.16: Consider a parallel plate waveguide of width b partially filled with a homogeneous dielectric of relative permittivity ϵ_r and thickness $a < b$ (Fig. 4.32). The dielectric interface is parallel to the guide walls. Find the characteristic equation for the guide propagation constant.

4.17: Consider a rectangular waveguide of dimension $a \times b$ operating in TM mode. Find the total power flowing in the waveguide and discuss the nature of this power when the operating frequency f is smaller and higher than the cutoff frequency f_{mnc}.

4.18: Consider a rectangular metallic waveguide of dimensions $a = 1$ cm and $b = 2.5$ cm.

(a) Identify the first three modes for the empty guide and find their cut-off frequencies.
(b) Sketch the electric field intensity distribution of the dominant mode over the waveguide cross section.
(c) How can we reduce the cut-off frequency of the dominant mode without changing the dimensions of the waveguide?
(d) If we reduce a to $a = 0.5$ cm, how much would the dominant cut-off frequency change? What about the next mode?

4.19: Find the reflected wave in a rectangular waveguide which contains a dielectric layer of thickness h, and of relative permittivity ϵ_r, placed at $0 \leq z \leq$. The incident wave is a TE_{10} mode described by $\mathbf{E} = E_0 \sin \frac{\pi x}{a} e^{-jk_z z}\widehat{y}$ where $k_z = \sqrt{k_0^2 - (\pi/a)^2}$.

4.20: Determine the phase velocity and the group velocity for the TE_{10} mode propagating in a 4×1 cm rectangular waveguide if the operating frequency is 1.8 times the cut off frequency.

4.21: Consider a grounded dielectric slab waveguide of thickness $a = 0.5$ cm and relative permittivity $\epsilon_r = 2.56$ operating in TM mode.

(a) Find the cutoff frequencies of the first six modes.
(b) What are the propagation constants of the propagating modes at f = 30 GHz?
(c) Calculate the ratio of the power outside the slab to the power inside for each propagating mode at 30 GHz.

4.22: Consider a thin film waveguide where the three regions $0 \leq x \leq a$, $x > a$ and $x < 0$ are occupied by three homogeneous dielectric materials n_1, n_2 and n_3, respectively, and $n_1 > n_2, n_3$. The structure is infinite in the y and z directions. Assume TE modes ($\mathbf{E} = \widehat{y}E_y$).

(a) Determine the fields for lossless propagation in the film and deduce the characteristic equation.
(b) By considering the z-directed power flows P_1, P_2 and P_3 in the three media where

$$P = \int S_z dx$$

show that

$$P = P_1 + P_2 + P_3 = \Gamma(a + \frac{1}{p_2} + \frac{1}{p_3})$$

where p_2 and p_3 are the attenuation factors in the n_2 and n_3 media and Γ is a factor proportional to the field intensity in the film.

4.23: A plane wave is normally incident in air upon a layer extending from $-\ell/2 \leq z \leq \ell/2$ in which the index of refraction varies as

$$n(z) = \frac{A}{A + (0.5 - z/\ell)}$$

where $A = n_1/(1 - n_1)$. Outside this layer, the refractive index is 1 (air) for $z > \ell/2$ and n_1 for $z < -\ell/2$. Calculate the magnitude of the magnitude and phase of the reflection coefficient at the top surface of the layer for $n_1 = 2$ and $k_0\ell = 0.2, 0.5, 1.0, 2.0, 4.0$, and 6.0.

4.24: The pahse velocity for some large scale sea waves is given by the relation $u_p = g/\omega$ where g is the gravitational acceleration and ω is the angular frequency of the wave. Find the group velocity to phase velocity ratio for this type of waves.

4.25: Using the spatial Fourier transform of the Maxwell's equations, find the dispersion relation for a progressive wave in a conducting medium.

4.26: A 100 MHz plane wave is propagating in a dispersive medium at a phase velocity of $c/3$ where c is the speed of light in vacuum. The dispersion of the medium is such that the phase velocity is proportional to square root of the wavelength. Find the group velocity.

4.27: Find the phase and group velocities for a nonmagnetic medium whose permittivity is given by

$$\epsilon(\omega) = 1 + \frac{\omega_p^2}{\omega_0^2 - \omega^2}$$

Consider only the cases of large and small frequencies (compared with ω_0).

4.28: Construct a one-dimensional wave packet ψ at time $t = 0$ in the case where the amplitude is given by the Gaussian function

$$A(k) = A_0 \exp\left[-(k - k_0)^2/(\Delta k)^2\right]$$

and A_0, k_0 and Δk are constants. Find the relation between the width of the packet Δx and the range of wave numbers Δk which contribute to the wave packet.

4.29: A surface electric current of density

$$\mathbf{K} = K_0 \cos\frac{\pi x}{a}\widehat{y} \quad -a/2 \leq x \leq a/2 \quad -b/2 \leq y \leq b/2$$

exist in plane $z = 0$. Find the electric field radiated by this current source in the far field.

4.30: In cylindrical coordinates, show that any electromagnetic field having the form

$$\mathbf{E}(\rho,\phi,z) = \mathbf{E}(\rho,\phi)e^{-jk_z z}, \quad \mathbf{H}(\rho,\phi,z) = \mathbf{H}(\rho,\phi)e^{-jk_z z}$$

can be expressed in terms of E_z and H_z alone, and deduce the source-free differential equations satisfied by E_z and H_z. The medium is assumed to be isotropic and homogeneous.

4.31: Using the result of Problem 24, write down the most general expressions for all the field components in a cylindrical region which includes the z axis ($\rho = 0$) but excludes infinity ($\rho = \infty$).

4.32: A TM_{01} wave is being propagated in a highly conducting hollow pipe of inside radius 1 cm. Its frequency is 1.5 times the cutoff frequency for this mode. The greatest peak value of electric field in the z direction is 1,000 V/m.

(a) What is the operating frequency?
(b) What is the maximun peak value of current density in Amperes per meter in the z-direction?

4.33: Find the vector wave function $\mathbf{M} = \nabla \times (\psi\widehat{z})$ which can be used to describe the magnetic field of the TEM mode ($E_z = 0, H_z = 0$) in a coaxial line of radii a and b with ($a \le \rho \le b$). Assume that the line is perfectly conducting and the medium is air.

4.34: Find the vector wave function $\mathbf{M} = \nabla \times (\psi\widehat{z})$ which can be used to represent the TE modes of the electric field in a waveguide of cross section corresponding to a quarter of a circular guide.

4.35: Find the Fourier-Legender series expansion of the function

$$f(\theta,\phi) = \begin{cases} 1 & 0 < \theta < \pi/2 \\ 0 & \pi/2 < \theta < \pi \end{cases}$$

4.36: Given that $R(r)$ satisfies

$$\frac{d}{dr}(r^2\frac{dR}{dr}) + [(kr)^2 - n(n+1)]R = 0$$

and writing

$$R(r) = \sqrt{\frac{\pi}{2kr}}B_{n+1/2}(kr)$$

verify that $B_{n+1/2}$ is a Bessel function of order $n + 1/2$, i.e. that it satisfies Bessel's equation of order $n + 1/2$.

4.37: The definition of $J_\nu(x)$ for any ν is

$$J_\nu(x) = \sum_{m=0}^{\infty}(-1)^m \frac{(x/2)^{\nu+2m}}{m!(m+\nu)!}$$

By manipulating the series and recognizing those for $\sin x$ and $\cos x$, show that

$$j_0(x) = \sqrt{\pi/2x}J_{1/2}(x) = \frac{\sin x}{x}$$
$$n_0(x) = -\sqrt{\pi/2x}J_{-1/2}(x) = -\frac{\cos x}{x} \tag{4.530}$$

Note: the duplication formula gives

$$(m + 1/2)! = \frac{(2m+1)!}{m!} 2^{-2m-1} \sqrt{\pi}$$

Part II
Scattering Theory

Chapter 5
Radar

Radar is an electronic system used to locate objects beyond the range of vision, and to determine their distance by projecting radio waves against them. The term *radar* stands for radio detection and ranging. A radar is used to measure the range or distance to some object or target, such as an aircraft, or a naval ship, that is located in the radiated beam of the transmitting antenna.

Radars not only indicate the presence and range of a distant object, called the target, but also determine its position in space, its size and shape, and its velocity and direction of motion. Although originally developed as an instrument of war, radar today is used extensively in many peacetime pursuits, which include air traffic control, detecting weather patterns, and tracking spacecraft.

All radar systems employ a high-frequency radio transmitter to send out a beam of electromagnetic waves, ranging in frequency from 300 MHz all the way through the microwave frequencies. Objects in the path of the beam reflect these waves back to the transmitter.

5.1 Historical Remarks

The basic concepts of radar are based on the laws of radio-wave reflection. The German engineer Christian Hulsmeyer was the first to propose the use of radio echoes in a detecting device designed to avoid collisions in marine navigation. A similar device was suggested in 1922 by the Italian inventor Guglielmo Marconi.

The first successful radio range-finding experiment occurred in 1924, when the British physicist Sir Edward Victor Appleton used radio echoes to determine the height of the ionosphere, an ionized layer of the upper atmosphere that reflects longer radio waves. The American physicists Gregory Breit and Merle Antony Tuve achieved independently the same measurements of the ionosphere in the following year, using the radio-pulse technique that was subsequently adopted in most radar systems. Development of the latter was impossible until electronic techniques and equipment were improved in the 1930s.

© Springer International Publishing Switzerland 2015

K. Barkeshli, *Advanced Electromagnetics and Scattering Theory*,
DOI 10.1007/978-3-319-11547-4_5

The first practical radar system was produced in 1935 by the British physicist Sir Robert Watson-Watt. His work gave the British a head start in this important technology, and by 1939 they had established a chain of radar stations along the south and east coasts of England to detect aggressors in the air or on the sea. In that same year two British scientists were responsible for the most important advance made in the technology of radar during World War II. The physicist Henry Boot and biophysicist John T. Randall invented an electron tube called the resonant-cavity magnetron. This type of tube is capable of generating high-frequency radio pulses with a large amount of power, thus permitting the development of microwave radar using lasers. Microwave radar, also called LIDAR (light detection and ranging), is used in the present day for communications and to measure atmospheric pollution.

The advanced radar systems developed in the 1930s played an important role in the Battle of Britain, an air battle from August through October 1940, in which Adolf Hitler's Luftwaffe failed to win control of the skies over England. Although the Germans had their own radar systems, throughout the rest of the war the British and the Americans were able to maintain technical superiority.

Over-the-horizen (OTH) radars were developed to detect military targets far beyond the optical horizen. They use 5–28 MHz radio waves, which reflect from the ionsphere, reaching up to 3,500 km in one "hop". Properties of the ocean surface are extracted from the minute amount of energy scattered by the sea surface back to the radar. The OTH radar can measure surface wind (Bragg-resonant wave) direction, radial surface currents (current vectors with two radars), Sea state (eg., rms wave height), surface wind speed dominant wave period dominant wave direction, non-directional (scalar) ocean wave spectrum and combined swell and wind-wave spectrum.

5.2 Operation

A block diagram of a typical radar system is shown in Fig. 5.1. Radar equipment consists of a transmitter, an antenna, a receiver, and an indicator. Radar transmitters and receivers are usually located in the same place. The transmitter broadcasts a beam of electromagnetic waves by means of an antenna, which concentrates the waves into a shaped beam pointing in the desired direction. When these waves strike an object in the path of the beam, some are reflected from the object, forming an echo signal (Fig. 5.2). The antenna collects the energy contained in the echo signal and delivers it to the receiver. Through an amplification process and computer processing, the radar receiver produces a visual signal on the screen of the indicator, essentially a computer display monitor.

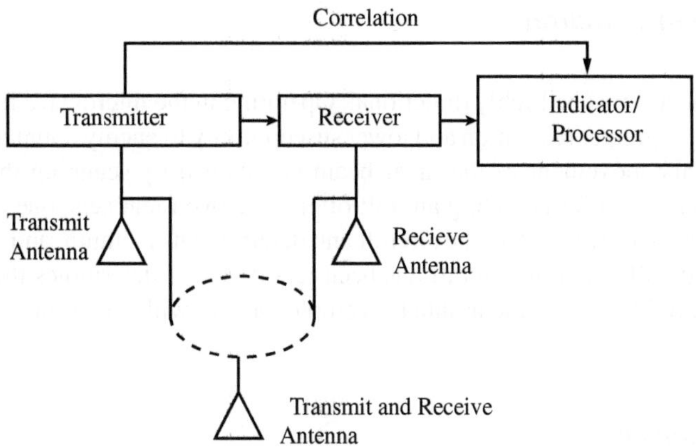

Fig. 5.1 The block diagram of a typical radar system

5.2.1 Transmitters

To operate radar successfully, the transmitter must emit a large burst of energy and receive, detect, and measure about 10^{-12} of the total radio energy, returned in the form of an echo.

One way to solve the problem of detecting the small echo in the presence of the enormously strong searching signal is by using the pulse system. A pulse of energy is transmitted for 0.1–5 μs; thereafter the transmitter is silent for a period of hundreds or thousands of microseconds. During the pulse, or broadcast, phase the receiver is isolated from the antenna by means of a TR (transmit-receive) switch; during the period between pulses the transmitter is disconnected from the antenna by means of an ATR (anti-TR) switch.

The pulse modulator, or pulser draws current continuously from a power source such as a generator and delivers pulses of the necessary voltage, power, duration, and spacing to the magnetron in the transmitter. The pulse must start and end suddenly, but the power and voltage should not vary appreciably during the pulse.

Continuous-wave radar broadcasts a continuous signal rather than pulses. Doppler radar, which is often used to measure the speed of an object, such as an automobile, transmits at a constant frequency.

Frequency-modulated (FM) radar broadcasts a continuous signal of uniformly changing frequency. During the time it takes a signal to be transmitted, reflected, and received, the transmitting frequency changes. The difference between the echo frequency and the transmitter frequency at the instant of echo reception is then measured and translated into the distance between transmitter and object. These systems are more accurate than the pulse type, although they operate over a shorter range.

5.2.2 Radar Antennas

Radar antennas must be highly directional. Operating in the microwave frequencies, antennas have greater resolution and lower susceptibility to enemy countermeasures. The necessary movement of the radar beam is obtained by scanning the antenna. Ground radars used for detecting aircraft often have two radar sets, one of which is scanned horizontally to detect the aircraft and determine its azimuth, and the other is scanned vertically after an aircraft has been reported, and determines the elevation of the aircraft. Many new radar antennas employ arrays with electronic steering.

5.2.3 Receivers

An ideal receiver must amplify and measure an weak signal at extremely high frequency. Because a mobile amplifier has not been devised that can satisfactorily perform this function directly, the signal is converted to an intermediate frequency (IF signal) of about 30 MHz by a superheterodyne circuit and amplified at this frequency. The radio frequency (RF signals) of the radar signal requires high precision oscillator and mixer. Suitable circuits have been developed, employing as oscillators high-power microwave tubes called klystrons. The intermediate frequency is amplified in a conventional fashion. The signal is then fed to a computer.

5.2.4 Computer Processing

Most modern radars convert received analog signals to digital signals by means of an analog-to-digital converter. The digital signals are processed by a high-speed computer to extract information about the target. First, the signal returns from the ground, where unwanted objects are removed by a moving target indicator (MTI) filter. Next, the signal is resolved into separate frequency components by means of a fast frequency transformer (FFT). Finally, after signals from multiple pulses are combined, target detection is determined by the constant false alarm rate CFAR) processor. The CFAR computer must balance detections against false alarms in an optimal manner.

5.2.5 Radar Displays

In modern radar displays, target detection, speed, and position may be overlaid onto maps showing terrain prominent features with high resolution. Recent advances in computer and high-speed electronics have contributed to the radar displays and processing.

5.3 Secondary-Radar System

The radar systems discussed above are known as primary systems, operating on the principle of a passive echo from the target. Another group of radar devices, known collectively as the secondary system, depends on a response from the target; most of these devices are used in navigation and in communication.

5.3.1 Transponder

A radar beacon, which is also called a racon or a transponder, is a secondary-radar set that sends out a pulse whenever it receives a pulse. Such beacons greatly extend the range of radar sets, because a transmitted pulse, even from a low-power transmitter, is far stronger than an echo. The radar transmitter that sends the initial pulse is called the interrogator, and the action of this pulse at the beacon is called triggering.

The simplest form of radar beacon sends out a single pulse of the same frequency that it receives, almost instantaneously, thus acting as a strong echo. Beacons, can also reply on a different frequency, or a calibrated delay can be incorporated at the beacon, so that it appears to be farther from the interrogator than it really is. Such a delay may be used in instrument landing systems, to measure the distance from an airport runway instead of from the beacon itself.

Beacons can also be designed to send back a coded response, thus ensuring that the navigator cannot mistake the blip that appears on the scope.

5.3.2 Radar Identification (IFF)

This is a type of coded radar beacon carried in aircraft for identification purposes in wartime, IFF being an abbreviation for Identification, Friend or Foe. IFF sets contain an emergency switch that, when turned on by a crew member of an aircraft in distress, immediately alerts the interrogating radar set and indicates the position of the aircraft.

5.4 Countermeasures

Two primary methods exist for so-called jamming of the enemy's radar: electronic jamming, transmitting on frequencies that interfere with enemy receivers; and mechanical jamming, dispersing material such as small strips of aluminum foil that produce echoes and interfere with the detection of genuine targets.

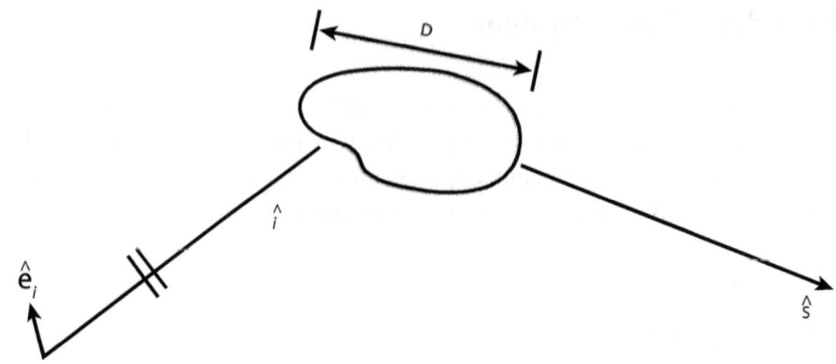

Fig. 5.2 A plane wave being scattered by a target

Other techniques include effective use of geometric shape of the target so that minimum energy is returned to the radar, and the use of radar absorping material (RAM) on the target body in order to dissipate much of the reflected energy.

5.5 Radar Cross Section

When the radar transmitter sends a short pulse of energy, some of this energy is reflected back towards the radar unit by the target. This energy is usually received by the same antenna as is used for transmitting. The elapsed time between the transmitted pulse and the received pulse is $2r/c$, where r is the range and c is the speed of light. By measuring this elapsed time, the range r is determined.

Consider an incident wave illuminating a target in free space as shown in Fig. 5.3. Let the incident wave be expressed as

$$\mathbf{E}^i = \mathbf{E}_0 e^{-j\mathbf{k}\cdot\mathbf{r}} = \hat{e}_i E_0 e^{-jk\hat{\imath}\cdot\mathbf{r}} \tag{5.1}$$

where k is the wavenumber, $\hat{\imath}$ is the unit vector in the direction of wave propagation and $\hat{e}_i$ is the polarization unit vector. The scattered field is defined as the difference between the total field in the presence of the target and that in its absence

$$\mathbf{E}^s = \mathbf{E}^t - \mathbf{E}^i \quad , \quad \mathbf{H}^s = \mathbf{H}^t - \mathbf{H}^i \tag{5.2}$$

The aim in the solution of a scattering problem is to find the scattered field $(\mathbf{E}^s, \mathbf{H}^s)$ ususally by solving a boundary value problem, and hence finding the total field $(\mathbf{E}^t, \mathbf{H}^t)$.

The region external to the scatterer can be divided into two subregions:

Region I: $R < 2D^2/\lambda$ where the scattered field has a relatively complicated amplitude and phase variation.
Region II: $R > 2D^2/\lambda$ where the scattered field has a spherical wave behavior.

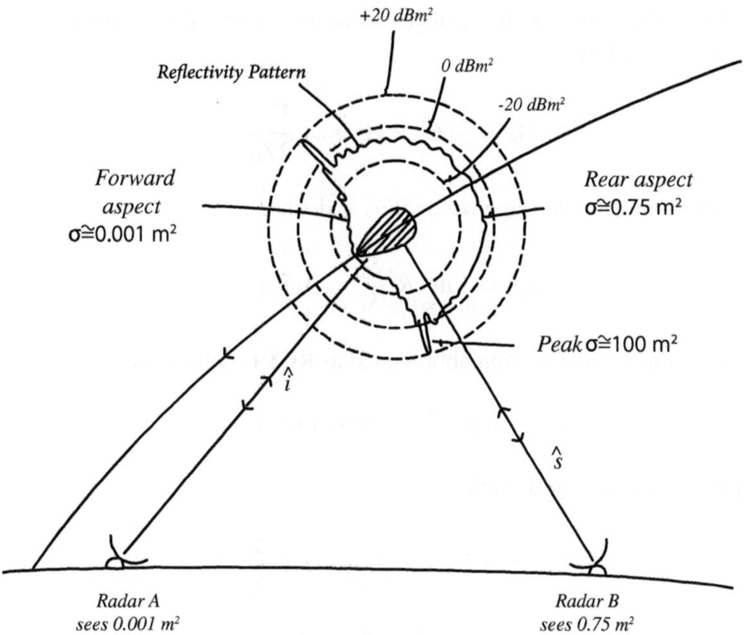

Fig. 5.3 A plane wave being scattered by a target

Thus, far from the scatterer, the filed intensity has the following form

$$\mathbf{E}^s = E_0 \frac{e^{-jk_0 R}}{R} \mathbf{f}(\widehat{\mathbf{s}}, \widehat{\imath}) \tag{5.3}$$

The vector function $\mathbf{f}(\widehat{\mathbf{s}}, \widehat{\imath})$ represents the amplitude, phase and polarization of the scattered wave in the far field in the direction $\widehat{s}$ with the object illuminated by a plane wave propagating in the direction $\widehat{\imath}$. It is called the *scattering amplitude* and is given by

$$\mathbf{f}(\widehat{s}, \widehat{\imath}) = -jk_0 Z_0 \int_V [\hat{r} \times \hat{r} \times \mathbf{J}(r')] e^{jk_0 \mathbf{r}'.\hat{r}} dv' \tag{5.4}$$

when use has been made of (2.71). The scattering amplitude **f** is a vector function of θ and ϕ and has $\hat{\theta}$ and $\hat{\phi}$ components only. In other words, the scattered field in the far-field is an outward-travelling spherical TEM wave. The scattered field is in general elliptically polarized.

The incident power flux density $\mathbf{S}_i$ at the target is given by

$$\mathbf{S}_i = \frac{1}{2}(\mathbf{E}^i \times \mathbf{H}^{i\star}) = \frac{|\mathbf{E}^i|^2}{2Z_0} \widehat{\imath} \tag{5.5}$$

while the scattered power flux density $\mathbf{S}_s$ at a distance R from the object in the direction $\widehat{s}$ is given by

$$\mathbf{S}_s = \frac{1}{2}(\mathbf{E}^s \times \mathbf{H}^{s\star}) = \frac{|\mathbf{E}^s|^2}{2Z_0}\widehat{s} \tag{5.6}$$

The *differential scattering cross section* is defined as

$$\sigma_d = \lim_{R\to\infty} R^2 \frac{S_s}{S_i} = |\mathbf{f}(\widehat{\mathbf{s}}, \widehat{\imath})|^2 \tag{5.7}$$

The *bistatic radar cross section* abbreviated as RCS is defined as

$$\sigma_{bi}(\widehat{s}, \widehat{\imath}) = 4\pi\sigma_d(\widehat{s}, \widehat{\imath}) \tag{5.8}$$

and in terms of power densities

$$\sigma_{bi}(\widehat{s}, \widehat{\imath}) = \lim_{R\to\infty} 4\pi R^2 \frac{S_s}{S_i} \tag{5.9}$$

The *backscattering* or *monostatic RCS* is defined as

$$\sigma_b(\widehat{\imath}) = \sigma_{bi}(-\widehat{\imath}, \widehat{\imath}) \tag{5.10}$$

The radar cross section σ_{bi} is the equivalent area which when multiplied by the incident power density, and which if it then scatters this power isotropically in all directions will produce the same incident returned power at the radar as the target actually does. The RCS is defined such that it is independent of the distance from the target. It is, however, a strong function of the angle of incidence as well as the observation angle, polarization, frequency, and material properties of the target.

The radar cross sections of a number of objects at microwave frequencies are listed in Table 5.1.

It is also instructive to use the following definition for the radar cross section

$$\sigma_{bi}(\widehat{s}, \widehat{\imath}) = \frac{P_s}{S_i} \tag{5.11}$$

with P_s being the total (time averaged) power radiated by a ficticious scatterer (at the target's position) that maintains the same field $\mathbf{E}^s$ in all directions as that maintained by the scatterer in the direction under consideration. Thus,

$$P_s = 4\pi R^2 S_s \tag{5.12}$$

where S_s is the scattered power density given by (5.6) and S_i is the incident power density. Clearly, the definitions in (5.9) and (5.11) are equivalent.

Table 5.1 Typical radar cross sections of a number of objects at microwave frequencies

Target	RCS, σ (m^2)
Small Van	200
Car	100
Jambo Jet	100
Regular Jet Plane	40
Large Bomber	40
Cabinned Motor Boat	10
Large Fighter Plane	6
Small Fighter Plane	2
Small 4 Crew Plane	2
An Adult Man	1
Cruise Missile	0.5
Bird	0.01
Fly	0.00001

The above definitions for the radar cross section hold for three dimensional scatterers. For two dimensional targets, we must define radar cross section per unit length, also referred to as the *radar echo width*. It is defined as

$$\sigma_{bi}^{2d} = \lim_{\rho \to \infty} 2\pi\rho \frac{S_s}{S_i} \tag{5.13}$$

where ρ is the radius in cylindrical system of coordinates.

Now consider the total observed scattered power at all angles surrounding the object. Then the *scattering cross section* is defined as

$$\sigma_s = \int_{4\pi} \sigma_d d\Omega = \int_{4\pi} |\mathbf{f}(\widehat{s}, \widehat{\imath})|^2 d\Omega \tag{5.14}$$

where $d\Omega$ denotes the differential solid angle. Equivalently, the scattering cross section can be expressed in terms of scattered fields as

$$\sigma_s = \frac{\int_{S_0} \Re e[\frac{1}{2}(\mathbf{E}^s \times \mathbf{H}^{s\star})] \cdot \mathbf{dS}}{|S_i|} \tag{5.15}$$

where S_0 is any closed surface surrounding the target.

A measure of the total power absorbed by the object can be given by the *absorption cross section*. The cross section of an object that would correspond to the total power flux entering the object is given by

$$\sigma_a = \frac{-\int_S \Re\mathrm{e}[\frac{1}{2}(\mathbf{E}^t \times \mathbf{H}^{t\star})] \cdot \mathbf{dS}}{|S_i|} \tag{5.16}$$

Alternatively, the absorption cross section σ_a can be expressed in terms of the total power loss inside the volume of the scatterer. Thus

$$\sigma_a = \frac{\int_v \mathbf{k}\epsilon_r''(\mathbf{r}')|\mathbf{E}^t(\mathbf{r}')|^2 dv'}{|E_i|^2} \tag{5.17}$$

The total or *extinction* cross section is the sum of the scattering and absorption cross sections

$$\sigma_t = \sigma_s + \sigma_a \tag{5.18}$$

The ratio of the scattering and extinction cross sections is referred to as the *scattering albedo*

$$W_0 = \frac{\sigma_s}{\sigma_t} = \frac{1}{\sigma_t} \int_{4\pi} |\mathbf{f}(\widehat{s}, \widehat{\imath})|^2 d\Omega \tag{5.19}$$

Also, the ratio of the scattering cross section and the geometrical cross section of a target is called the *absorption efficiency* of the target

$$Q_s = \frac{\sigma_s}{\sigma_g} \tag{5.20}$$

5.6 Radar Equation

Consider a radar system illuminating a target of radar cross section σ located in the far field of the transmitting and receiving antennas

$$R_1 > 2D_t^2/\lambda, \quad R_2 > 2D_r^2/\lambda \tag{5.21}$$

where D_t and D_r denote the maximum dimensions of antennas. The incident power density at the target is given by

$$S_i = \frac{1}{4\pi R_1^2} G_t(\widehat{\imath}) P_t \tag{5.22}$$

where G_t is the transmitter antenna gain. The returned power density at the receiver antenna can be expressed as

$$S_r = \sigma_{bi}(\widehat{s}, \widehat{\imath})\frac{S_i}{4\pi R_2^2} \tag{5.23}$$

The received power can be written as

$$P_r = A_r(\widehat{s})S_r \tag{5.24}$$

where A_r is the effective area of the receiving antenna

$$A_r(\widehat{s}) = \frac{\lambda^2}{4\pi}G_r(-\widehat{s}) \tag{5.25}$$

and G_r is the receiver antenna gain. Thus, for a bistatic radar system, we have

$$\frac{P_r}{P_t} = \frac{\lambda^2 G_t(\widehat{\imath})G_r(-\widehat{s})\sigma_{bi}(\widehat{s}, \widehat{\imath})}{(4\pi)^3 R_1^2 R_2^2} \tag{5.26}$$

and if the radar operates in monostatic mode ($\widehat{s} = -\widehat{\imath}$ and $R_1 = R_2 = R$)

$$\frac{P_r}{P_t} = \frac{\lambda^2 G_t^2(\widehat{\imath})\sigma_b(-\widehat{\imath}, \widehat{\imath})}{(4\pi)^3 R^4} \tag{5.27}$$

Example 5.1 A radar system of peak power $P_t = 10^5$ W and antenna gain $G = 40$ dB is operating at 3 GHz. If the minimum detectable power is 10 pW, find the maximum range at which a target of cross section $\sigma = 2\,\text{m}^2$ can be detected.

From the radar equation, we have

$$R_{max} = \left[\frac{\lambda^2 P_t G_t^2 \sigma_b}{(4\pi)^3 P_r}\right]^{1/4}$$

Substituting for the specified values, we find that

$$R_{max} = \left[\frac{(0.1)^2(10^5)(10^4)^2(2)}{(4\pi)^3(10^{-12})}\right]^{1/4}$$

And the maximum range is found to be R_{max}. □

5.7 Doppler Effect

Doppler effect is the apparent variation in frequency of any emitted wave, such as a wave of light or sound, as the source of the wave approaches or moves away, relative to an observer.[1]

The frequency of the returned signal from a moving target is different from the transmitted frequency, and this change in frequency may be used to measure the velocity of the target in the radial direction.

In order to describe the Doppler effect, assume that we have a plane wave described by

$$\Phi(z,t) = A\cos(\omega t - \beta z) \tag{5.28}$$

where $\beta = \omega/c$. This is a plane wave moving at the velocity c in the (z,t) frame. Let us assume that there is (z',t) frame which is moving in the positive z direction with a velocity $v \ll c$. If the two frames coincide at $t = 0$ such that $z = z'$, we then have

$$z' = z - vt \tag{5.29}$$

To an observer in the moving frame, the plane wave looks like

$$\begin{aligned}\Phi(z',t) &= A\cos(\omega t - \beta z' - \beta v t)\\ &= A\cos[\omega(1 - \frac{v}{c})t - \beta z']\\ &= A\cos(\omega' t - \beta z)\end{aligned} \tag{5.30}$$

where

$$\omega' = \omega(1 - \frac{v}{c}) \tag{5.31}$$

Therefore, the frequency of the wave appears lower to the moving observer who is moving in the same direction as the travelling wave. If the observer is travelling in the opposite direction to the wave, then, the observer notices a rise in frequency of the wave. In other words,

$$\omega' = \omega(1 \mp \frac{2v}{c}) \tag{5.32}$$

The minus sign applies to a receding target, while the plus sign applies to an approaching target.

For a wave travelling in a $\widehat{c}$ direction, and an observer travelling with a velocity $\mathbf{v}$, the Doppler shift is given by the formula

[1] After Christian J. Doppler, the Austrian physicist, who first stated the physical principle in 1842.

$$\Delta\Omega = -\omega\frac{\mathbf{v}\cdot\widehat{c}}{c} \tag{5.33}$$

The difference in frequncy bears the same ratio to the transmitted frequency as the target velocity bears to the speed of light.

The Doppler effect is used by the weather radar to detect wind velocity and by the police radar to detect the velocity of moving vehicles. Since for usual speeds, the ratio v/c is very small, the Doppler shift is small unless the radar frequency is very large. Therefore, most Doppler radars work in the microwave region.

Example 5.2 Radar guns are used to detect speeding motorists. Here a gun transmits waves at a given frequency toward an oncoming car. Reflected waves return to the gun at a different frequency, depending on how fast the car being tracked is moving. A device in the gun compares the transmission frequency to the received frequency to determine the speed of the car.

A police radar operates at 3 GHz. Consider a car moving toward the radar at 179 km/h. Calculate the frequency shift measured by the radar.

The frequency of the police radar is shifted by exactly 1 kHz. If a radar receiver is so arranged that it rejects echoes that have the same frequency as the transmitter and amplifies only those echoes that have different frequencies it shows only moving targets. Such a receiver can pick out vehicles moving over terrain in darkness. □

5.8 Radar Clutter

Radar clutter is defined as the unwanted reflective waves from irrelevant target Clutter constitutes unwanted signals, echoes, or images which interfere with observation of desired signals. The clutter may be caused by everything from buildings to trees to ecean waves. Such interference can hinder radar operator looking for particular targets or particular features of the environment.

The exact meaning of the clutter depends on the actual application a radar system is used for. For example, for a surveillance radar, unwanted echo from ground, sea, rain and other atmospheric activities, as well as birds and bugs constitute the clutter. On the other hand, for a remote sensing radar, the ground and sea characteristics would be the target and the echoes coming from them cannot be considered as clutter.

Radar clutter can be devided into three categories:

- point clutters such as structures, birds, and automobiles,
- surface clutters such as ground and sea clutter, and
- volume clutter such as rain and snow clutter.

Surface Clutter

The ground and see surfaces as targets have radar cross sections. However, because of their distributed area, it is more convenient to define a new parameter as radar cross section per unit area. This is a dimensionless parameter and is called *normalized*

Table 5.2 The mean values of γ for various train types

Train	RMS height, σ_h, m	γ, dB
Mountain	100	−5
Urban	10	−5
Jungle valley	10	−10
Valley	10	−12
Farm & desert	3	−15
Forest	1	−20
Smooth areas	0.3	−25

radar cross section (NRCS) and is denoted by σ°. This parameter is a function of the wavelength, angle of incidence, roughness, and the texture of the surface.

Ground Clutter

Radar land clutter constitutes the unwanted radar echoes returned from the earths surface that compete against and interfere with the desired echoes returned by targets such as aircraft and other moving and stationary targets. Land surface variablity includes roughness, material, moisture, and vegetation cover.

A simple model for the land clutter is the constant γ model. It follows the general behavior

$$\sigma^\circ(\alpha) = \gamma \sin^n \alpha, \quad -1 < n < 3 \tag{5.34}$$

where α is the grazing (or depression) angle (the incidence angle relative to the horizen), γ is a constant and is related to the scattering power of the surface under consideration. The typical values of γ for various train types are given in Table 5.2.

This model matches the measurements at large grazing angles away from the normal incidence. The negative values of n correspond to cases of high vegetation.

Sea Clutter

Sea clutter is the result of the interaction of the radiated electromagnetic wave with ocean waves. Sea clutter is a self-generated interface and can be characterized as a distributed, non-directional source. Radars operating in a maritime environment have a serious limitation imposed on their performance by unwanted sea echoes. Under rough wind conditions, sharp-tipped waves can reflect microwave energy back to the radar. Sea clutter can be of modest to large reflectivity and extend to far range. It further complicates near-surface velocity analysis by returning a mix of both the wind and wave motion. Breaking waves give a stronger signal than non-breaking waves on the radar.

The main contribution to sea clutter is Bragg-resonant scattering by surface gravity waves. This behavior is produced by scattering from ocean waves having a wavelength half that of the radar wavelength, and moving radially to and away from the radar. For example, an over-the-horizen radar operating at 20 MHz sees only the 7.5 m-long ocean waves that travel exactly toward and away from the radar. The spectrum of the sea echo at High frequencies exhibits a surprisingly simple structure: two sharp Doppler-shifted lines, one positively shifted echo from the Bragg-resonant waves travelling toward the radar, and a negatively shifted line from those receding from the radar. The exact shifts and relative magnitudes of these lines permit us to extract information about surface winds and currents. Bragg resonant scattering also occurs at harmonics of the principle wavelength. These result in second-order peaks in the spectrum.

Since the ocean surface consists of random collection of sea waves and the detailed configuration of the ocean surface varies in an stochastic manner, additional scattering processes will be in operation. These include scattering from lateral waves and interaction between crossing sea waves. If these crossing sea waves generate a new sea wave with a wavelength equal to one-half of radar wavelength. Then Bragg-resonance scattering will occur. Other directions produce intermediate ratios, permitting extraction of the direction, although with a left-right ambiguity with respect to the radar radial.

Radially travelling gravity waves Doppler-shift the backscattered radar energy by an amount proportional to their phase velocity. If surface currents transport the gravity waves, an extra Doppler shift is added that is proportional to the radial component of the surface current.

Because of the similar dependance of σ° to the incidence angle for various surfaces, attempts have been made to establish a simple model with the minimum number of parameters to predict the behavior of σ°. We will present two such models here.

The first model is (5.34) which is used for both the ground and the sea. As mentioned before, the constant γ model is generally valid for large grazing angles. However, at small grazing angles, the measured values of σ° are actually lower than the predicted values of the model due to the propagation path loss. Also, close to the normal incidence, the measured values are higher than that predicted by the model because of the semi-specular reflection from the surface. To measure the NRCS for surface clutter close to the normal incidence the radar should have a very narrow pulse or antenna beamwidth. Thus, the measurements should be carried out from very high altitude.

The above model is also used to predict σ° for the sea surface. In this case, γ is related to the *Beaufort wind constant*, K_B, and the radar wavelength. The mean value of γ averaged over all sea wind directions is given by the following equation

$$10 \log \gamma = 6K_B - 10 \log \lambda - 64 \tag{5.35}$$

where λ is the wavelength in meters and K_B is the wind constant and is an indication of the state of the sea surface. Table 5.3 lists γ at various sea states.

Another useful empirical model for the ground and sea surface clutter is given by

$$\sigma^\circ = \begin{cases} C\frac{\sin^n \alpha}{\cos^u \alpha}, & 0^\circ \leq \alpha \leq \alpha_G \\ \sigma^\circ(\alpha_G), & \alpha_G \leq \alpha \leq 90^\circ \end{cases} \tag{5.36}$$

Table 5.3 The mean values of γ for various sea states

Sea state no.	Wind constant K_B	λ_γ, dB	RMS height, σ_h, m
0	1	−58	0.003
1	2	−52	0.027
2	3	−46	0.09
3	4	−40	0.21
4	5	−34	0.42
5	6	−28	0.72
6	7	−22	1.14
7	8	−16	1.70
8	9	−10	2.43
9	10	−4	3.33

where

$$\begin{cases} -1 \le n \le 3, & 0 \le u \le 1 \quad \text{for land} \\ 1 \le n \le 2, & 1 \le u \le 5 \quad \text{for sea} \end{cases} \tag{5.37}$$

This model describes the behavior of σ° by four parameters. The parameters u and n control the dependence of σ° to the angle close to the normal incidence and away from normal, respectively. Also, C and α_G are level parameters that control its relative size. The latter parameters are usually chosen to be

$$C = 0.01, \quad \alpha_G = 85^\circ \tag{5.38}$$

The empirical model gives constant σ° for $85^\circ \le \alpha \le 90^\circ$, as the exact shape of σ° close to the normal incidence is hard to predict.

5.8.1 Clutter Statistics

Sea clutter has been traditionally treated as a stochastic process merely because of its random-looking waveform. The models given in the previous section, are actually the average NRCS as a function of grazing angle. If one plots the σ° itself as a function of angle, the data would be scattered in a range of values ±3 dB around the given curve.

For many years, in radars with low resolution capabilities, clutter echoes were considered as having a Gaussian distributed disturbance. In modern radar systems, operating with high resolution capabilities, the statistics of the clutter have been observed to deviate from the normality. The amplitude statistic of clutter have been modelled by *Rayleigh, log-normal, contaminated-normal, Weibull, log-Weibull and K-distributions*.

There is some evidence that sea clutter is actually chaotic and can be modelled by a system of five coupled non-linear differntial equations.

5.9 Non-Meteorological Echoes

In closing this chapter, we review some of the major non-meteorological radar echos and clutter encountered in practice.

Stationary targets such as hills, trees and towers are not the only sources of clutter. Returns from insects in the atmospheric boundary layer are common occurences. Since, on average, the bugs are advected along with the mean winds, they provide excellent tracers to diagnose low level winds, and a fantastic background against which to track such otherwise low-reflectively phenomena as gust fronts, thin lines, the sea breeze.

Birds can be extremely troublesome in radar meteorology. Pulse resolution volumes are typically several hundred meters on a side, and it only takes one bird per volume to return a large, moving radar echo. This problem becomes very important when migration seasons set in.

When wind shear conditions are just right, the dry thermal plumes of rising warm, moist air in the lowest few hundred meters of the atmosphere will often organize into long, band like rolls. Insects and birds find thier way into the thermal updrafts and collectively return strong echoes.

Under the proper atmospheric conditions, the refractive index of air can change with height in such a way as to bend the transmitted beam back down towards the surface; it then hits the ground, and returns along its curved path to the radar. This is called anomalous propagation (AP). AP is often observed in cold weather, when the associated temperature inversion bends the radar beam down.

Chapter 6
Canonical Scattering Problems

In this chapter, we will discuss scattering from objects whose boundaries are conformal to the major systems of coordinates. Because of the formal elegance of the solutions to these boundary value problems and the discussions that they lead to, they are often called the canonical electromagnetic scattering problems.

6.1 The Circular Cylinder

We start our discussion of canonical scattering problems from the circular cylinder of infinite length. We consider both conducting and dielectric cylinders.

6.1.1 The Conducting Cylinder

Consider a perfectly conducting cylinder of radius a with the z axis as its axis of symmetry illuminated by a plane wave. We may consider two different incident polarizations defined with respect to the axis: the E-polarization or transverse magnetic to z (TM_z) and the H-polarization or transverse electric to z (TE_z). We treat each polarization separately.

6.1.1.1 Transverse Magnetic Polarization

Let the cylinder be illuminated by a plane wave with the electric field polarized in the z direction as shown in Fig. 6.1. Without loss of generality, assume that the wave is propagating in the x direction. Therefore

$$\mathbf{E}^i = \widehat{z} E_0 e^{-jkx} \tag{6.1}$$

© Springer International Publishing Switzerland 2015

K. Barkeshli, *Advanced Electromagnetics and Scattering Theory*,
DOI 10.1007/978-3-319-11547-4_6

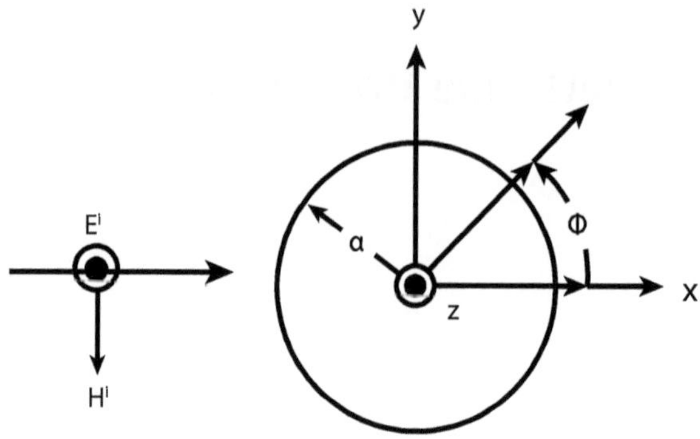

Fig. 6.1 TM_z scattering of a plane wave by a perfectly conducting cylinder

The presence of the cylinder causes a diffraction of the incident field. The diffracted field can be written as

$$\mathbf{E}^t = \mathbf{E}^i + \mathbf{E}^s \tag{6.2}$$

The task is to find the scattered field $\mathbf{E}^s$ in the external region $\rho \geq a$ by solving the wave equation and matching the boundary conditions on the surface of the conductor. For this purpose we expand the incident field in terms of cylindrical harmonics using (C.13) in Appendix C as

$$\mathbf{E}^i = \hat{z} E_0 \sum_{n=-\infty}^{\infty} j^{-n} J_n(k\rho) e^{jn\phi} \tag{6.3}$$

The scattered field is in turn written as

$$\mathbf{E}^s = \hat{z} E_0 \sum_{n=-\infty}^{\infty} j^{-n} a_n H_n^{(2)}(k\rho) e^{jn\phi} \tag{6.4}$$

where we have used the Hankel function of the second kind in order to allow the proper behavior far from the cylinder. Enforcing the Dirichlet boundary condition for the tangential component of the total electric field on the surface of the cylinder

$$E_z^t = 0, \quad \rho = a \tag{6.5}$$

we have

$$E_z^t = E_0 \sum_{n=-\infty}^{\infty} j^{-n} [J_n(ka) + a_n H_n^{(2)}(ka)] e^{jn\phi} = 0 \tag{6.6}$$

and the unknown coefficients are found to be

$$a_n = -\frac{J_n(ka)}{H_n^{(2)}(ka)} \tag{6.7}$$

The scattered field is, therefore, written as

$$E^s = -\widehat{z}E_0 \sum_{n=-\infty}^{\infty} j^{-n} \frac{J_n(ka)}{H_n^{(2)}(ka)} H_n^{(2)}(k\rho)e^{jn\phi} \tag{6.8}$$

and the diffracted (total) field is given by

$$E^t = \widehat{z}E_0 \sum_{n=-\infty}^{\infty} j^{-n}[\frac{H_n^{(2)}(ka)J_n(k\rho) - J_n(ka)H_n^{(2)}(k\rho)}{H_n^{(2)}(ka)}]e^{jn\phi} \tag{6.9}$$

The variation of the electric field intensity E_z^t along the x axis for a conducting cylinder of size $ka = 2$ is shown in Fig. 6.2. The standing wave pattern in the incident part of the x axis results from the interference of the small wavelets radiated from the illuminated parts of the cylinder. Also, notice that the shadow region is not very "dark" at this frequency. In addition, as physically expected, the total field E_z^t

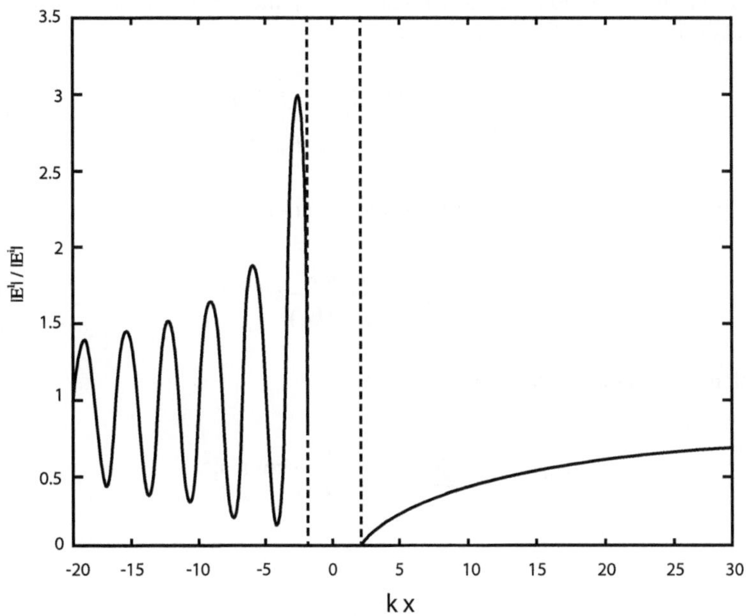

Fig. 6.2 The magnitude of the diffracted electric field along the x-axis by a perfectly conducting cylinder of size $k_0a = 2$ illuminated by a TM_z plane wave

approaches the incident field E_z^i at sufficiently large distances.

$$\lim_{kx \to \infty} |\frac{E_z}{E_z^i}| = 1 \tag{6.10}$$

The extent of the region in which E_z^t is significantly less than E_z^i gives an idea of the size of the shadow region.

Using the Maxwell's equations, the magnetic field intensity is given by

$$\mathbf{H} = -\frac{1}{j\omega\mu}\left(\widehat{\rho}\frac{1}{\rho}\frac{\partial E_z}{\partial \phi} - \widehat{\phi}\frac{\partial E_z}{\partial \rho}\right) \tag{6.11}$$

whose components are explicitly written as

$$H_\rho^s = -\frac{E_0}{j\omega\mu}\frac{1}{\rho}\sum n j^{-(n-1)}\frac{J_n(ka)}{H_n^{(2)}(ka)}H_n^{(2)}(k\rho)e^{jn\phi} \tag{6.12}$$

$$H_\phi^s = -\frac{kE_0}{j\omega\mu}\sum j^{-n}\frac{J_n(ka)}{H_n^{(2)}(ka)}H_n^{(2)'}(k\rho)e^{jn\phi} \tag{6.13}$$

6.1.1.2 Far-Zone Scattered Field

In order to find the far-zone scattered field, we use the large argument expansion of the Hankel function (4.377) to find

$$E_z^s \sim -E_0\sqrt{\frac{2j}{\pi k}}\frac{e^{-jk\rho}}{\sqrt{\rho}}\sum_{n=-\infty}^{\infty}\frac{J_n(ka)}{H_n^{(2)}(ka)}e^{jn\phi} \tag{6.14}$$

Therefore, the radar scattering echo width (5.13) is given by

$$\sigma_{2d} = \frac{4}{k}|\sum_{n=-\infty}^{\infty}\frac{J_n(ka)}{H_n^{(2)}(ka)}e^{jn\phi}|^2 \tag{6.15}$$

Equivalently, the echo width can be written as

$$\sigma_{2d} = \frac{2\lambda}{\pi}|\sum_{n=0}^{\infty}\epsilon_n\frac{J_n(ka)}{H_n^{(2)}(ka)}\cos n\phi|^2 \tag{6.16}$$

where

$$\epsilon_n = \frac{2}{1+\delta_{0n}} \tag{6.17}$$

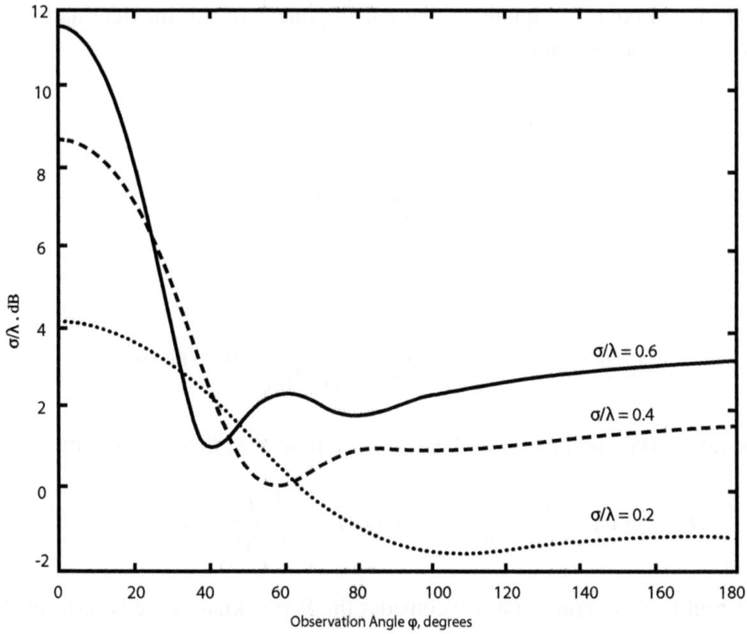

Fig. 6.3 Radar echo widths per unit wavelength for perfectly conducting cylinders of various radii illuminated by a TM_z plane wave

is the Neumann number. It is customary to express σ in terms of echo width per unit wavelength. Thus

$$\frac{\sigma_{2d}}{\lambda} = \frac{2}{\pi} |\sum_{n=0}^{\infty} \epsilon_n \frac{J_n(ka)}{H_n^{(2)}(ka)} \cos n\phi|^2 \tag{6.18}$$

Figure 6.3 shows the TM_z radar echo widths per unit wavelength for perfectly conducting cylinders of various radii.

6.1.1.3 Induced Surface Current

We may attribute the electromagnetic fields scattered by a conducting target to the electric current induced on its surface.

$$\mathbf{K} = \widehat{n} \times \mathbf{H}^t, \quad \rho = a \tag{6.19}$$

where $\widehat{n} = \widehat{\rho}$ is unit vector normal to the cylinder surface. Thus

$$\mathbf{K} = \widehat{z} H_\phi^t, \quad \rho = a \tag{6.20}$$

From the Maxwell's equations, the ϕ-component of the incident and scattered magnetic fields are found to be

$$
\begin{aligned}
H_\phi^i &= \frac{1}{j\omega\mu}\frac{\partial E_z^i}{\partial\rho}\widehat{\phi} \\
&= \frac{kE_0}{j\omega\mu}\sum_n j^{-n} J_n'(k\rho)e^{jn\phi}
\end{aligned} \tag{6.21}
$$

and

$$
H_\phi^s(\rho = a) = -\frac{kE_0}{j\omega\mu}\sum_n j^{-n}\frac{J_n(ka)}{H_n^{(2)}(ka)}H_n^{(2)'}(ka)e^{jn\phi} \tag{6.22}
$$

and the tangential component of the diffracted magnetic field is given by

$$
H_\phi^t(\rho = a) = \frac{kE_0}{j\omega\mu}\sum_n j^{-n}[J_n'(ka) - \frac{J_n(ka)}{H_n^{(2)}(ka)}H_n^{(2)'}(ka)]e^{jn\phi} \tag{6.23}
$$

To simplify this expression, we consider the Wronskian of the Bessel functions

$$
W\{J_n(ka), H_n^{(2)}(ka)\} = -\frac{2j}{\pi ka} \tag{6.24}
$$

to find

$$
H_\phi^t(\rho = a) = \frac{2E_0}{\pi a\omega\mu}\sum_n j^{-n}\frac{e^{jn\phi}}{H_n^{(2)}(ka)} \tag{6.25}
$$

Using (6.20), we obtain the induced surface current density as

$$
\mathbf{K} = \widehat{z}\frac{2E_0}{\pi a\omega\mu}\sum_{n=0}^{\infty}\epsilon_n j^{-n}\frac{\cos n\phi}{H_n^{(2)}(ka)} \tag{6.26}
$$

6.1.1.4 Small Radius Approximation

When the frequency of operation is sufficiently low, the radius of the cylinder is much smaller than the wavelength ($ka \ll 1$). Therefore, at the low frequency limit, we retain the dominant $n = 0$ term and obtain

$$
\mathbf{K} \simeq \widehat{z}\frac{2E_0}{\pi a\omega\mu}\frac{1}{H_0^{(2)}(ka)} \tag{6.27}
$$

Using the small argument approximation for the Hankel function, we have

$$\begin{aligned} H_0^{(2)}(ka) &= J_0(ka) - jN_0(ka) \\ &\simeq 1 - j\frac{2}{\pi}\ln(\frac{\gamma ka}{2}) \simeq -j\frac{2}{\pi}\ln(\frac{\gamma ka}{2}) \end{aligned} \tag{6.28}$$

where use has been made of (4.257) and (4.260). Thus, the induced electric current takes the form

$$\mathbf{K} = \hat{z}j\frac{E_0}{a\omega\mu}\frac{1}{\ln(\frac{\gamma ka}{2})} \tag{6.29}$$

It is noted that the induced current **K** for the thin circular wire is 90° out of phase with respect to $\mathbf{E}^i$.

Also, taking the $n = 0$, the scattering echo width in this case is given by

$$\frac{\sigma_{2d}}{\lambda} = \frac{\pi}{2}|\frac{1}{\ln(\frac{\gamma ka}{2})}|^2 \tag{6.30}$$

which is independent of ϕ.

6.1.1.5 Transverse Electric Polarization

We now turn to the H-polarization (Fig. 6.4). In this case, the incident magnetic field is given by

$$\mathbf{H}^i = \hat{z}H_0 e^{-jkx} = \hat{z}H_0 \sum_{n=-\infty}^{\infty} j^{-n}J_n(k\rho)e^{jn\phi} \tag{6.31}$$

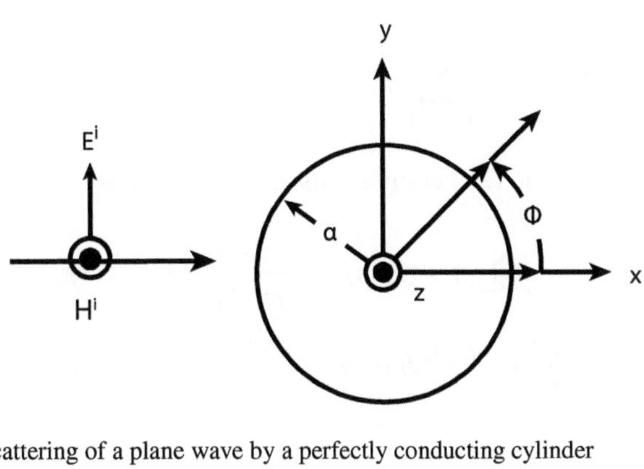

Fig. 6.4 TH_z scattering of a plane wave by a perfectly conducting cylinder

The diffracted field is the sum of the incident and scattered fields

$$\mathbf{H}^t = \mathbf{H}^i + \mathbf{H}^s \tag{6.32}$$

Expanding the scattered field in terms of cylindrical harmonics

$$\mathbf{H}^s = \hat{z} H_0 \sum_{n=-\infty}^{\infty} j^{-n} b_n H_n^{(2)}(k\rho) e^{jn\phi} \tag{6.33}$$

the total magnetic field is written as

$$\mathbf{H}^t = \hat{z} H_0 \sum_{n=-\infty}^{\infty} j^{-n} [J_n(k\rho) + b_n H_n^{(2)}(k\rho)] e^{jn\phi} \tag{6.34}$$

From Faraday's law, the electric field is found to be

$$\mathbf{E} = \frac{1}{j\omega\epsilon} [\hat{\rho} \frac{1}{\rho} \frac{\partial H_z}{\partial \phi} - \hat{\phi} \frac{\partial H_z}{\partial \rho}] \tag{6.35}$$

The appropriate boundary condition in this case is the homogeneous Neumann boundary condition

$$\frac{\partial H_z^t}{\partial \rho} = 0, \quad \rho = a \tag{6.36}$$

Substituting from (6.34), we find

$$b_n = -\frac{J_n'(ka)}{H_n^{(2)'}(ka)} \tag{6.37}$$

Thus, the scattered magnetic field is given by

$$H_z^s = -H_0 \sum_{n=-\infty}^{\infty} j^{-n} \frac{J_n'(ka)}{H_n^{(2)'}(ka)} H_n^{(2)}(k\rho) e^{jn\phi} \tag{6.38}$$

Also, the scattered electric field components are expressed as

$$E_\phi^s = -\frac{kH_0}{j\omega\epsilon} \sum j^{-n} [J_n'(k\rho) - \frac{J_n'(ka)}{H_n^{(2)'}(ka)} H_n^{(2)'}(k\rho)] e^{jn\phi} \tag{6.39}$$

$$E_\rho^s = \frac{1}{\rho} \sum j^{-(n-1)} [J_n(k\rho) - \frac{J_n'(ka)}{H_n^{(2)'}(ka)} H_n^{(2)}(k\rho)] e^{jn\phi} \tag{6.40}$$

6.1.1.6 Far-Zone Scattered Field

In order to find the far-zone scattered field, we again use the large argument approximation for the Hankel function ((4.265) and find that)

$$H_z^s \sim -H_0\sqrt{\frac{2j}{\pi k}}\frac{e^{-jk\rho}}{\sqrt{\rho}}\sum_{n=-\infty}^{\infty}\frac{J_n'(ka)}{H_n^{(2)'}(ka)}e^{jn\phi} \tag{6.41}$$

The scattering echo width is given by

$$\sigma_{2d} = \frac{2\lambda}{\pi}|\sum_{n=0}^{\infty}\epsilon_n\frac{J_n'(ka)}{H_n^{(2)'}(ka)}\cos n\phi|^2 \tag{6.42}$$

The echo width per wavelength is expressed as

$$\frac{\sigma_{2d}}{\lambda} = \frac{2}{\pi}|\sum_{n=0}^{\infty}\epsilon_n\frac{J_n'(ka)}{H_n^{(2)'}(ka)}\cos n\phi|^2 \tag{6.43}$$

Figure 6.5 depicts the TE_z echo widths per unit wavelength for perfectly conducting cylinders of various radii.

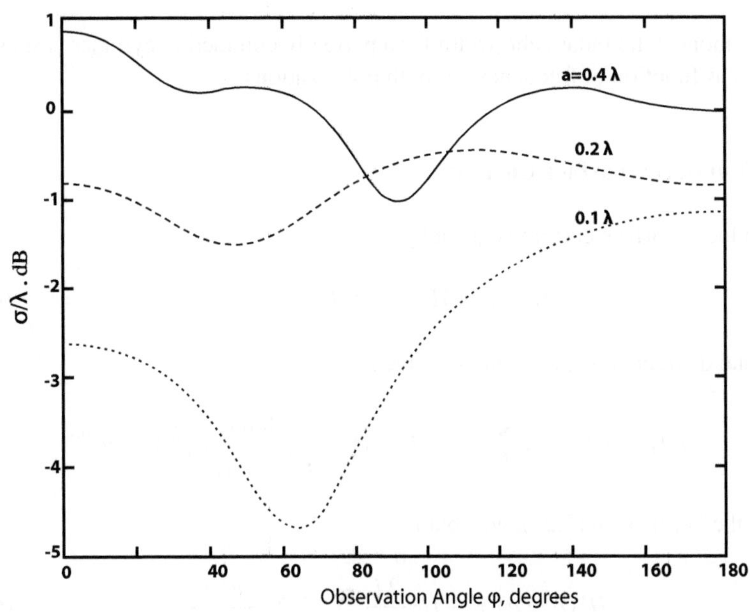

Fig. 6.5 Bistatic radar echo widths per unit wavelength for perfectly conducting cylinders of various radii illuminated by a TE_z plane wave

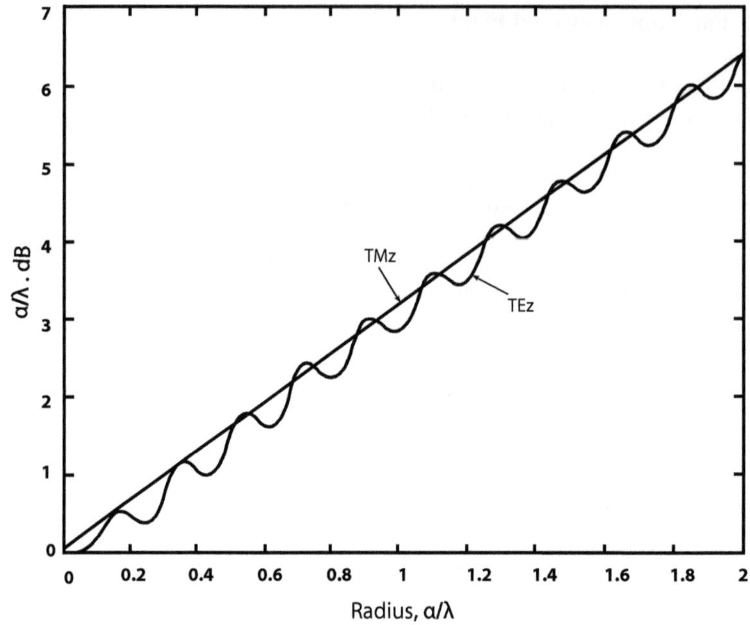

Fig. 6.6 The monostatic radar echo widths per unit wavelength for a perfectly conducting cylinder as function of frequency for the principal polarizations

The monostatic radar echo width for a perfectly conducting cylinder is shown in Fig. 6.6 as function of frequency for both polarizations.

6.1.1.7 Induced Surface Current

The induced surface current is given by

$$\mathbf{K} = \widehat{n} \times \mathbf{H}^t = -\widehat{\phi} H_z^t, \quad \rho = a \tag{6.44}$$

The total diffracted magnetic field is given by

$$H_z^t(\rho = a) = H_0 \sum j^{-n} [J_n(ka) - \frac{J_n'(ka)}{H_n^{(2)'}(ka)} H_n^{(2)}(ka)] e^{jn\phi} \tag{6.45}$$

Using the Wronskian (6.24), we obtain

$$H_z^t(\rho = a) = -H_0 \frac{2j}{\pi ka} \sum j^{-n} \frac{e^{jn\phi}}{H_n^{(2)'}(ka)} \tag{6.46}$$

Thus, the current is given by

$$\mathbf{K} = \widehat{\phi}\frac{2j}{\pi ka}H_0 \sum_{n=-\infty}^{\infty} j^{-n}\frac{e^{jn\phi}}{H_n^{(2)\prime}(ka)} \tag{6.47}$$

It is noted that this circulating current is induced by the E_ϕ component.

6.1.1.8 Small Radius Approximation

In the low frequency limit, the leading terms in the induced current are those corresponding to the $n = 0$ and $n = \pm 1$. Therefore,

$$\mathbf{K} = \widehat{\phi}\frac{2j}{\pi ka}H_0[\frac{1}{H_0^{(2)\prime}(ka)} + j^{-1}\frac{e^{j\phi}}{H_1^{(2)\prime}(ka)} + j\frac{e^{-j\phi}}{H_{-1}^{(2)\prime}(ka)}] \tag{6.48}$$

Using the small argument approximations for the Bessel functions, we find that

$$\mathbf{K} \simeq -\widehat{\phi}H_0[1 - j2ka\cos\phi] \tag{6.49}$$

The radar echo width for the thin cylinder is found to be

$$\frac{\sigma_{2d}}{\lambda} = \frac{\pi}{8}(ka)^4(1 - 2\cos\phi)^2 \tag{6.50}$$

It is noted that the echo width is a function of ϕ in this case and it is maximum in the backscattering direction at $\phi = \pi$. There is a subsidiary maximum at $\phi = 0$ in the forward direction and nulls at $\phi = \pm 60°$ as shown in Fig. 6.7.

6.1.1.9 Line Source Excitation

Consider a conducting circular cylinder of radius a illuminated by an electric line source positioned at $\boldsymbol{\rho}'$.

$$\mathbf{E}^i = -\widehat{z}\frac{k^2 I}{4\omega\epsilon}H_0^{(2)}(k|\boldsymbol{\rho} - \boldsymbol{\rho}'|) \tag{6.51}$$

From the addition theorem of the Hankel function (4.410), we may write

$$\mathbf{E}^i = -\widehat{z}\frac{k^2 I}{4\omega\epsilon}\sum_{n=-\infty}^{\infty}\begin{cases} J_n(k\rho)H_n^{(2)}(k\rho')e^{jn(\phi-\phi')} & \rho \le \rho' \\ J_n(k\rho')H_n^{(2)}(k\rho)e^{jn(\phi-\phi')} & \rho \ge \rho' \end{cases} \tag{6.52}$$

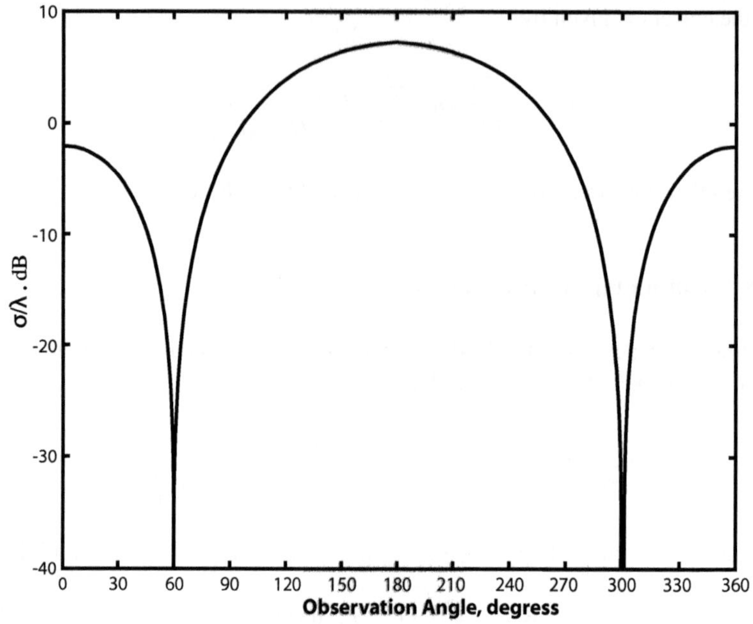

Fig. 6.7 Radar echo widths per unit wavelength for an electrically narrow perfectly conducting cylinder illuminated by a TE_z plane wave propagating along the x-axis

Nothing that the scattered field exists only outside the cylinder, we write

$$\mathbf{E}^s = -\hat{z}\frac{k^2 I}{4\omega\epsilon}\sum_{n=-\infty}^{\infty} C_n H_n^{(2)}(k\rho)e^{jn(\phi-\phi')} \quad \rho > a \tag{6.53}$$

In order to find the unknown coefficients, C_n, we enforce the boundary condition on the tangential component of the diffracted electric field E^t to find

$$E_z^t(\rho = a) = -\frac{k^2 I}{4\omega\epsilon}[J_n(ka)H_n^{(2)}(k\rho') + C_n H_n^{(2)}(ka)]e^{jn(\phi-\phi')} \tag{6.54}$$

and, therefore

$$C_n = -H_n^{(2)}(k\rho')\frac{J_n(ka)}{H_n^{(2)}(ka)} \tag{6.55}$$

Hence, the scattered field is written as

$$\mathbf{E}^s = \frac{k^2 I}{4\omega\epsilon}\hat{z}\sum_{n=-\infty}^{\infty} H_n^{(2)}(k\rho')\frac{J_n(ka)}{H_n^{(2)}(ka)}H_n^{(2)}(k\rho)e^{jn(\phi-\phi')} \tag{6.56}$$

The induced current is given by the tangential magnetic field over the cylinder

$$\mathbf{K} = \hat{\rho} \times \mathbf{H}^t|_{\rho=a} = -\hat{z}\frac{I}{2\pi a}\sum_{n=-\infty}^{\infty}\frac{H_n^{(2)}(k\rho')}{H_n^{(2)}(ka)}e^{jn(\phi-\phi')} \tag{6.57}$$

Far from the cylinder, the scattered field takes the form

$$\mathbf{E}^s \simeq \hat{z}\frac{k^2 I}{4\omega\epsilon}\sqrt{\frac{2j}{\pi k\rho}}e^{-jk\rho}\sum_{n=-\infty}^{\infty} j^n \frac{J_n(ka)}{H_n^{(2)}(ka)}H_n^{(2)}(k\rho') \tag{6.58}$$

Therefore, the radiation pattern of the line source in the presence of the conducting cylinder is given by

$$E_z^t \simeq -\frac{k^2 I}{4\omega\epsilon}\sqrt{\frac{2j}{\pi k}}\frac{e^{-jk\rho}}{\sqrt{\rho}}\sum_{n=-\infty}^{\infty} j^n[J_n(k\rho') - \frac{J_n(ka)}{H_n^{(2)}}(ka)H_n^{(2)}(k\rho')]e^{jn(\phi-\phi')} \tag{6.59}$$

Figures 6.8 and 6.9 show the normalized radiation patterns of a line source in the vicinity of a cylinder of radius 5λ, positioned at 5.25λ and 5.5λ from the center of the cylinder, respectively.

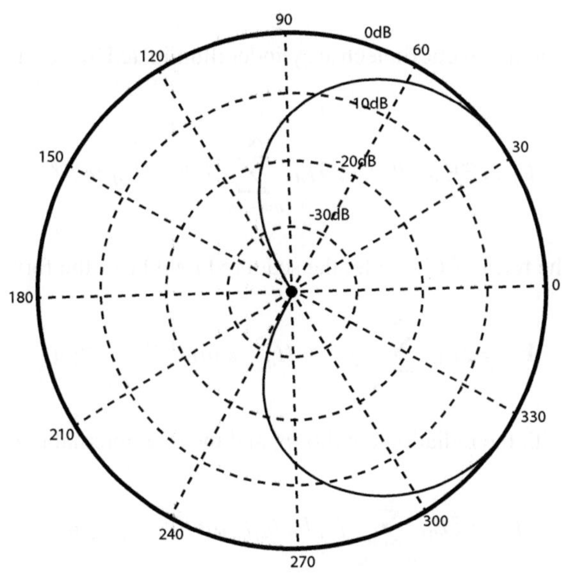

Fig. 6.8 Normalized radiation pattern of a line source in the vicinity of a conducting cylinder. Cylinder radius $a = 5\lambda$, $\rho' = 5.25\lambda$, $\phi' = 0$

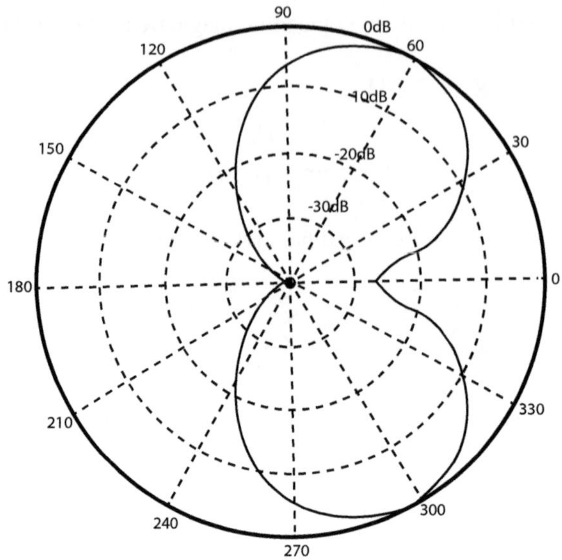

Fig. 6.9 Normalized radiation pattern of a line source in the vicinity of a conducting cylinder. Cylinder radius $a = 5\lambda$, $\rho' = 5.5\lambda$, $\phi' = 0$

6.1.2 *The Homogeneous Dielectric Cylinder*

Consider now a nonmagnetic dielectric cylinder illuminated by a transverse magnetic plane wave

$$\mathbf{E}_i = \hat{z}E_0 e^{-jk_0x} = \hat{z}E_0 \sum_{n=-\infty}^{\infty} j^{-n} J_n(k_0\rho)e^{jn\phi} \tag{6.60}$$

Guided by the form of $\mathbf{E}^i$, we let the scattered field be of the form

$$\mathbf{E}^s = \hat{z}E_0 \sum_{n=-\infty}^{\infty} j_{-n} a_n H_n^{(2)}(k_0\rho)e^{jn\phi}, \quad \rho \geq a \tag{6.61}$$

in accordance with the radiation condition, and the total internal field given

$$\mathbf{E}^t = \hat{z}E_0 \sum_{n=-\infty}^{\infty} j_{-n} b_n J_n(k_1\rho)e^{jn\phi}, \quad \rho \leq a \tag{6.62}$$

since the field must be finite at $\rho = 0$. In the above, k_1 is the wavenumber inside the dielectric cylinder.

From the continuity of the tangential electric field at the dielectric interface $\rho = a$, we find that

$$J_n(k_0a) + a_n H_n^{(2)}(k_0a) = b_n J_n(k_1a) \tag{6.63}$$

and from the continuity of the tangential magnetic field

$$J_n(k_0a) + a_n H_n^{(2)}(k_0a) = b_n J_n(k_1a) \tag{6.64}$$

Solving the above system of equations, we find

$$a_n = -\frac{J_n(k_0a) - \gamma_e J_n'(k_0a)}{H_n^{(2)}(k_0a) - \gamma_e H_n^{(2)'}(k_0a)} \tag{6.65}$$

where

$$\Gamma_e = \frac{Z_0}{Z_1}\frac{J_n(k_1a)}{J_n'(k_1a)} \tag{6.66}$$

so that

$$\mathbf{E}^s = -\widehat{z}\sum_{n=-\infty}^{\infty} j^{-n}\frac{J_n(k_0a) - \Gamma_e J_n'(k_0a)}{H_n^{(2)}(k_0a) - \Gamma_e H_n^{(2)'}(k_0a)} H_n^{(2)}(k_0\rho)e^{jn\phi} \tag{6.67}$$

The magnitude of the diffracted electric field along the x-axis by a dielectric cylinder of size $k_0a = 2$ and relative permitivity $\epsilon_r = 2.5$ is shown in Fig. 6.10. This should be compared with the result shown in Fig. 6.2 for a perfectly conducting cylinder.

Consider the same cylinder illuminated by a transverse electric plane wave

$$\mathbf{H}^i = \widehat{z}H_0 e^{-jk_0x} = \widehat{z}H_0\sum_{n=-\infty}^{\infty} j^{-n}J_n(k_0\rho)e^{jn\phi} \tag{6.68}$$

with the ϕ-component of the incident electric field given by

$$E_\phi^i = -\frac{jZ_0}{k_0}\widehat{\phi}\cdot\nabla\times\mathbf{H}^i = jZ_0\sum_{n=-\infty}^{\infty} j^{-n}J_n'(k_0\rho)e^{jn\phi} \tag{6.69}$$

We let

$$\mathbf{H}^s = \widehat{z}H_0\sum_{n=-\infty}^{\infty} j^{-n}a_n H_n^{(2)}(k_0\rho)e^{jn\phi}, \quad \rho \geq a \tag{6.70}$$

and

$$\mathbf{H}^t = \widehat{z}H_0\sum_{n=-\infty}^{\infty} j^{-n}b_n J_n(k_1\rho)e^{jn\phi}, \quad \rho \leq a \tag{6.71}$$

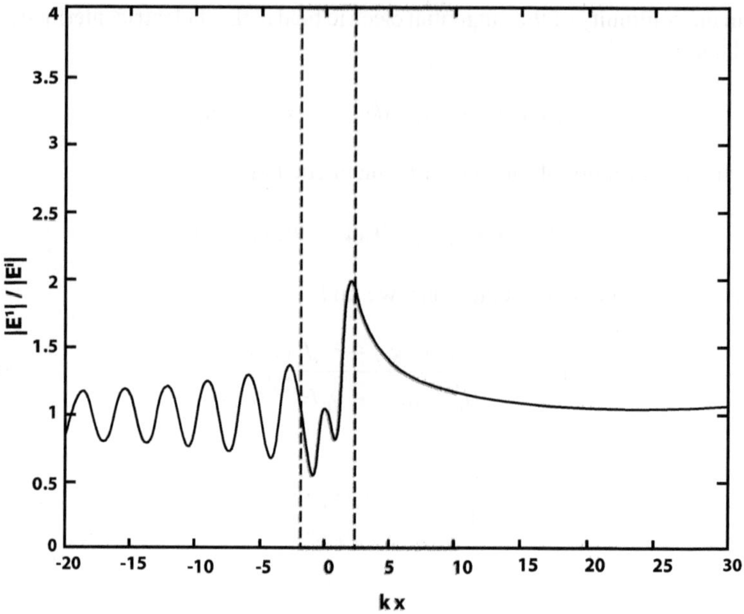

Fig. 6.10 The magnitude of the diffracted electric field along the x-axis by a dielectric cylinder of size $k_0 a = 2$ and relative permitivity $\epsilon_r = 2.5$ illuminated by a TM_z plane wave

From the continuity of the tangential magnetic field H_z at the interface $\rho = a$, we now have

$$J_n(k_0 a) + a_n H_n^{(2)}(k_0 a) = b_n J_n(k_1 a) \tag{6.72}$$

and from the continuity of the tangential electric field E_ϕ^t

$$J_n'(k_0 a) + a_n H_n^{(2)'}(k_0 a) = \frac{Z_1}{Z_0} b_n J_n'(k_1 a) \tag{6.73}$$

Solving the above system of equations, we find that

$$a_n = -\frac{J_n(k_0 a) - \Gamma_h J_n'(k_0 a)}{H_n^{(2)}(k_0 a) - \Gamma_h H_n^{(2)'}(k_0 a)} \tag{6.74}$$

where

$$\Gamma_h = \frac{Z_0}{Z_1} \frac{J_n(k_1 a)}{J_n'(k_1 a)} \tag{6.75}$$

so that

$$\mathbf{H}^s = -\hat{z} \sum_{n=-\infty}^{\infty} j^{-n} \frac{J_n(k_0 a) - \Gamma_h J_n'(k_0 a)}{H_n^{(2)}(k_0 a) - \Gamma_h H_n^{(2)'}(k_0 a)} H_n^{(2)}(k_0 \rho) e^{jn\phi} \tag{6.76}$$

6.2 The Conducting Wedge

In this section, we consider the important problem of radiation of a line source in the presence of a conducting wedge.

Assume that a perfectly conducting wedge of internal angle 2α whose edge is aligned with the z-axis is illuminated with an electric line source

$$\mathbf{J} = \hat{z} I \delta(\rho - \rho') \tag{6.77}$$

as shown in Fig. 6.11. The electric field is given by

$$\mathbf{E}^i = -\hat{z} \frac{k_0^2 I}{4\omega\epsilon} H_0^{(2)}(k_0 |\rho - \rho'|) \tag{6.78}$$

We express the incident field in terms of the cylindrical harmonics by using the addition theorem for the Hankle functions (4.298). The scattered field may also be expressed in a similar fashion. The total diffracted field has only a z component, since this is sufficient to satisfy the boundary condition

$$E_z^t = 0, \quad \phi = \alpha, 2\pi - \alpha \tag{6.79}$$

This boundary condition requires that the ϕ variations be represented by standing wave functions. The total field must also satisfy reciprocity as the observation and the source points are interchanged. In addition, $\mathbf{E}^t$ should be continuous at $\rho = \rho'$.

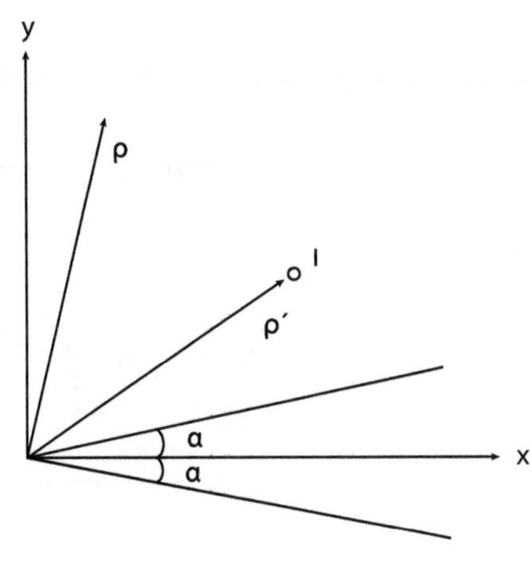

Fig. 6.11 A conducting wedge of internal angle 2α illuminated by a linear source

Therefore, in the angular range of interest, $\alpha \leq \phi \leq 2\pi - \alpha$, we write

$$E_z^t = \sum_\nu \begin{cases} a_\nu J_\nu(k\rho) H_\nu^{(2)}(k\rho') \sin\nu(\phi-\alpha) \sin\nu(\phi'-\alpha) & \rho \leq \rho' \\ a_\nu J_\nu(k\rho') H_\nu^{(2)}(k\rho) \sin\nu(\phi-\alpha) \sin\nu(\phi'-\alpha) & \rho \geq \rho' \end{cases} \tag{6.80}$$

Clearly, the boundary condition on the $\phi = \alpha$ face is automatically satisfied. In order to satisfy the boundary condition on the $\phi = 2\pi - \alpha$ face, we must choose

$$\nu = \frac{n\pi}{\psi_0} \tag{6.81}$$

where $\psi_0 = 2(\pi - \alpha)$ is the external wedge angle. It should be noted that n can only take positive integer values. This will be shown later when we discuss the proper behavior of electromagnetic fields in the vicinity of the edge.

The magnetic field can be found from Faraday's law. In particular, the ϕ-component of the diffracted magnetic field is given by

$$H_\phi^t = \frac{k_0}{j\omega\mu} \sum_{n=1}^{\infty} \begin{cases} a_\nu J_\nu'(k_0\rho) H_\nu^{(2)}(k_0\rho') \sin\nu(\phi-\alpha) \sin\nu(\phi'-\alpha) & \rho \leq \rho' \\ a_\nu J_\nu(k_0\rho') H_\nu^{(2)'}(k_0\rho) \sin\nu(\phi-\alpha) \sin\nu(\phi'-\alpha) & \rho \geq \rho' \end{cases} \tag{6.82}$$

In order to find a_ν, we consider the boundary condition that must be satisfied by the magnetic field at the source

$$\begin{aligned} \mathbf{K} &= \widehat{n} \times [\mathbf{H}^t(\rho'_+) - \mathbf{H}^t(\rho'_-)] \\ &= \widehat{z}[H_\phi^t(\rho'_+) - H_\phi^t(\rho'_-)] \end{aligned} \tag{6.83}$$

Clearly, $\mathbf{K}$ is a concentrated current source. From the definition of the surface current (1.44), we have

$$\mathbf{K} = \widehat{z} \lim_{\Delta\to 0} \int_{-\Delta/2}^{\Delta/2} I\delta(\rho - \rho')d\rho \tag{6.84}$$

where the integration is carried out over the source. Noting that

$$\delta(\boldsymbol{\rho} - \boldsymbol{\rho}') = \frac{\delta(\rho-\rho')\delta(\phi-\phi')}{\rho'} \tag{6.85}$$

we have

$$\begin{aligned} \mathbf{K} &= \widehat{z} \lim_{\Delta\to 0} \int_{-\Delta/2}^{\Delta/2} \frac{I\delta(\rho-\rho')\delta(\phi-\phi')}{\rho} d\rho \\ &= \frac{I\delta(\phi-\phi')}{\rho'} \end{aligned} \tag{6.86}$$

substituting in (6.83), we have

$$K_z = -\frac{2}{\pi\omega\mu\rho'}\sum_\nu a_\nu \sin\nu(\phi' - \alpha)\sin\nu(\phi - \alpha) \tag{6.87}$$

However, the Fourier representation of $\delta(\phi - \phi')$ is given by

$$\delta(\phi - \phi') = \frac{1}{\pi - \alpha}\sum_\nu \sin\nu(\phi' - \alpha)\sin\nu(\phi - \alpha) \tag{6.88}$$

Combining this with (6.86) and comparing the result with (6.87), we conclude that

$$a_\nu = -\frac{\omega\mu\pi I}{2(\pi - \alpha)} \tag{6.89}$$

which is independent of n. The electric field is, therefore, given by

$$\mathbf{E}^t = -\hat{z}\frac{\omega\mu\pi I}{2(\pi - \alpha)}\sum_{n=1}^{\infty}\left\{\begin{matrix} J_\nu(k\rho)H_\nu^{(2)}(k\rho') \\ J_\nu(k\rho')H_\nu^{(2)}(k\rho)\end{matrix}\right\}\sin\nu(\phi - \alpha)\sin\nu(\phi' - \alpha) \tag{6.90}$$

where $\nu = \dfrac{n\pi}{\psi_0}$.

In the far field, $k_0\rho \gg 1$, the electric field takes the form

$$E_z^t = -\frac{\omega\mu\pi I}{\psi_0}\sqrt{\frac{2j}{\pi k}}\frac{e^{-jk\rho}}{\sqrt{\rho}}\sum_{n=1}^{\infty} j^\nu J_\nu(k\rho')\sin\nu(\phi' - \alpha)\sin\nu(\phi - \alpha) \tag{6.91}$$

where we used the large argument approximation for the Hankel function (4.265). Equation (6.91) is the far field of a line source radiating in the vicinity of the conducting wedge.

Figure 6.12 shows the far field radiation pattern of a line source in the presence of a conducting wedge.

In order to find the diffracted field by a wedge illuminated by a plane wave, we let the line source recede to infinity ($k\rho' \to \infty$).

$$\begin{aligned} E_z^i &= -\frac{k_0^2 I}{4\omega\epsilon}H_0^{(2)}(k_0|\boldsymbol{\rho} - \boldsymbol{\rho}'|) \\ &\simeq -\frac{k_0^2 I}{4\omega\epsilon}\sqrt{\frac{2j}{\pi k_0}}\frac{e^{-jk_0[\rho' - \rho\cos(\phi - \phi')]}}{\sqrt{\rho'}} \end{aligned} \tag{6.92}$$

This represents a plane wave incident at angle $\phi_0 = \phi'$ with TM$_z$ polarization. That is

$$E_z^i = E_0 e^{jk\rho\cos(\phi - \phi_0)} \tag{6.93}$$

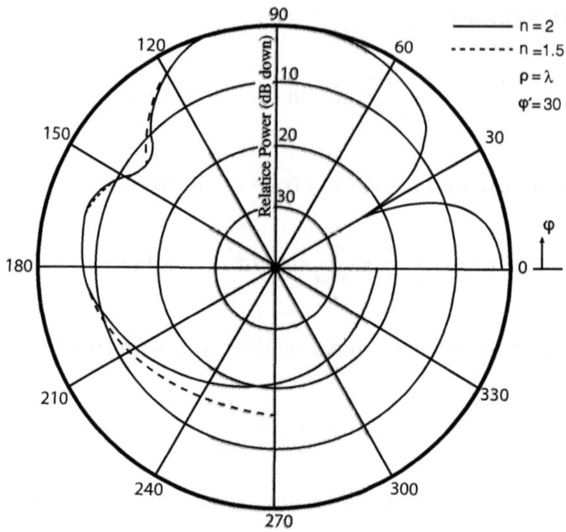

Fig. 6.12 The normalized radiation pattern of a line source in the vicinity of a perfectly conducting wedge

where

$$E_0 = -\frac{k_0^2 I}{4\omega\epsilon}\sqrt{\frac{2j}{\pi k_0}}\frac{e^{-jk_0\rho'}}{\sqrt{\rho'}} \tag{6.94}$$

Therefore, the diffracted electric field is given by

$$\mathbf{E}^t = -\hat{z}\frac{4\pi E_0}{\psi_0}\sum_{n=1}^{\infty} j^\nu J_\nu(k\rho)\sin\nu(\phi_0-\alpha)\sin\nu(\phi-\alpha) \tag{6.95}$$

with $\nu = \dfrac{n\pi}{\psi_0}$. Also, the azimuthal component of the magnetic field is found to be

$$H_\phi^t = -j\frac{4\pi E_0 Y_0}{\psi_0}\sum_{n=1}^{\infty} j^\nu J_\nu'(k\rho)\sin\nu(\phi_0-\alpha)\sin\nu(\phi-\alpha) \tag{6.96}$$

6.2.1 The Half Plane

A half plane is a wedge with an interior angle $2\alpha = 0$ (Fig. 6.13). Thus,

$$\nu = n/2, \quad n = 1, 2, \ldots \quad a_\nu = -\frac{\omega\mu I}{2} \tag{6.97}$$

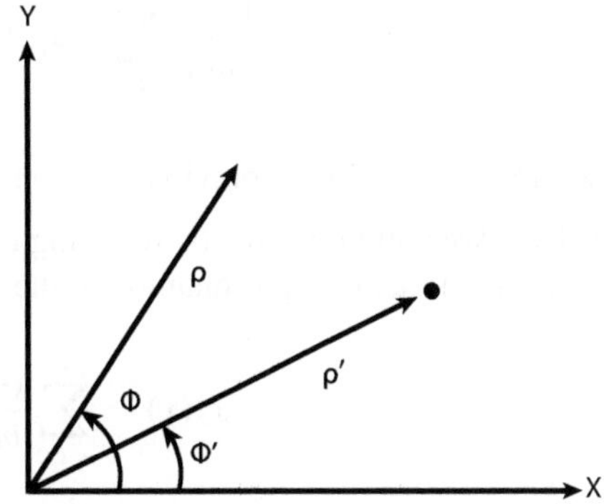

Fig. 6.13 A half plane illuminated by a line source

Substituting in (6.90), the diffracted field in the presence of a line source is given by

$$E_z^t = -\frac{\omega\mu I}{2}\sum_{n=1}^{\infty}\left\{\begin{array}{ll} J_{n/2}(k\rho)H_{n/2}^{(2)}(k\rho') & \rho \le \rho' \\ J_{n/2}(k\rho')H_{n/2}^{(2)}(k\rho) & \rho \ge \rho' \end{array}\right\} \sin\frac{n\phi}{2}\sin\frac{n\phi'}{2} \tag{6.98}$$

For a plane wave excitation, we have from (6.95)

$$E_z^t = 2E_0\sum_{n=1}^{\infty} j^{n/2} J_{n/2}(k\rho')\sin\frac{n\phi'}{2}\sin\frac{n\phi}{2} \tag{6.99}$$

6.2.2 The Edge Condition

The behavior of the electric and magnetic field components in the neighborhood of a conducting wedge depends on the wedge angle. Boukamp (1946) observed that an infinite number of solutions can be constructed for a perfectly conducting half plane, all of which satisfy the Maxwell's equations and boundary conditions. Miexner (1949, 1972) was able to eliminate the bogus solutions by imposing and edge condition from a physical point of view.

The edge condition states that the power density in the vicinity of an edge, or any geometrical singularity, must be integrable so the local energy remains finite. That is,

$$\Re\text{e}\int_S \mathbf{E}\times\mathbf{H}^\star\cdot\mathbf{ds} \to 0 \tag{6.100}$$

as the surface S enclosing the edge shrinks to the edge.

Consider the diffracted fields (6.95) and (6.96) in the presence of a conducting wedge illuminated by a TM_z plane wave.

$$E_z^t = -\frac{4\pi E_0}{\psi_0}\sum_{\nu} j^\nu J_\nu(k\rho)\sin\nu(\phi_0-\alpha)\sin\nu(\phi-\alpha) \tag{6.101}$$

$$H_\phi^t = -j\frac{4\pi E_0}{Z_0\psi_0}\sum_\nu j^\nu J_\nu'(k\rho)\sin\nu(\phi_0-\alpha)\sin\nu(\phi-\alpha) \tag{6.102}$$

where $\nu = \dfrac{n\pi}{\psi_0}$. In the previous section, we noted that n can only take on positive values. We will prove this by resorting to energy considerations.

From the series representation of the Bessel function

$$J_\nu(x) = \sum_{m=0}^{\infty}\frac{(-1)^m(x/2)^{2m+\nu}}{m!\Gamma(m+\nu+1)} \tag{6.103}$$

we may approximate the function close to the edge ($k\rho \ll 1$) using (4.257) as

$$J_\nu \sim \frac{(k\rho/2)^\nu}{\Gamma(\nu+1)} \tag{6.104}$$

Taking the dominant term of the summation, we have

$$E_z^t \sim \rho^\nu \sin\nu(\phi-\alpha) \tag{6.105}$$

and similarly

$$H_\phi^t \sim \rho^{\nu-1}\sin\nu(\phi-\alpha) \tag{6.106}$$

$$H_\rho^t \sim \rho^{\nu-1}\cos\nu(\phi-\alpha) \tag{6.107}$$

$$J_z \sim \rho^{\nu-1}\sin\nu(\phi_0-\alpha) \tag{6.108}$$

The total power (per unit length of the edge) entering the region within the radius ρ_0 is given by

$$\begin{aligned} P &= \int_\alpha^{2\pi-\alpha}\left(\frac{1}{2}\mathbf{E}\times\mathbf{H}^\star\right)\cdot(-\widehat{\rho})\rho_0 d\phi \\ &= \frac{1}{2}\int_\alpha^{2\pi-\alpha} E_z H_\phi^\star \rho_0 d\phi \end{aligned} \tag{6.109}$$

Using the edge field behavior of the fields, we find that

$$P = f(\alpha,\phi_0)\rho_0^{2\nu} \tag{6.110}$$

By enforcing the requirement that the power must remain finite that

$$\lim_{\rho_0\to 0} P < \infty \tag{6.111}$$

we conclude that $\nu > 0$, which is compatible with our choice of solutions. This condition automatically rejects solutions of the form

$$E_z^t = \sum_{n=1}^{\infty} B_\nu J_{-\nu_n}(k\rho) \sin \nu_n(\phi - \alpha), \quad \nu_n = \frac{n\pi}{\psi_0} \tag{6.112}$$

since this form renders the local power infinite.

Similarly, we have for the TE_z polarization

$$H_z^t \sim B + C\rho^\nu \cos \nu(\phi - \alpha) \tag{6.113}$$

$$E_\rho^t \sim \rho^{\nu-1} \sin \nu(\phi - \alpha) \tag{6.114}$$

$$E_\phi^t \sim \rho^{\nu-1} \cos \nu(\phi - \alpha) \tag{6.115}$$

$$J_\rho \sim B + C\rho^\nu \tag{6.116}$$

We conclude that except under symmetry conditions,

- The field components perpendicular to a sharp edge, the charge density, and the current density parallel to the edge, become infinite like $r^{\nu-1}$ where $\frac{1}{2} < \nu < 1$.
- The field components parallel to the edge and the current density perpendicular to it remain finite. In particular, the electric field parallel to the edge must vanish like r^ν.

These results are depicted in Fig. 6.14.

In particular, in the neighborhood of a diffracting half plane ($\nu = 1/2$), the edge fields behave as ($k\rho \ll 1$)

$$\mathbf{E} \sim \widehat{z} A\rho^{1/2} \sin \phi/2 \tag{6.117}$$

$$\mathbf{H} \sim \frac{A}{2j\omega\mu} \rho^{-1/2} (\widehat{\phi} \sin \phi/2 - \widehat{\rho} \cos \phi/2) \tag{6.118}$$

for the TM_z polarization, and

$$\mathbf{H} \sim \widehat{z}(B + C\rho^{1/2} \cos \phi/2) \tag{6.119}$$

$$\mathbf{E} \sim -\frac{C}{2j\omega\epsilon} \rho^{-1/2} (\widehat{\phi} \cos \phi/2 + \widehat{\rho} \sin \phi/2) \tag{6.120}$$

for the TE_z polarization, where A, B and C are complex constants determined by the excitation. Also, the excited current behaves like

Example 6.1 The induced surface current on a half plane illuminated by a TM_z plane wave can be shown to be

$$\mathbf{K} = \widehat{z} \frac{2E_0}{j\omega\mu x} \sum_n \frac{n}{2} j^{n/2} J_{n/2}(kx) \sin \frac{n\phi_0}{2} [1 - (-1)^n]$$

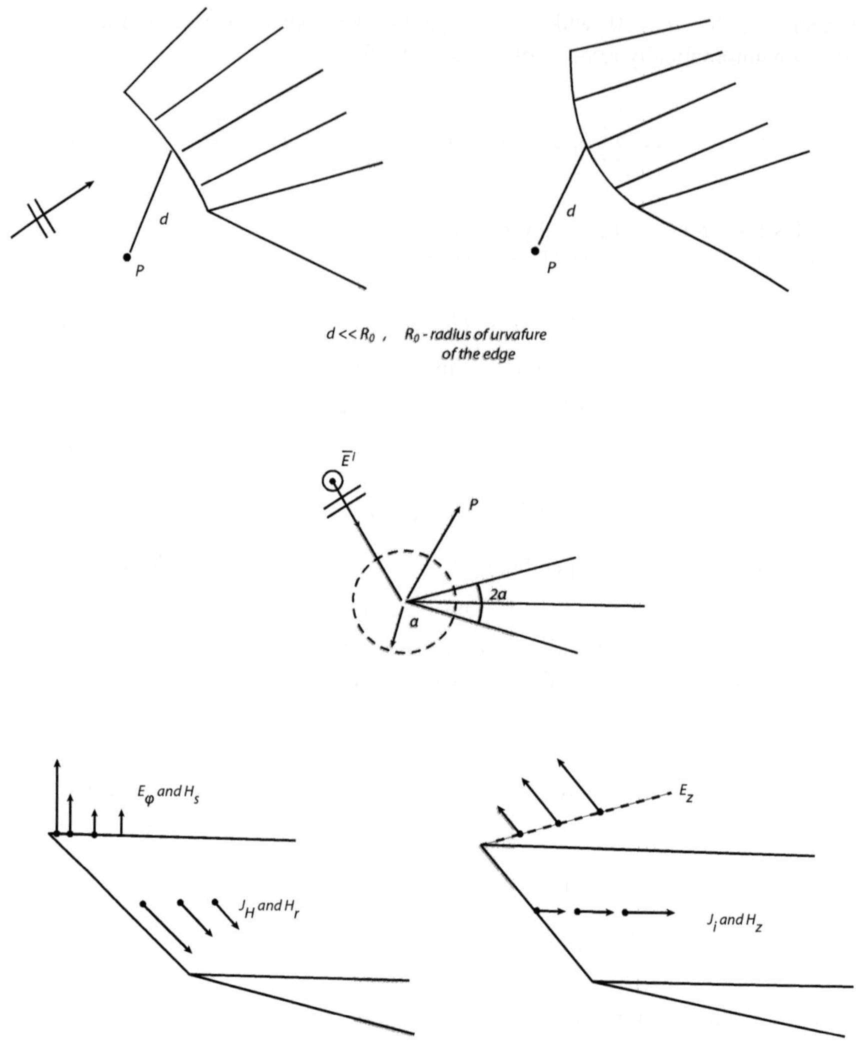

Fig. 6.14 The fields near the edge of a wedge

where x is the distance from the edge of the wedge. This result is clearly compatible with the edge condition. □

The knowledge of the singularity of fields near an edge makes it possible to estimate the extent of the region in which electric breakdown may occur, an important factor in high-voltage applications.

The singularity may be incorporated into the numerical algorithm used to solve the problem. This may increase the speed of convergence of the numerical process. When the singularity is not incorporated into the algorithm, the accuracy of the numerical solution may be checked by verifying whether fields and sources behave as predicted in the vicinity of singular points (or zones). The same checks apply to analytical solutions as well.

6.3 The Sphere

Sphere is one of the few geometries for which "exact" solution exists. The scattering behavior of a sphere can be classified into three regions:

Low frequency (Rayleigh) region,
Intermediate (resonance) frequency region (Mie scattering), and
High frequency (optical) region.

Because of its symmetry, the sphere is often used as a reference scatterer to measure the scattering properties (such as radar cross section) of other targets.

We will use eigenfunction expansion to solve the wave equation subject to appropriate boundary conditions as we specialize the results to each frequency regions.

6.3.1 Low Frequency Scattering

Low frequency scattering characterizes the scattering of electromagnetic waves by particles much smaller than a wavelength. It is, therefore, a quasi-static solution which is closely related to the field distribution of a dielectric sphere in a uniform static field. In this case, the Helmholtz wave equation reduces to the Lapalace's equation.

Consider a dielectric sphere of radius a and dielectric constant ϵ_r illuminated by a $\widehat{z}$-polarized field propagating in the $+\widehat{x}$ direction. We have

$$\mathbf{E}^i = \widehat{z} E_0 e^{-jkx} \tag{6.121}$$

Under low frequency approximation ($kr \ll 1$), the incident field represents a $\widehat{z}$-directed uniform static electric field

$$\mathbf{E}^i = \widehat{z} E_0 = \mathbf{E}_0 \tag{6.122}$$

The potential field ψ satisfies the Laplace's equation in spherical coordinates. The solutions with azimuthal symmetry to the separated equations are

$$R_n = C_1 r^n + C_2 r^{-(n+1)} \tag{6.123}$$

$$H_n = P_n(\cos\theta) \tag{6.124}$$

where $P_n(\theta)$ are the Legendre polynomials.

The general solution is

$$\Phi(r,\theta) = \sum_{n=0}^{\infty} P_n(\cos\theta)[C_{1n} r^n + C_{2n} r^{-(n+1)}] \tag{6.125}$$

Inside the sphere, the field should be finite at $r = 0$. This implies that

$$\Phi_{in} = \sum_{n=0}^{\infty} A_n r^n P_n(\cos\theta) \quad (6.126)$$

Outside the sphere, the potential should be such that it produces the uniform incident electric field far from the sphere. That is

$$\Phi^i = -E_0 z = -E_0 r \cos\theta \quad (6.127)$$

In the vicinity of the sphere, the potential is supplemented by the *scattered* potential given by $r^{-(n+1)}$ terms

$$\Phi_{out} = -E_0 r \cos\theta + \sum_{n=0}^{\infty} B_n r^{-(n+1)} p_n(\cos\theta) \quad (6.128)$$

Note that r^n terms must be excluded so fields stay finite.

We now impose the boundary conditions. The potential ψ has to be continuous across the boundary in the absence of surface currents.

$$\sum_{n=0}^{\infty} A_n a^n p_n(\cos\theta) = -E_0 a \cos\theta + \sum_{n=0}^{\infty} B_n a^{-(n+1)} P_n(\cos\theta) \quad (6.129)$$

while the normal electric flux should also be continuous across the boundary,

$$\epsilon \sum_{n=0}^{\infty} A_n n a^{n-1} P_n(\cos\theta) = -\epsilon_0 E_0 \cos\theta - \epsilon_0 \sum_{n=0}^{\infty} B_n (n+1) a^{-(n+2)} P_n(\cos\theta) \quad (6.130)$$

Using the orthogonality of the Legendre polynomials (4.384), we find

$$A_0 = B_0 = 0 \quad (6.131)$$

$$A_n = B_n = 0 \quad n > 1 \quad (6.132)$$

and

$$A_1 a = B_1 a^{-2} - E_0 a \quad (6.133)$$

$$\epsilon A_1 = -2\epsilon_0 B_1 a^{-3} - \epsilon_0 E_0 \quad (6.134)$$

implying that

$$A_1 = -\frac{3E_0}{\epsilon_r + 2} \tag{6.135}$$

$$B_1 = \frac{\epsilon_r - 1}{\epsilon_r + 2} E_0 a^3 \tag{6.136}$$

Substituting into the expressions for the potential Φ,

$$\Phi_{in} = -(\frac{3}{\epsilon_r + 2}) E_0 r \cos\theta \tag{6.137}$$

$$\Phi_{out} = -E_0 r \cos\theta + (\frac{\epsilon_r - 1}{\epsilon_r + 2}) E_0 \frac{a^3}{r^2} \cos\theta \tag{6.138}$$

Note that the internal potential is a function of $r \cos\theta$, that is a function of z

$$\Phi_{in} = -(\frac{3}{\epsilon_r + 2}) E_0 z \tag{6.139}$$

The internal field is given by

$$\mathbf{E}_{in} = \frac{3}{\epsilon_r + 2} \mathbf{E}_0 \tag{6.140}$$

The field inside the sphere is therefore uniform, parallel to the incident field, and lower in intensity than the incident field.

The external potential can be written as the sum of two incident and *scattered* potentials

$$\Phi_{out} = -E_0 z + (\frac{\epsilon_r - 1}{\epsilon_r + 2}) E_0 a^3 \frac{\cos\theta}{r^2} \tag{6.141}$$

The second term on the right hand side exhibits the familiar potential field in a dipole. The external electric field can thus be expressed as

$$\mathbf{E}_{out} = -\nabla\Phi_{out} = E_0 \widehat{z} + (\frac{\epsilon_r - 1}{\epsilon_r + 2}) E_0 a^3 \left(-\nabla(\frac{\cos\theta}{r^2})\right) \tag{6.142}$$

The electric field lines are shown in Fig. 6.15.

The above identification of the scattered potential as an equivalent dipole potential motivates the calculation of the polarization vector (dipole moment per unit volume) inside the sphere defined as

$$\mathbf{P} = \epsilon_0 \chi_e \mathbf{E}_{in} = 3\epsilon_0 \left(\frac{\epsilon_r - 1}{\epsilon_r + 2}\right) \mathbf{E}_0 \tag{6.143}$$

The total dipole moment is defined as

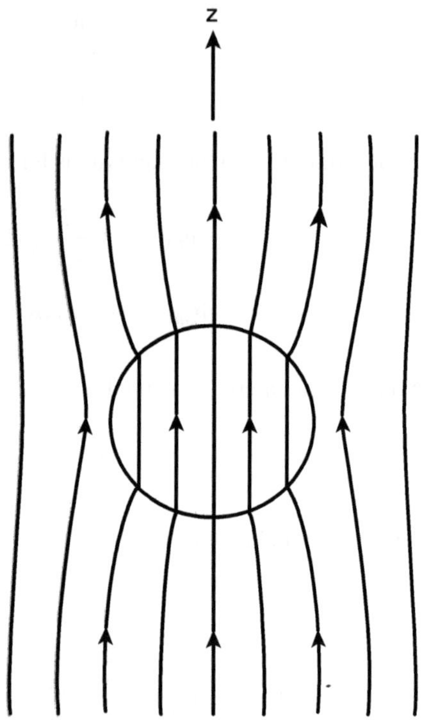

Fig. 6.15 The field lines for a dielectric sphere immersed in a z-directed quasi-static electric field

$$\mathbf{p} = \int \mathbf{P}(\mathbf{r}')dv' = 4\pi\epsilon_0 a^3 \frac{\epsilon_r - 1}{\epsilon_r + 2}\mathbf{E}_0 \tag{6.144}$$

The vector $\mathbf{p}$ is the equivalent dipole moment of the sphere and appears in the dipole term in Φ_{out}. Thus, the dielectric sphere introduces a disturbance which, as far as the exterior region is concerned, can be represented by a dipole located at the center of the sphere, of moment $\mathbf{p}$.

The electric flux density inside the sphere is given by

$$\mathbf{D}_{in} = \epsilon\mathbf{E}_{in} = \frac{3\epsilon_r}{\epsilon_r + 2}\mathbf{D}_0 \tag{6.145}$$

6.3.1.1 Rayleigh Scattering

Continuing with our interpretation of dipole action of the sphere, we may pass on to the "time varying" case by introducing the equivalent volume current density $\mathbf{J}_{eq}$ given by

$$\begin{aligned}\mathbf{J}_{eq} &= j\omega\epsilon_0(\epsilon_r - 1)\mathbf{E}_{in} \\ &= j\omega\mathbf{P}\end{aligned} \tag{6.146}$$

where **P** is the polarization vector defined in (6.143). The current moment is given by

$$\mathbf{p}_i = \int \mathbf{J}_{eq} dv = j\omega \int \mathbf{P} dv = j\omega \mathbf{p} \tag{6.147}$$

Using the volumetric equivalence principle, the scattered field far from the sphere is given by

$$\mathbf{E}^s = -jk_0 Z_0 \frac{e^{-jk_0 r}}{4\pi r} \mathbf{N}_t \tag{6.148}$$

where $\mathbf{N}_t = \mathbf{N} - N_r \widehat{r}$ is the scattering amplitude and

$$\mathbf{N} = \int_v \mathbf{J}_{eq}(\mathbf{r}') e^{jk\mathbf{r}' \cdot \widehat{r}} dv' \tag{6.149}$$

is the schelkunoff's vector. In the low frequency limit $e^{jk\mathbf{r}' \cdot \widehat{r}} \simeq 1$ and we have

$$\mathbf{N} = \int_v \mathbf{J}_{eq}(\mathbf{r}') dv' = j\omega \mathbf{p} \tag{6.150}$$

where **p** is the dipole moment of the sphere given by (6.144) and the transverse component of **N** (to $\widehat{r}$) is given by

$$\begin{aligned} \mathbf{N}_t &= -j\omega p \sin\theta \widehat{\theta} \\ &= -j\omega 4\pi\epsilon_0 a^3 \left(\frac{\epsilon_r - 1}{\epsilon_r + 2}\right) E_0 \sin\theta \widehat{\theta} \end{aligned} \tag{6.151}$$

Thus, the scattered electric field is obtained as

$$E^s_\theta = -\left(\frac{\epsilon_r - 1}{\epsilon_r + 2}\right) (ka)^2 E_0 (\frac{a}{r}) e^{jkr} \sin\theta \tag{6.152}$$

The scattering cross section is thus given by

$$\sigma = 4\pi \left(\frac{\epsilon_r - 1}{\epsilon_r + 2}\right)^2 k^4 a^6 \sin^2\theta \tag{6.153}$$

Clearly, the radar cross section is proportional to the sixth power of the radius a and inversely proportional to the fourth power of the wavelength. Thus, in the Rayleigh regime, the high-frequency waves are scattered more than the low-frequency waves. The blue color of the sky can be explained by noting that the blue portion of a light spectrum scatters more than the red portion. The air molecules can contribute to this scattering mechanism.

6.3.2 Mie Scattering

Mie scattering refers to the rigorous solution of plane wave scattering by a sphere in free space. This problem was one of the first problems to receive successful theoretical treatment. Unlike the cylindrical coordinate system, the spherical coordinate system does not possess an axis with constant direction. Therefore, the usual definition of transverse magnetic and electric polarizations used in studying scattering from cylindrical targets can not be used here.

The Hertz potentials discussed before, used the Lorenz Gauge (1.152) and (1.119). To solve the problem of scattering from a sphere, a new set of potentials, called the Debye[1] potentials need to be introduced. Debye potentials are obtained under a different gauge condition.

These potentials can be derived by examining (1.103) in the time harmonic case

$$\nabla^2 \mathbf{A} - \mu\epsilon \frac{\partial^2 \mathbf{A}}{\partial t^2} = -\mu \mathbf{J} + \nabla(\nabla \cdot \mathbf{A}) + j\omega\mu\epsilon \nabla \Phi \tag{6.154}$$

Clearly, using the Lorenz's gauge, does not reduce (6.154) to a simple scalar equation in a spherical coordinate system. Let us assume that the $\mathbf{A}$ has only a radial component

$$\mathbf{A} = \widehat{r} A_r \tag{6.155}$$

Accordingly, $\mathbf{J}$ would also have a radial component. The spherical components of (6.154) are as follows

$$\frac{1}{r^2 \sin\theta} \frac{\partial}{\partial \theta}(\sin\theta \frac{\partial A_r}{\partial \theta}) + \frac{1}{r^2 \sin^2\theta} \frac{\partial^2 A_r}{\partial \phi^2} + k^2 A_r = j\omega\mu\epsilon \frac{\partial \Phi}{\partial r} - \mu J_r \tag{6.156}$$

and

$$-\frac{1}{r} \frac{\partial^2 A_r}{\partial r \partial \theta} = \frac{j\omega\mu\epsilon}{r} \frac{\partial \Phi}{\partial \theta} \tag{6.157}$$

$$-\frac{1}{r\sin\theta} \frac{\partial^2 A_r}{\partial r \partial \phi} = \frac{j\omega\mu\epsilon}{r\sin\theta} \frac{\partial \Phi}{\partial \phi} \tag{6.158}$$

If we use the *Debye gauge*

$$\frac{\partial A_r}{\partial r} = -j\omega\mu\epsilon\Phi \tag{6.159}$$

then the ϕ-component (6.157) is automatically satisfied, and (6.156) reduces to

$$\frac{1}{r^2} A_r + \frac{1}{r^2 \sin\theta} \frac{\partial}{\partial \theta}(\sin\theta \frac{\partial A_r}{\partial \theta}) + \frac{1}{r^2 \sin^2\theta} \frac{\partial^2 A_r}{\partial \phi^2} + k^2 A_r = -\mu J_r \tag{6.160}$$

---

[1] After Debye, physicist.

In order to reduce this equation to the scalar wave equation in the spherical coordinate system, we employ the *Debye transformation*

$$j\omega\mu\epsilon r\pi_{er}^{D} = A_r \tag{6.161}$$

Then (6.160) becomes

$$\nabla^2\pi_{er}^{D} + k^2\pi_{er}^{D} = -\frac{J_r}{j\omega\epsilon r} \tag{6.162}$$

In other words, π_{er}^{D} satisfies the scalar inhomogeneous wave equation in spherical coordinate system. The potential π_{er}^{D} is known as the *electric Debye Hertz potential.* The field components are thus given by

$$\begin{aligned}\mathbf{E} &= \nabla\times\nabla\times(r\pi_{er}^{D}\widehat{r}) - \frac{J_r}{j\omega\epsilon}\widehat{r} \\ \mathbf{H} &= j\omega\epsilon\nabla\times(r\pi_{er}^{D}\widehat{r})\end{aligned} \tag{6.163}$$

From the above analysis, we see that only the electric field has a radial component and the magnetic field is transverse to r. These fields are thus called E-modes or TM_r modes. By duality, we may define the *magnetic Debye Hertz potential* satisfying

$$\nabla^2\pi_{mr}^{D} + k^2\pi_{mr}^{D} = -\frac{J_{mr}}{j\omega\mu r} \tag{6.164}$$

producing the fields

$$\begin{aligned}\mathbf{E} &= -j\omega\mu\nabla\times(r\pi_{mr}^{D}\widehat{r}) \\ \mathbf{H} &= \nabla\times\nabla\times(r\pi_{mr}^{D}\widehat{r}) - \frac{J_{mr}}{j\omega\mu}\widehat{r}\end{aligned} \tag{6.165}$$

which are called H-modes or TE_r modes.

In view of (6.162) and (6.164), the Debye Hertz potentials are given in free space by

$$\pi_{er}^{D} = \frac{1}{4\pi j\omega\epsilon}\int\limits_V \frac{J_r(\mathbf{r}')}{r'}\frac{e^{-jk|\mathbf{r}-\mathbf{r}'|}}{|\mathbf{r}-\mathbf{r}'|}dv' \tag{6.166}$$

$$\pi_{mr}^{D} = \frac{1}{4\pi j\omega\mu}\int\limits_V \frac{J_{mr}(\mathbf{r}')}{r'}\frac{e^{-jk|\mathbf{r}-\mathbf{r}'|}}{|\mathbf{r}-\mathbf{r}'|}dv' \tag{6.167}$$

The total fields are the superposition of those given by (6.163) and (6.165). In component form, they are given by

$$E_r = \frac{\partial^2}{\partial r^2}(r\pi_{er}^{D}) + k^2 r\pi_{er}^{D}$$

$$
\begin{aligned}
E_\theta &= \frac{1}{r}\frac{\partial^2}{\partial r \partial \theta}(r\pi_{er}^D) - j\omega\mu\frac{1}{\sin\theta}\frac{\partial \pi_{mr}^D}{\partial \phi} \\
E_\phi &= \frac{1}{r\sin\theta}\frac{\partial^2}{\partial r \partial \phi}(r\pi_{er}^D) + j\omega\mu\frac{\partial \pi_{mr}^D}{\partial \theta} \\
H_r &= \frac{\partial^2}{\partial r^2}(r\pi_{mr}^D) + k^2 r\pi_{mr}^D \\
H_\theta &= j\omega\epsilon\frac{1}{\sin\theta}\frac{\partial \pi_{er}^D}{\partial \phi} + \frac{1}{r}\frac{\partial^2}{\partial r \partial \theta}(r\pi_{mr}^D) \\
H_\phi &= -j\omega\epsilon\frac{\partial \pi_{er}^D}{\partial \phi} + \frac{1}{r\sin\theta}\frac{\partial^2}{\partial r \partial \phi}(r\pi_{mr}^D)
\end{aligned}
\tag{6.168}
$$

As the vector and scalar potentials A and Φ are related by the Lorenz's gauge

$$
\frac{\partial A_r}{\partial r} + j\omega\mu\epsilon\Phi = 0 \tag{6.169}
$$

Combining this with (6.161), we conclude that

$$
\Phi = -\frac{\partial(r\pi_{er}^D)}{r} \tag{6.170}
$$

This should be compared with (1.119) where the Hertz vector and Φ are related by $\Phi = -\nabla \cdot \pi$.

Now, consider a dielectric sphere of radius a in free space illuminated by plane wave propagating in the z-direction

$$
\mathbf{E}^i = \widehat{x}E_0 e^{-jk_0 z} \tag{6.171}
$$

as shown in Fig. 6.16. The incident magnetic field is given by

$$
\mathbf{H}^i = \widehat{y}Y_0 E_0 e^{-jk_0 z} \tag{6.172}
$$

We will first find the Debye Hertz potentials for the incident field. The radial component of the incident electric field is given by

$$
\begin{aligned}
E_r^i = E_0\widehat{x}\cdot\mathbf{r}e^{-jk_0 z} &= E_0 \sin\theta\cos\phi e^{-jk_0 r\cos\theta} \\
&= E_0\cos\phi\frac{1}{jk_0 r}\frac{\partial}{\partial\theta}(e^{-jk_0 r\cos\theta})
\end{aligned}
\tag{6.173}
$$

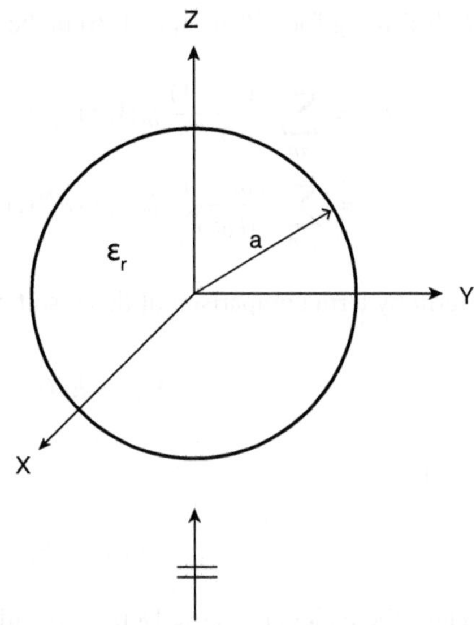

Fig. 6.16 A dielectric sphere of radius a and relative permitivity ϵ_r illuminated by a plane wave

Using the spherical wave transformation for the plane wave (4.405), we have

$$E_r^i = E_0 \cos\phi \frac{1}{jk_0 r} \frac{\partial}{\partial\theta} \sum_{n=0}^{\infty} j^{-n}(2n+1) j_n(k_0 r) P_n(\cos\theta)$$
$$= jE_0 \frac{\cos\phi}{(k_0 r)^2} \sum_{n=1}^{\infty} j^{-n}(2n+1) \widehat{j_n}(k_0 r) P_n^1(\cos\theta) \tag{6.174}$$

where

$$P_n^1(\cos\theta) = -\frac{\partial}{\partial\theta} P_n(\cos\theta) \tag{6.175}$$

and $\widehat{j_n}(k_0 r) = k_0 r j_n(k_0 r)$. The latter definition is due to Schellkunoff. On the other hand, the incident electric potential satisfies the Helmholtz equation (6.162). Thus, it can be expressed in terms of spherical harmonics

$$\pi_{er}^{Di} = \sum_{m=0}^{\infty} \sum_{n=0}^{\infty} j_n(k_0 r) P_n^m(\cos\theta)[A_{mn}^e \cos m\phi + B_{mn}^e \sin m\phi] \tag{6.176}$$

Also, E_r^i must satisfy

$$E_r^i = \frac{\partial^2}{\partial r^2}(r\pi_{er}^{Di}) + k_0^2 r \pi_{er}^{Di} \tag{6.177}$$

Substituting for π_{er}^{Di} from (6.176) in the above equation, we obtain

$$E_r^i = \sum_{mn} \frac{n(n+1)}{r} j_n(k_0 r) P_n^m(\cos\theta)[A_{mn}^e \cos m\phi + B_{mn}^e \sin m\phi]$$
$$= \sum_{mn} \frac{n(n+1)}{(k_0 r)^2} \widehat{j}_n(k_0 r) P_n^m(\cos\theta)[A_{mn}^e \cos m\phi + B_{mn}^e \sin m\phi] \quad (6.178)$$

Term by term comparison of this result with (6.175) gives

$$A_{1n}^e = E_0(-j)^{n-1} \frac{2n+1}{k_0 n(n+1)} \quad (6.179)$$

while

$$A_{mn}^e = B_{mn}^e = 0, \quad m \neq 1 \quad (6.180)$$

Thus, the incident electric Debye potential is given by

$$\pi_{er}^{Di} = \frac{E_0 \cos\phi}{k_0^2 r} \sum_{n=1}^{\infty} (-j)^{n-1} \frac{2n+1}{n(n+1)} \widehat{j}_n(k_0 r) P_n^1(\cos\theta) \quad (6.181)$$

By a dual process, we also find that

$$\pi_{mr}^{Di} = \frac{Y_0 E_0 \sin\phi}{k_0^2 r} \sum_{n=1}^{\infty} (-j)^{n-1} \frac{2n+1}{n(n+1)} \widehat{j}_n(k_0 r) P_n^1(\cos\theta) \quad (6.182)$$

The scattered fields can be obtained from the scattered Debye potential which must satisfy the radiation condition. Thus

$$\pi_{er}^{Ds} = \frac{E_0 \cos\phi}{k_0^2 r} \sum_{n=1}^{\infty} a_n \widehat{h}_n^{(2)}(k_0 r) P_n^1(\cos\theta) \quad (6.183)$$

$$\pi_{mr}^{Ds} = \frac{Y_0 E_0 \sin\phi}{k_0^2 r} \sum_{n=1}^{\infty} b_n \widehat{h}_n^{(2)}(k_0 r) P_n^1(\cos\theta) \quad (6.184)$$

The total potentials are expressed as

$$\pi_{er}^{D} = \pi_{er}^{Di} + \pi_{er}^{Ds}$$
$$= \frac{E_0 \cos\phi}{k_0^2 r} \sum_{n=1}^{\infty} [(-j)^{n-1} \frac{2n+1}{n(n+1)} \widehat{j}_n(k_0 r) + a_n \widehat{h}_n^{(2)}(k_0 r)] P_n^1(\cos\theta) \quad (6.185)$$
$$\pi_{mr}^{D} = \pi_{mr}^{Di} + \pi_{mr}^{Ds}$$

$$= \frac{Y_0 E_0 \sin\phi}{k_0^2 r} \sum_{n=1}^{\infty} [(-j)^{n-1} \frac{2n+1}{n(n+1)} \widehat{j}_n(k_0 r) + a_n \widehat{h}_n^{(2)}(k_0 r)] P_n^1(\cos\theta)$$

The external diffracted fields can be expressed in terms of π_{er}^D and π_{mr}^D. The internal fields are given by

$$\pi_{er}^{Din} = \frac{E_0 \cos\phi}{k_1^2 r} \sum_{n=1}^{\infty} c_n \widehat{j}_n(k_1 r) P_n^1(\cos\theta) \tag{6.186}$$

$$\pi_{mr}^{Din} = \frac{Y_1 E_0 \sin\phi}{k_1^2 r} \sum_{n=1}^{\infty} d_n \widehat{j}_n(k_1 r) P_n^1(\cos\theta) \tag{6.187}$$

where $Y_1 = \sqrt{\epsilon_1/\mu_1}$ and $k_1 = \omega\sqrt{\epsilon_1 \mu_1}$ are the intrinsic admittance and the wavenumber inside the sphere, respectively.

In order to find the coefficients a_n, b_n, c_n, and d_n, we apply the boundary conditions on the surface of the sphere. The boundary conditions are the continuity of tangential components of the electric and magnetic fields at $r = a$. However, from (6.168), we observe that these boundary conditions involve a combination of π_{er}^D and π_{mr}^D. In order to simplify the boundary conditions, we notice that if we consider the linear combination $\partial(\sin\theta E_\theta)/\partial\theta + \partial E_\phi/\partial\phi$, then $\partial(r\pi_{er}^D)/\partial r$ must be continuous at the boundary surface. Similarly, taking $\partial E_\theta/\partial\phi - \partial(\sin\theta E_\phi)/\partial\theta$, we find that $\mu\pi_{mr}^D$ must be continuous at $r = a$. Thus, we write

$$\frac{\partial(r\pi_{er}^D)}{\partial r} = \frac{\partial(r\pi_{er}^{Din})}{\partial r} \tag{6.188}$$

$$\mu_1 \pi_{mr}^D = \mu_0 \pi_{mr}^{Din} \tag{6.189}$$

Similar considerations for the tangential components of the magnetic field give the boundary conditions that $\partial(r\pi_{mr}^D)/\partial r$ and $\epsilon\pi_{er}^D$ must be continuous at the surface of the sphere. Hence, we have

$$\frac{\partial(r\pi_{mr}^D)}{\partial r} = \frac{\partial(r\pi_{mr}^{Din})}{\partial r} \tag{6.190}$$

$$\epsilon_1 \pi_{er}^D = \epsilon_0 \pi_{er}^{Din} \tag{6.191}$$

Substituting for the total potentials in the last two systems of equations and solving for the unknown coefficients, we find that

$$a_n = \frac{j^{-n}(2n+1)}{n(n+1)} \frac{\sqrt{\epsilon_0\mu_1}\widehat{j}_n(\alpha)\widehat{j}'_n(\beta) - \sqrt{\epsilon_1\mu_0}\widehat{j}'_n(\alpha)\widehat{j}_n(\beta)}{\sqrt{\epsilon_1\mu_0}\widehat{h}_n^{(2)'}(\alpha)\widehat{j}_n(\beta) - \sqrt{\epsilon_0\mu_1}\widehat{h}_n^{(2)}(\alpha)\widehat{j}'_n(\beta)} \tag{6.192}$$

$$b_n = \frac{j^{-n}(2n+1)}{n(n+1)} \frac{\sqrt{\epsilon_0\mu_1}\widehat{j}'_n(\alpha)\widehat{j}_n(\beta) - \sqrt{\epsilon_1\mu_0}\widehat{j}_n(\alpha)\widehat{j}'_n(\beta)}{\sqrt{\epsilon_1\mu_0}\widehat{h}_n^{(2)}(\alpha)\widehat{j}'_n(\beta) - \sqrt{\epsilon_0\mu_1}\widehat{h}_n^{(2)'}(\alpha)\widehat{j}_n(\beta)} \tag{6.193}$$

$$c_n = \frac{j^{-n}(2n+1)}{n(n+1)} \frac{-j\sqrt{\epsilon_1\mu_0}}{\sqrt{\epsilon_1\mu_0}\widehat{h}_n^{(2)'}(\alpha)\widehat{j}_n(\beta) - \sqrt{\epsilon_0\mu_1}\widehat{h}_n^{(2)}(\alpha)\widehat{j'}_n(\beta)} \tag{6.194}$$

$$d_n = \frac{j^{-n}(2n+1)}{n(n+1)} \frac{j\sqrt{\epsilon_0\mu_1}}{\sqrt{\epsilon_1\mu_0}\widehat{h}_n^{(2)}(\alpha)\widehat{j'}_n(\beta) - \sqrt{\epsilon_0\mu_1}\widehat{h}_n^{(2)'}(\alpha)\widehat{j}_n(\beta)} \tag{6.195}$$

where $\alpha = k_0 a$ and $\beta = k_1 a$. For small spheres, $k_1 a \ll 1$, only the $n = 1$ term dominate with

$$a_n = -(ka)^3 \frac{\epsilon_1 - \epsilon}{\epsilon_1 + 2\epsilon}, \quad b_n = -(ka)^3 \frac{\mu_1 - \mu}{\mu_1 + 2\mu} \tag{6.196}$$

and the Mie scattering reduces to Rayleigh scattering.

The above analysis may be extended to a perfectly conducting sphere by letting $\epsilon_1 \to \infty$ and $\mu_1 = 0$. The normalized radar cross section of a conducting sphere as function of frequency is shown in Fig. 6.17. In the low frequency region, the Rayleigh scattering is valid. In the high frequency regime the radar cross section approaches the optical geometric cross section, while in the intermediate frequencies the resonant Mie scattering dominates.

For a plane wave incidence on a conducting sphere, the far zone scattered fields do not experience depolarization in the forward and backward direction.

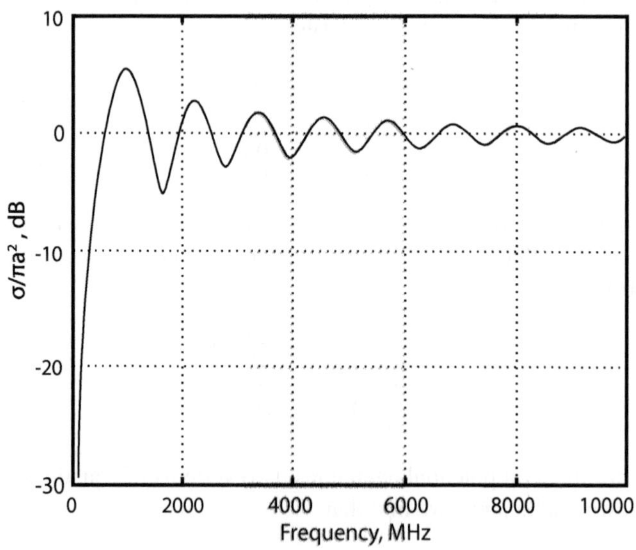

Fig. 6.17 The monostatic radar cross section of a perfectly conducting sphere of radius 5 cm as a function of frequency

Exercises

6.1: Consider an infinitely long conducting circular cylinder illuminated by a TM_z plane wave at $\phi_0 = 180°$.

(a) Using eigen-function expansion, find the bistatic radar echo width.
(b) Plot the bistatic echo width for $ka = 1$, $ka = 5$ and $ka = 10$.
(c) Express the result for an electrically thin wire ($ka \ll 1$).

6.2: A perfectly conducting circular cylinder of radius a is coated with a nonmagnetic dielectric of radius $b > a$. It is illuminated by a TM_z plane wave $\mathbf{E}^i = \widehat{z}e^{-jk_0x}$. Find the scattered fields in the region $\rho \geq b$.

6.3: A homogeneous dielectric cylinder of radius a, permitivity ϵ_1 and permeability μ_1 has its axis parallel to the z axis and is immersed in a medium with permittivity ϵ and permeability μ. If the cylinder is illuminated by the plane wave

$$\mathbf{H}^i = \widehat{z}e^{-jkx}$$

find the scattered field.

6.4: Using (6.153), find the scattering cross section σ_s, absorption cross section σ_a, scattering albedo W_0, and absorption efficiency Q_s of a dielectric sphere in the Rayleigh region.

Chapter 7
Approximate Methods

In Chap. 6, we discussed a number of scattering problems for which analytical solutions in the form of eigenfunction expansions were available. For this reason, these were referred to as canonical problems. However, when the geometry of the scatterer is complex, these analytical methods are not applicable. In such cases, we have to resort to approximate and/or numerical methods.

In this chapter, we will represent approximate methods for calculating the scattered field from dielectric scatterers as well as electrically large conducting objects.

For low-contrast dielectric targets, the Rayleigh-Debye method is presented. This approximation is also known as the Born approximation. We will also review the higher order Born approximations and give conditions under which these approximations are convergent. Electromagnetic scattering from dielectric objects whose permittivity fluctuate around an average value will also be discussed.

For large conducting targets, we will present the physical optics approximation which is valid when the typical dimensions of the target and its radii of curvature are large compared to the operating wavelength. The physical optics approximation belongs to the class of asymptotic methods which is applicable for higher frequencies.

7.1 Rayleigh-Debye Approximation

Consider a low-contrast dielectric scatterer for which the relative permittivity is close to unity ($\epsilon_r \simeq 1$) illuminated by an incident plane wave in free space

$$\mathbf{E^i} = \mathbf{E}_0 e^{-jk_0 \mathbf{r}\cdot\widehat{\imath}} \tag{7.1}$$

In order to calculate the scattered field, we may use the volumetric field equivalence principle (Sect. 5.1). The equivalent volumetric current is given by (3.56) and is expressed as

$$\mathbf{J_e} = j\omega\epsilon_0(\epsilon_r - 1)\mathbf{E} \tag{7.2}$$

© Springer International Publishing Switzerland 2015

K. Barkeshli, *Advanced Electromagnetics and Scattering Theory*,
DOI 10.1007/978-3-319-11547-4_7

This current is radiating in free space and we may use (2.71) to find the scattered field. Hence, we have

$$\mathbf{E}^{\mathbf{s}} = -jkZ_0 \frac{e^{-jkr}}{4\pi r} (-\widehat{r} \times \widehat{r} \times \mathbf{N}) \tag{7.3}$$

where

$$\mathbf{N} = \int_V \mathbf{J_e}(\mathbf{r}') e^{jk\mathbf{r}' \cdot \widehat{r}} dv' \tag{7.4}$$

The electric current $\mathbf{J_e}$ is given in terms of the electric field inside the scatterer which is presently an unknown quantity. We will now consider the lowest order approximation for the scattering amplitude in (7.3).

For a low-contrast dielectric object, the internal field may be approximated by the incident field. Thus,

$$\mathbf{E}^{[0]} = \mathbf{E}^{i} + \mathbf{E}^{s} \simeq \mathbf{E}^{\mathbf{i}} \tag{7.5}$$

is the zeroth order approximation for the total internal field and the equivalent current is given by

$$\mathbf{J_e} = j\omega\epsilon_0(\epsilon_r - 1)\mathbf{E}_0 e^{-jk_0 \mathbf{r} \cdot \widehat{\iota}} \tag{7.6}$$

The above approximation is known as the *Rayleigh-Debye* approximation and is valid if

$$(\epsilon_r - 1)kD \ll 1 \tag{7.7}$$

where D is the maximum dimension of the scatterer. Substituting (7.6) in (7.3), we find the first order approximation for the scattered field as

$$\mathbf{E}^{\mathrm{s}[1]} = \frac{k_0^2}{4\pi} (-\widehat{r} \times \widehat{r} \times \mathbf{E}_0) \int_V \chi_e(\mathbf{r}') e^{-jk_0 \widehat{\xi} \cdot \mathbf{r}'} dv' \tag{7.8}$$

where $\chi_e = \epsilon_r - 1$ is the electric susceptibility of the target and

$$\widehat{\xi} = (\widehat{\iota} - \widehat{r}), \quad |\widehat{\iota}| = 2\sin(\vartheta/2) \tag{7.9}$$

and ϑ is the angle between $\widehat{\iota}$ and $\widehat{r}$. Clearly, $\widehat{\xi}$ depends on the incident and observation angle. The above expression for the scattered field is known as *the first order Born approximation*.

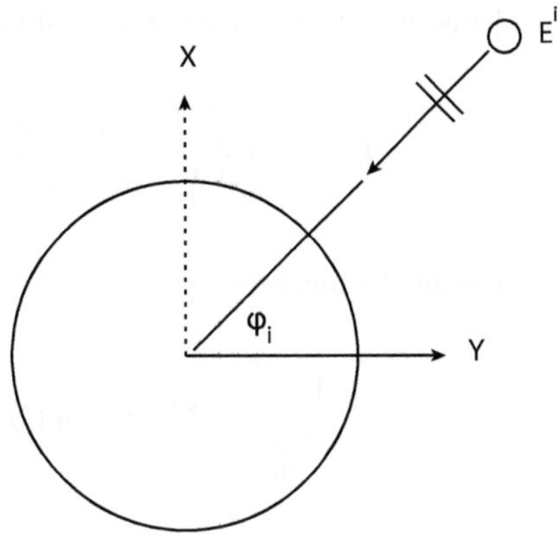

Fig. 7.1 A low-contrast dielectric cylinder of radius a illuminated by a TM_z plane wave

It is seen that the scattered field is proportional to the Fourier transform of χ_e evaluated at the wave number k_ξ.

For two-dimensional problems, the expression (7.7) reduces to

$$k_0 D\sqrt{\epsilon_r - 1} < \pi \tag{7.10}$$

Example 7.1 Consider a low-contrast dielectric cylinder of infinite extent and radius a illuminated by a transverse magnetic plane wave (Fig. 7.1)

$$\mathbf{E}^{\mathbf{i}}(\mathbf{r}) = \widehat{z}e^{-jk_0\widehat{\iota}\cdot\rho} \tag{7.11}$$

This is a two-dimensional problem. Using the Born approximation, we write

$$\mathbf{J_e} = j\omega\epsilon_0\widehat{z}\chi_e e^{-jk_0\widehat{\iota}\cdot\rho}$$

The scattered field is given by

$$\mathbf{E}^{\mathbf{s}} = \nabla\nabla\cdot\pi^s + k_0^2\pi^s$$

where

$$\pi^s = -j\frac{Z_0}{k_0}\widehat{z}\int_S \mathbf{J_e}(\rho')\left[\frac{1}{j}H_0^{(2)}(k_0|\rho-\rho'|)\right]ds'$$

Since $\mathbf{J_e}$ is z-directed and independent of z, the first term on the right hand side of the above will be zero and we find that

$$E_z^s = -j\frac{k_0^2}{4}\chi\int_S e^{-jk_0\widehat{\iota}\cdot\rho'}e^{-jk_0|\rho-\rho'|}ds'$$

Using the asymptotic expansion of the Hankel function, we find

$$E_z^s = -j\frac{k_0^2}{4}\sqrt{\frac{2j}{\pi k_0}}\frac{e^{-jk_0\rho}}{\sqrt{\rho}}\chi_e\int_0^a \rho' \int_0^{2\pi} e^{-jk_\xi \rho' \cos\phi'} d\phi' d\rho'$$

Employing the identities

$$\frac{1}{2\pi}\int_0^{2\pi} e^{-jk\rho} d\phi = J_0(k\rho), \quad \int x J_0(x) dx = x J_1(x)$$

we obtain

$$E_z^s = -j\frac{k_0^2}{4}\sqrt{\frac{2j}{\pi k_0}}\frac{e^{-jk_0\rho}}{\sqrt{\rho}}\chi_e 2\pi k_\xi a^2 J_1(k_\xi a)/k_\xi a$$

where $k_\xi = 2k_0 \sin(\phi_s - \phi_i)$.

For a dielectric cylinder (7.10) reduces to

$$\frac{a}{\lambda_0}\sqrt{\chi_e} < 0.25$$

The echo width of a low-contrast cylinder of radius $a = 0.5\lambda$ and $\epsilon_r = 1.1$ is shown in Fig. 7.2. □

If we write the first order approximation for the total internal field as

$$\mathbf{E}^{[1]} \simeq \mathbf{E}^{\mathbf{i}} + \mathbf{E}^{s[0]} \tag{7.12}$$

the second order scattered field is given by

$$\mathbf{E}^{\mathbf{s}[2]} = -jkZ_0\frac{e^{-jkr}}{4\pi r}\left(-\widehat{r}\times\widehat{r}\times\int_V j\omega\epsilon_r\chi_e(\mathbf{r}')(\mathbf{E}^{\mathbf{i}} + \mathbf{E}^{\mathbf{s}[0]})(\mathbf{r}')e^{jk\mathbf{r}'\cdot\widehat{r}}dv'\right) \tag{7.13}$$

The Born approximation of order $i+1$ for the scattered field is similarly expressed as

$$\mathbf{E}^{\mathbf{s}[i+1]} = -jkZ_0\frac{e^{-jkr}}{4\pi r}\left(-\widehat{r}\times\widehat{r}\times\int_V j\omega\epsilon_r\chi_e(\mathbf{r}')(\mathbf{E}^{\mathbf{i}} + \mathbf{E}^{\mathbf{s}[i]})(\mathbf{r}')e^{jk\mathbf{r}'\cdot\widehat{r}}dv'\right) \tag{7.14}$$

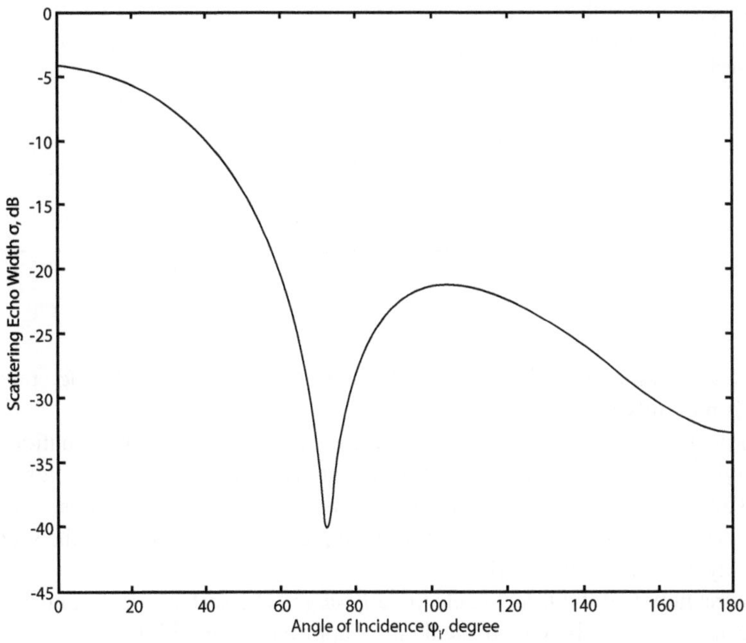

Fig. 7.2 The echo width of a dielectric cylinder of radius $a = 0.5\lambda$ and $\epsilon_r = 1.1$ illuminated by a TM_z plane wave

7.2 Physical Optics Approximation

Consider a perfectly conducting object illuminated by a plane wave in free space. By the surface physical equivalence (Sect. 5.2), the scattered field can be attributed to the induced surface electric current **K** over the target radiating in free space. Using (5.3) and (5.4), the scattered electric field far from the target can be expressed as

$$\begin{aligned}\mathbf{E}^s &= E_0 \frac{e^{-jk_0R}}{R}\mathbf{f}(\widehat{s},\widehat{\imath}) \\ &= -jk_0Z_0E_0\frac{e^{-jk_0R}}{R}\int_S [-\widehat{s}\times\widehat{s}\times\mathbf{K}(\mathbf{r}')]e^{jk_0\mathbf{r}'\cdot\widehat{s}}ds'\end{aligned} \quad (7.15)$$

where S is the surface of the conductor. The bistatic radar cross section is given by

$$\begin{aligned}\sigma_{bi}(\widehat{s},\widehat{\imath}) &= 4\pi|\mathbf{f}(\widehat{s},\widehat{\imath})|^2 \\ &= \frac{k_0^2Z_0^2}{4\pi}\left|\int_S [-\widehat{s}\times\widehat{s}\times\mathbf{K}(\mathbf{r}')]e^{jk_0\mathbf{r}'\cdot\widehat{s}}ds'\right|^2\end{aligned} \quad (7.16)$$

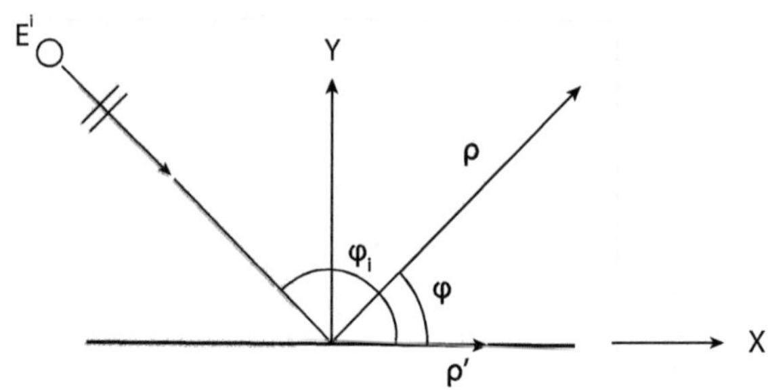

Fig. 7.3 A perfectly conducting strip of width w illuminated by a TM_z plane wave

The above formulation relies on an evaluation of the equivalent electric current **K** which is presently an unknown quantity.

In order to calculate the induced current, we may set up an integral equation for **K** by applying boundary conditions for the tangential electric or magnetic fields over S. The resulting integral equation can then be solved by numerical methods. However, if we can estimate the surface current by an approximate method, then the scattered fields may be calculated directly.

Assume now that the target is electrically large. The surface electric current can now be approximated by the physical optics approximation (3.79)

$$\mathbf{K} \simeq \mathbf{K}_{p0} = \begin{cases} 2\widehat{n} \times \mathbf{H}^i & \text{in the lit region} \\ 0 & \text{in the shadow region} \end{cases} \tag{7.17}$$

The scattered electric field and the bistatic radar cross section can be computed directly from (7.15) and (7.16).

Example 7.2 Find the physical optics scattering from a thin conducting strip of width w located on the xz-plane and illuminated by a TM_z plane wave.

Consider the strip shown in Fig. 7.3 illuminated by an E-polarized plane wave at the incidence angle ϕ_i

$$\begin{aligned} \mathbf{E^i} &= \widehat{z} E_0 e^{jk_0(x\cos\phi_i + y\sin\phi_i)} \\ \mathbf{H}^i &= \frac{E_0}{Z_0}(-\widehat{x}\sin\phi_i + \widehat{y}\cos\phi_i) e^{jk_0(x\cos\phi_i + y\sin\phi_i)} \end{aligned}$$

Using the physical optics approximation for the surface current, we have

$$\mathbf{K}_{p0} = 2\widehat{n} \times \mathbf{H}^i, \quad y = 0$$

or, equivalently

$$\mathbf{K}_{p0} = \widehat{z}\frac{2E_0}{Z_0}\sin\phi_i e^{jk_0\cos\phi_i}$$

The scattered electric field is given by

$$\mathbf{E}^{\mathbf{s}} = \nabla\nabla \cdot \pi^s + k_0^2 \pi^s$$

where

$$\pi^s = -j\frac{Z_0}{K_0}\int_{-w/2}^{w/2} \mathbf{K}_{p0}(\rho')[\frac{1}{j}H_0^{(2)}(k_0|\rho - \rho'|)]d\ell' \tag{7.18}$$

Since π^s is z-directed and independent of z, the scattered electric field is given by

$$\mathbf{E}^{\mathbf{s}} = \frac{k_0 Z_0}{4}\int_{-w/2}^{w/2} \mathbf{K}_{p0}(x')H_0^{(2)}(k_0|\rho - \rho'|)dx'$$

where ϕ is the observation angle.

Substituting for the surface current and using the large argument expansion for the Hankel function, the scattered field far from the strip is found to be

$$\begin{aligned} E_z^s &= -\frac{k_0 E_0}{2}\sqrt{\frac{2j}{\pi k_0}}\frac{e^{-jk_0\rho}}{\sqrt{\rho}}\sin\phi_i \int_{-w/2}^{w/2} e^{jk_0x'(\cos\phi_i + \cos\phi)}dx' \\ &= -\frac{k_0 w E_0}{2}\sqrt{\frac{2j}{\pi k_0}}\frac{e^{-jk_0\rho}}{\sqrt{\rho}}\sin\phi_i \mathrm{sinc}[\frac{k_0 w}{2}(\cos\phi + \cos\phi_i)] \end{aligned}$$

The bistatic echo width is thus given by

$$\sigma^{2d} = \lim_{\rho\to\infty} 2\pi\rho\frac{|\mathbf{E}^{\mathbf{s}}|^2}{|\mathbf{E_i}|^2}$$

or

$$\sigma^{2d}/\lambda = (k_0 w)^2 \sin^2\phi_i \mathrm{sinc}^2[\frac{k_0 w}{2}(\cos\phi + \cos\phi_i)]$$

For the monostatic echo width $\phi = \phi_i$ and we obtain

$$\sigma^{2d}/\lambda = (k_0 w)^2 \sin^2\phi \mathrm{sinc}^2[k_0 w \cos\phi]$$

The bistatic echo width for a 2λ wide conducting strip illuminated by a plane wave at an incidence angle of 60° is shown in Fig. 7.4. It is observed that the maximum echo width occurs at the specular direction of 120°. The monostatic echo width of the same strip is shown in Fig. 7.5. □

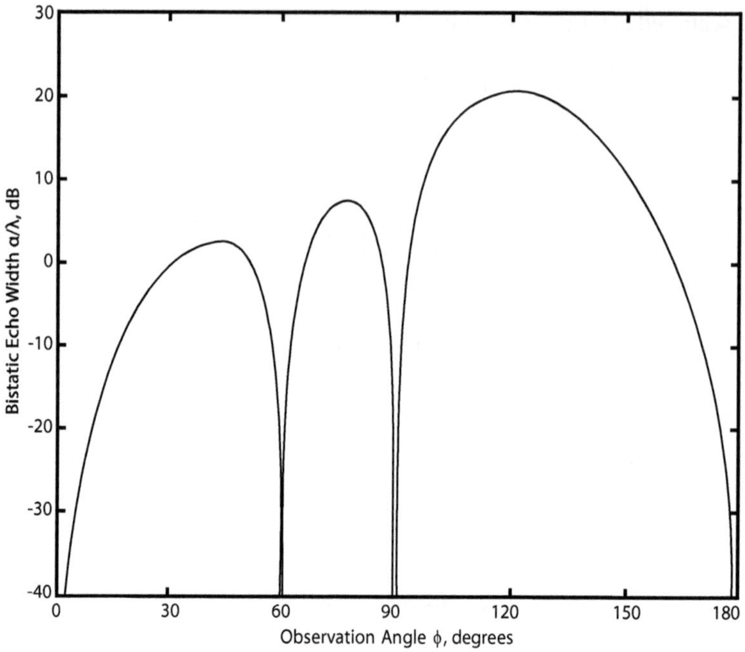

Fig. 7.4 The TM_z bistatic echo width of a 2λ wide conducting strip illuminated by at $\phi_i = 60°$

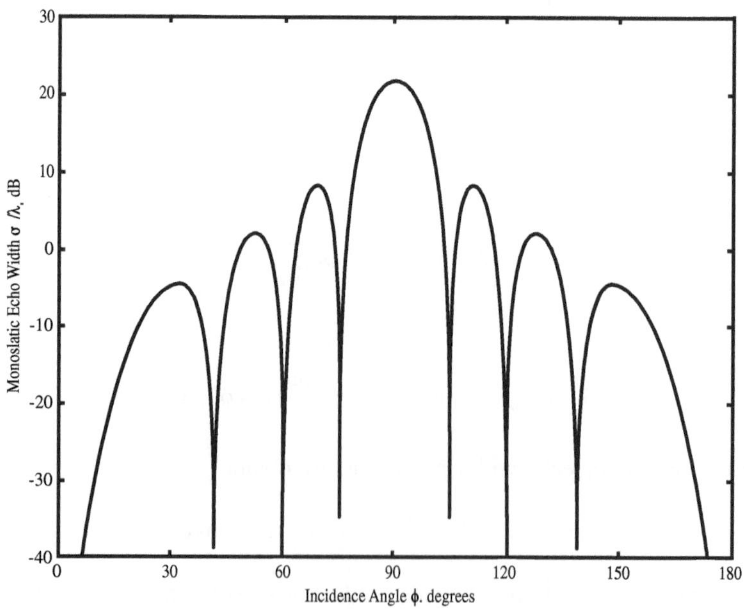

Fig. 7.5 The TM_z monostatic echo width of a 2λ wide conducting strip

Fig. 7.6 A perfectly conducting right triangular plate illuminated by a plane wave

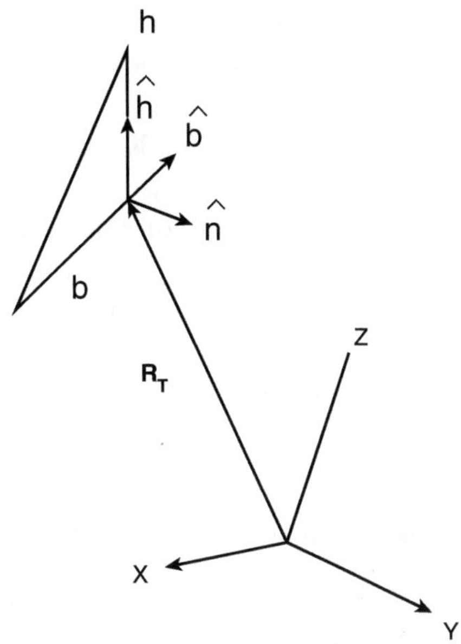

It is clear from the above example that the physical optics current does not satisfy the edge condition. This is because the underlying assumption for this approximation is that the strip is infinitely wide. For this reason, the edge on backscattering echo width is null.

7.2.1 Scattering from a Right Triangular Plate

Consider a perfectly conducting right triangular plate illuminated by a plane wave as shown in Fig. 7.6. This is one of the shapes for which the physical optics scattering can be calculated in closed form.

In order to calculate the scattered field for this plate, we define a global system of coordinates denoted by $(\widehat{x}, \widehat{y}, \widehat{z})$ and a local system of coordinates for the triangular plate denoted by $(\widehat{b}, \widehat{h}, \widehat{n})$ where $\widehat{n}$ is the unit normal to the plate.

Let the incident magnetic field be given by

$$\mathbf{H}^i = \widehat{h}^i \frac{E_0}{Z_0} e^{jk_0 \widehat{\iota} \cdot \mathbf{r}} \tag{7.19}$$

where $\widehat{h^i}$ denotes the polarization of the incident magnetic field while E_0^i and Z_0 are the amplitude of the incident electric field and the characteristic impedance of the free space, respectively. Also, $\hat{\iota}$ is the unit vector in the direction of the incident wave. The physical optics current is given by

$$\mathbf{K}_{p0} = 2\hat{n} \times \hat{h}^i \frac{E_0}{Z_0} e^{jk_0\hat{\iota}\cdot\mathbf{r}'} \tag{7.20}$$

where $\mathbf{r}'$ is located on the triangular plate. The far-zone scattered field is given by

$$\mathbf{E}^s = \mathbf{f}(\hat{s},\hat{\iota})\frac{e^{-jk_0 r}}{r} \tag{7.21}$$

where the scattering amplitude $\mathbf{f}^s$ is expressed as

$$\mathbf{f}^s = \hat{e}^s f \tag{7.22}$$

The unit vector $\hat{e}^s$ in the direction of scattered field is given by

$$\hat{e}^s = \{(\hat{n} \times \hat{h}^i) - [(\hat{n} \times \hat{h}^i)\cdot\hat{s}]\hat{s}\} \tag{7.23}$$

and

$$f(\hat{s},\hat{\iota}) = -\frac{jk_0}{2\pi} E_0^i\left(\frac{hb}{2}\right)e^{-j(\beta-\psi)} \begin{cases} \frac{1}{j\alpha}[e^{j\alpha}\mathrm{sinc}(\alpha+\beta) - \mathrm{sinc}\beta] & \\ \frac{1}{j\beta}[e^{j\beta} - \mathrm{sinc}\beta] & \alpha \to 0 \\ 1 & \alpha \to 0, \beta \to 0 \end{cases} \tag{7.24}$$

In the above α and β are given by

$$\alpha = \frac{k_0 h}{2}(\hat{s}+\hat{\iota})\cdot\hat{h}, \quad \beta = \frac{k_0 b}{2}(\hat{s}+\hat{\iota})\cdot\hat{b} \tag{7.25}$$

and ψ is the phase reference relative to the origin

$$\psi = k_0(\mathbf{r}+\mathbf{r}^i)\cdot\mathbf{R}_t \tag{7.26}$$

7.2.2 Scattering from a Convex Target

The physical optics scattering from a large perfectly conducting convex target can be calculated by discretizing the surface of the object to right triangular patches.

Let us assume that the surface has been divided to N such patches. If we denote the scattered electric field from each of these patches by $\mathbf{E}_n^s$, then total scattered field is given by superposition as

$$\mathbf{E^s} = \sum_{n=1}^{N} \mathbf{E}_n^s \tag{7.27}$$

Then, the radar cross section is given by

$$\begin{aligned}\sigma_{bi}(\widehat{s},\widehat{\iota}) &= \lim_{r\to\infty} 4\pi r^2 \frac{|\mathbf{E^s}|^2}{|\mathbf{E^i}|^2} \\ &= \lim_{r\to\infty} 4\pi r^2 \frac{|\Sigma_{n=1}^{N}\mathbf{E}_n^s|^2}{|\mathbf{E^i}|^2}\end{aligned} \tag{7.28}$$

In order to find the lit and the shadow regions of the target, we may simply use the following test (Fig. 7.7)

If $n_j \cdot \widehat{\iota} < 0$, then patch j is in the lit region, and

If $n_j \cdot \widehat{\iota} > 0$, then patch j is in the shadow region.

Example 7.3 We will apply the above method to estimate the physical optics radar cross section of a perfectly conducting cube of side $a = 1\,\text{m}$ at 10 GHz. At this frequency, the physical optics approximation is valid since $a \simeq 33\lambda$.

The cube can be divided into 12 right triangular patches. Using (7.24), (7.27), (7.28), the RCS is given in Fig. 7.8. □

If the target under consideration is not convex, the physical optics approximation can still be used to obtain an estimate of the RCS.

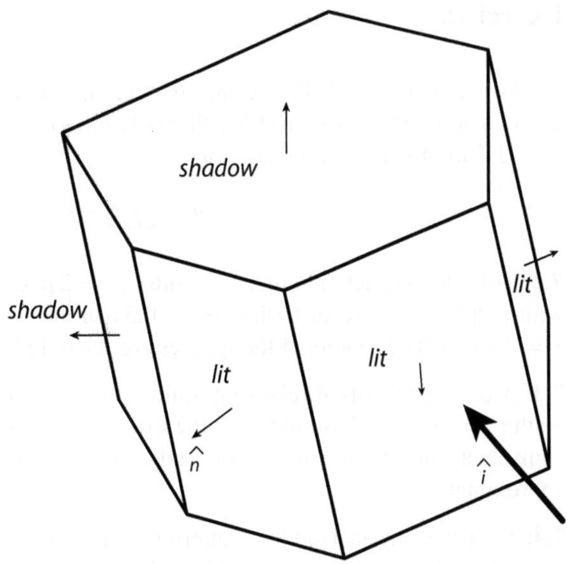

Fig. 7.7 Determination of the lit and shadow regions for a convex target with respect to the incident wave. In the lit region $\widehat{n} \times \widehat{\iota} < 0$, and in the shadow region $\widehat{n} \times \widehat{\iota} > 0$

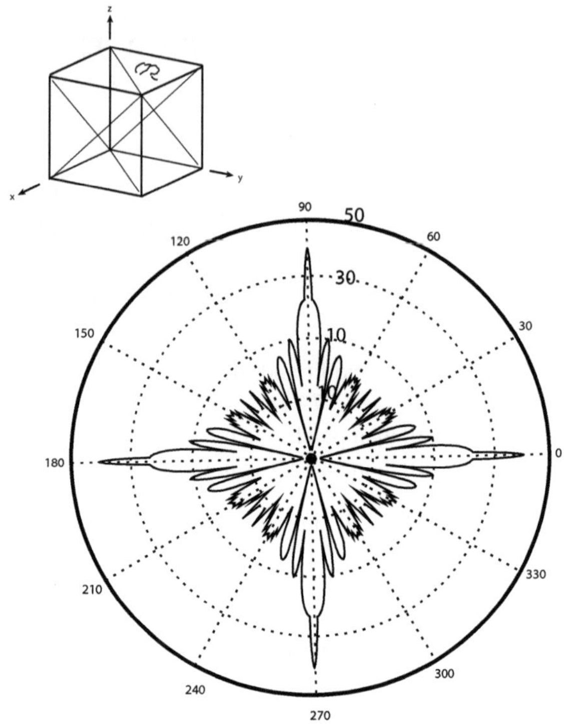

Fig. 7.8 The physical optics RCS from a perfectly conducting cube of side $a = 1\,\mathrm{m}$ at 10 GHz (azimuthal cut at $\theta = 90°$)

Exercises

7.1: Using the Rayleigh-Debye approximation, calculate the bistatic radar cross section of a homogeneous dielectric sphere of radius $a = \lambda/20$ and relative permittivity $\epsilon_r = 1.2$ illuminated by a plane wave

$$\mathbf{E}^{\mathbf{i}} = \hat{z}E_0 e^{-jkx}$$

7.2: Consider a spherical object of radius $a_1 = 2\,\mu\mathrm{m}$ and refractive index $n_1 = 1.0$ with a spherical core of radius $a_2 = 0.5\,\mu\mathrm{m}$ and refractive index $n_2 = 1.02$ at $\lambda = 0.6\,\mu\mathrm{m}$. The centers of the spheres are separated by a distance $d = 1\,\mu\mathrm{m}$.

7.3: A dielectric strip of relative permittivity ϵ_r, width w and thickness t is located on the $y = 0$ plane. Assuming that the strip has $(\epsilon_r - 1)kw \ll 1$, find the volumetric equivalent current and monostatic radar cross section using the first order Born approximation.

7.4: Find the physical optics scattering from a thin conducting strip of width w located on the xz-plane and illuminated by a TE_z plane wave. Plot the monostatic echo width as a function of incidence angle.

7.5: Using the physical optics approximation, find the backscattering radar cross section of a rectangular perfectly conducting thin plate illuminated by a plane wave propagating from the direction (θ, ϕ). Show that the monostatic cross section of the plate at normal incidences is

$$\sigma_b = 4\pi A^2$$

where A is the area of the plate.

7.6: Consider physical optics scattering from a thin circular conducting plate of radius a.
(a) Find the induced physical optics current on the plate.
(b) Find the scattered electric and magnetic fields.
(c) Calculate the bistatic and monostatic radar cross sections.
[Hint: Use the integral presentation of the Bessel function

$$J_0(x) = \frac{1}{2\pi}\int_0^{2\pi} e^{jx\sin\phi}d\phi$$

and the identity $\int_x J_0(x)dx = xJ_1(x)$.]

7.7: Using the physical optics approximation, find the backscattering cross section of a finite conducting cylinder of radius a and length ℓ, symmetrically positioned along the z-axis, and illuminated by a plane wave propagating from $(\theta = \theta_0, \phi = 0)$.

7.8: A plane wave is incident on an infinite cone of half angle θ_0 along its axis. Find the physical optics monostatic radar cross section (Fig. 7.9).

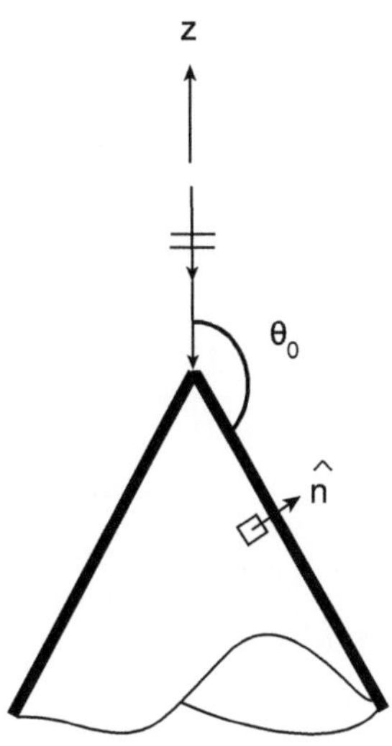

Fig. 7.9 An infinitely long perfectly conducting cone illuminated by a plane wave

Chapter 8
Integral Equation Method

The key to the solution of scattering or radiation problems is a knowledge of the physical or equivalent current distributions on the volume or the surface of the scatterer.

From the uniqueness principle, if the tangential electromagnetic fields (or equivalent currents) on a closed surface are known, all other electromagnetic quantities such as scattered fields, far field pattern, etc., are known.

8.1 Types of Integral Equations

In order to find the induced current on the scatterer, we construct an integral equation based on an appropriate boundary condition.

The types and characteristics of integral equations can be discussed in the context of applied mathematics. In this context, one would usually classify the integral equations to those of first or second type depending on the appearance of the unknown function in the equation. They can also be classified as *Fredholm* or *Volterra* equations depending on the limits of the integral operators involved. These equations are different in that their underlying differential analogs differ in their boundary conditions and have different types of eigenvalues and eigenfunctions. A detailed mathematical discussion of integral equations is beyond the scope of this work. However, we may classify the integral equations often encountered in scattering theory based on the specific boundary relations upon which they are constructed.

If the condition enforces a relation on the electric field, the integral equation is referred to as the *electric field integral equation* (*EFIE*). If the condition enforces a relation on the magnetic field, the resulting integral equation is referred to as the *magnetic field integral equation* (*MFIE*).

Also, depending on whether the equivalent current introduced is a surface current or a volumetric current, the formulation is referred to as *surface integral* or *volume*

© Springer International Publishing Switzerland 2015
K. Barkeshli, *Advanced Electromagnetics and Scattering Theory*,
DOI 10.1007/978-3-319-11547-4_8

integral formulation. The surface integral equation is in general suitable for perfectly conducting bodies or antennas while the volume integral equation method is suitable for volume scattering problems.

8.2 Perfectly Conducting Scatterers

Consider a perfectly conducting scatterer in free space characterized by (μ_0, ϵ_0) illuminated by an incident field $(\mathbf{E}^i, \mathbf{H}^i)$ as shown in Fig. 8.1. The incident field is radiated by impressed sources $(\mathbf{J}^i, \mathbf{M}^i)$ and is taken to be the field that would exist in space if the body were not present. The total external field is designated $(\mathbf{E}, \mathbf{H})$ and the internal field is identically zero.

Applying the physical equivalence priniple for scattering problems, we have

$$\mathbf{K}_e = \widehat{n} \times \mathbf{H} \tag{8.1}$$

where $\mathbf{K}_e$ is the physical equivalent current on S. The surface current $\mathbf{K}_e$, in turn, radiates the scattered field $(\mathbf{E}^s, \mathbf{H}^s)$ in unbounded space such that $\mathbf{E}^s = \mathbf{E} - \mathbf{E}^i$ and $\mathbf{H}^s = \mathbf{H} - \mathbf{H}^i$.

8.2.1 Electric Field Integral Equation (EFIE)

In general, the electric field can be expressed in terms of the Hertz vector potential as

$$\mathbf{E}^s = \nabla(\nabla \cdot \pi^s) + k_0^2 \pi^s \tag{8.2}$$

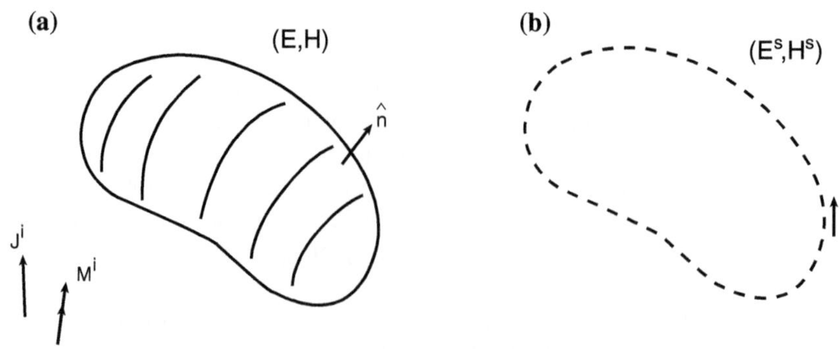

Fig. 8.1 **a** Scattering from a perfectly conducting body. **b** The physical equivalence

where

$$\pi^s(\mathbf{r}) = -j\frac{Z_0}{k_0}\oint_S \mathbf{K}_e(\mathbf{r}')G_0(\mathbf{r};\mathbf{r}')ds' \tag{8.3}$$

and G_0 is the free space Green's function. Thus,

$$\mathbf{E}^s = -jk_0Z_0\left(1+\frac{\nabla\nabla}{k_0^2}\right)\oint_S \mathbf{K}_e(\mathbf{r}')G_0(\mathbf{r};\mathbf{r}')ds' \tag{8.4}$$

In order to find the equation for $\mathbf{K}_e$, we apply the boundary condition on the tangential component of the electric field at the surface of the conductor

$$\hat{n}\times\mathbf{E} = \hat{n}\times(\mathbf{E}^i+\mathbf{E}^s) = 0, \quad \mathbf{r}\in S \tag{8.5}$$

Substituting for the scattered field, we obtain

$$\hat{n}\times\mathbf{E}^i(\mathbf{r}) = \hat{n}\times\left\{jk_0Z_0\oint_S \mathbf{K}_e(\mathbf{r}')\left(1+\frac{\nabla\nabla}{k_0^2}\right)G_0(\mathbf{r};\mathbf{r}')ds'\right\} \quad \mathbf{r}\in S \tag{8.6}$$

Which is the desired electric field integral equation for the induced electric current on the surface of a conductor. Note that S could be an open or closed surface. For open surfaces, $\mathbf{K}_e$ is the equivalent current density that represents the vector sum of the physical current densities on the opposite sides of the surface.

Once the electric current density $\mathbf{K}_e$ is found by solving (8.6), the scattered field may be calculated using the radiation integral (8.4).

8.2.2 Magnetic Field Integral Equation (MFIE)

The scattered magnetic field is given by

$$\mathbf{H}^s = j\frac{k_0}{Z_0}\nabla\times\pi^s = \nabla\times\oint_S \mathbf{K}_e(\mathbf{r}')G_0(\mathbf{r};\mathbf{r}')ds' \tag{8.7}$$

Since the Green's function is singular, the order of differentiation and integration cannot be interchanged. Using the vector identity

$$\nabla\times(\mathbf{K}_eG_0) = G_0\nabla\times(\mathbf{K}_e(\mathbf{r}')) + (\nabla G_0)\times\mathbf{K}_e \tag{8.8}$$

and noting the fact that ∇ operates on unprimed coordinates, (8.7) may be written as

$$\mathbf{H}^s = \oint_S \mathbf{K}_e(\mathbf{r}') \times \left[\nabla' G_0(\mathbf{r};\mathbf{r}')\right] ds' \tag{8.9}$$

On the surface of the scatterer, we have the boundary condition

$$\mathbf{K}_e = \widehat{n} \times \mathbf{H} = \widehat{n} \times (\mathbf{H}^i + \mathbf{H}^s) \quad \mathbf{r} \to \mathbf{r}_{s+} \tag{8.10}$$

where $\mathbf{r} \to \mathbf{r}_{s+}$ indicates that the conducting surface S is approached by r from the outside. Substituting from (8.9), the magnetic field integral equation is obtained as

$$\widehat{n} \times \mathbf{H}^i(\mathbf{r}) = \mathbf{K}_e(\mathbf{r}) - \lim_{\mathbf{r}\to S} \left\{ \widehat{n} \times \oint_S \mathbf{K}_e(\mathbf{r}') \times \left[\nabla' G_0(\mathbf{r};\mathbf{r}')\right] ds' \right\} \tag{8.11}$$

Note that because of the boundary condition used in the derivation of (8.11), the MFIE is valid only for closed surfaces and the current density $\mathbf{K}_e$ in the MFIE is the actual current density induced on the surface of the conductor.

If the surface is approached by r from the inside, we have

$$\widehat{n} \times \mathbf{H} = \widehat{n} \times (\mathbf{H}^i + \mathbf{H}^s) = 0 \quad \mathbf{r} \to \mathbf{r}_{s-} \tag{8.12}$$

since the total electromagnetic fields inside the perfectly conducting medium are zero. Combining (8.10) and (8.12) and substituting from (8.9) for $\mathbf{H}^s$, we find that

$$\mathbf{K}_e - \lim_{\mathbf{r}\to S} \widehat{n} \times \left[\oint_S \mathbf{K}_e(\mathbf{r}') \times \nabla' G_0(\mathbf{r};\mathbf{r}') ds' \right] = 2\widehat{n} \times \mathbf{H}^i \tag{8.13}$$

The limiting process appearing in MFIE must be carefully carried out since the tangential magnetic field is discontinuous at S and the surface integral should be interpreted as the *Cauchy principle value*. The first term in the MFIE (8.13) gives the physical optics approximation

$$\mathbf{K}_e \simeq 2\widehat{n} \times \mathbf{H}^i \tag{8.14}$$

Thus, the MFIE should be useful for a large object with a smooth surface whose radius of curvature is large compared with the wavelength. In this case, the physical optics is a good approximation and the second term provides a correction.

Once the current density $\mathbf{K}_e$ is found from (8.11), the scattered electromagnetic fields may be calculated from radiation integrals (8.4) and (8.9).

It should be noted that both EFIE and MFIE integral equations fail to have a unique solution at frequencies corresponding to the resonances of a cavity whose interior is bounded by the conducting surface S.

8.3 Two-Dimensional Problems

We consider applications of the above formulations to two-dimensional problems involving infinite cylindrical scatterers.

Consider an infinitely long perfectly conducting cylinder as shown in Fig. 8.2 with its axis parallel to the z-axis. We treat both principle polarizations and first employ the EFIE formulation.

In the TM_z polarization (E-polarization), the incident electric field is polarized along the cylinder axis

$$\mathbf{E}^i = \hat{z} E_0 e^{jk_0(x\cos\phi_i + y\sin\phi_i)} \tag{8.15}$$

Because of symmetry, the scattered field and the induced current are also z-directed. The Hertz potential is given by

$$\pi^s = -j\frac{Z_0}{k_0}\hat{z}\oint_S \mathbf{K}(\rho')G_0(\rho;\rho')ds' \tag{8.16}$$

where S is the conducting cylindrical surface and G_0 is the free space Green's function. Since π^s is z-directed and independent of z, the first term on the right hand side of (8.2) is zero. Thus, the scattered electric field is given by

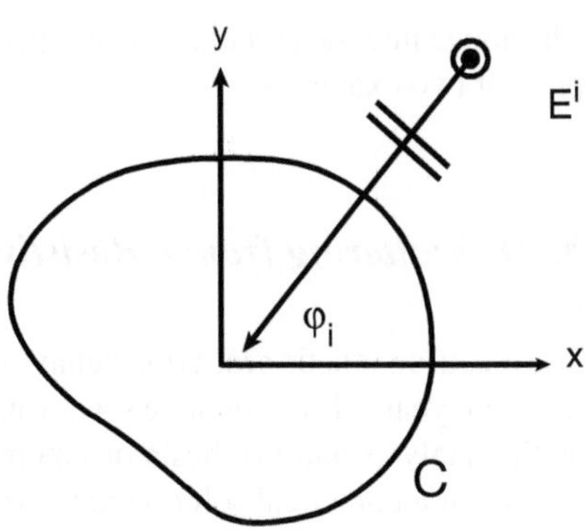

Fig. 8.2 A two-dimensional perfectly conducting cylinder illuminated by an E-polarized plane wave

$$\mathbf{E}^s = -jk_0 Z_0 \hat{z} \oint_S K_z(\mathbf{r}') G(\mathbf{r}; \mathbf{r}') ds' \tag{8.17}$$

Because the cylinder is infinite in the z direction, we may write the last equation as

$$E_z^s = -jk_0 Z_0 \oint_C K_z(\rho') \int_{-\infty}^{\infty} \frac{e^{-jk|\mathbf{r}-\mathbf{r}'|}}{4\pi|\mathbf{r}-\mathbf{r}'|} dz' d\ell' \tag{8.18}$$

We recognize the infinite integral on z' as

$$\int_{-\infty}^{\infty} \frac{e^{-jk\sqrt{(\rho-\rho')^2+z'^2}}}{4\pi\sqrt{(\rho-\rho')^2+z'^2}} dz' = \frac{1}{4j} H_0^{(2)}(k|\rho-\rho'|) \tag{8.19}$$

as the two-dimensional free space Green's function, and thus

$$E_z^s(\rho) = -\frac{k_0 Z_0}{4} \oint_C K_z(\rho') H_0^{(2)}(k|\rho-\rho'|) d\ell' \tag{8.20}$$

Imposing the boundary condition

$$E_z^i + E_z^s = 0 \quad \text{on } S \tag{8.21}$$

we obtain the desired integral equation for the induced surface current as

$$\frac{k_0 Z_0}{4} \oint_C K_z(\rho') H_0^{(2)}(k|\rho-\rho'|) d\ell' = E_z^i(\rho), \quad \rho \in C \tag{8.22}$$

The above integral equation can be specialized to a particular geometry. We will now present two examples.

8.3.1 Scattering from a Resistive Strip

The electromagnetic scattering behavior of thin strips has been studied in some detail in recent years. These include scattering from conducting and resistive strips as well as the analysis and synthesis of tapered strips. These studies have focused on flat strips, but a numerical solution method for thin dielectric slabs of uniform thickness and arbitrary shape was also given as early as 1965.

The problem of scattering from thin resistive strips serves as a useful and relatively simple example in scattering theory for illustrating fundamental concepts and practical methods which can be extended to more complex problems. As discussed in Sect. 5.1, the resistive sheet model can be used to simulate dielectric layers whose thickness τ is small compared to the wavelength.

In this section, we present a general analysis of scattering from thin resistive strips of constant curvature using the integral equation formulation. The integral equations are specialized to the perfectly conducting strip and are solved analytically for two limiting cases. For the electrically wide strips, the well known physical optics solutions are obtained through an exercise in Fourier transform theory. For narrow conducting strips, on the other hand, a quasi-static solution is derived based on the low frequency approximation of the equations. The analysis is then extended to the circular cylindrical resistive strip problem.

8.3.1.1 Flat Strips

Using the resistive sheet boundary condition (3.60), integral equations may be derived for computing the current induced on thin strips for a given excitation (see Fig. 3.9).

Consider the E-polarized plane wave

$$\mathbf{E}^i = \widehat{z} e^{jk_o(x\cos\phi_o + y\sin\phi_o)} \tag{8.23}$$

incident on a thin flat dielectric layer of thickness τ, width w and relative permittivity ϵ_r coincident with the x-axis. The resistive sheet boundary condition (3.60)

$$\mathbf{E} - (\widehat{n}\cdot\mathbf{E})\,\widehat{n} = Z_s\mathbf{K} \tag{8.24}$$

states that the dielectric layer may be replaced by a resistive strip of surface resisitivity

$$Z_s = \frac{Z_o}{jk_o\tau(\epsilon_r - 1)}$$

This excitation generates a z-directed current K_z on the strip, giving rise to the scattered field

$$E_z^s(\rho) = -jk_oZ_o\int_{-w/2}^{w/2} K_z(x')G_s(\rho; x')dx' \tag{8.25}$$

where G_s is the two dimensional free space Green's function given by (8.19).

Imposing the condition (8.24) on the total tangential electric field over the strip, an integral equation for K_z is obtained as

$$\begin{aligned} Y_0 E_0 e^{jk_0 x \cos\phi_0} &= \eta_s K_z(x) \\ &+ \frac{k_o}{4} \int_{-w/2}^{w/2} K_z(x') H_0^{(2)}\left(k_o|x - x'|\right) dx' \end{aligned} \tag{8.26}$$

where $\eta_s = Z_s/Z_0$ is the normalized surface resitivity.

Consider now the H-polarized plane wave

$$\mathbf{H}^i = \hat{z}\, \mathbf{H}_0 e^{jk_0(x\cos\phi_0 + y\sin\phi_0)} \tag{8.27}$$

incident on the resistive strip. This excitation generates an x-directed current K_x responsible for the scattered field given by

$$E^s(\rho) = -jk_0 Z_0 \left(1 + \frac{1}{k_0^2}\frac{\partial^2}{\partial x^2}\right) \int_{-w/2}^{w/2} K_x(x') G_s(\rho; x'\hat{x}) dx' \tag{8.28}$$

Again, by imposing the resistive boundary condition (8.24), the integral equation satisfied by the current density K_x is obtained as

$$\begin{aligned} \sin\phi_0 H_0 e^{jk_0 x\cos\phi_0} &= \eta_s K_x(x) \\ &+ \frac{k_0}{4}\left(1 + \frac{1}{k_0^2}\frac{\partial^2}{\partial x^2}\right) \int_{-w/2}^{w/2} K_x(x') H_0^{(2)}(k_0|x - x'|) dx' \end{aligned} \tag{8.29}$$

Equations (8.26) and (8.29) are integral equations to be solved for the unknown current densities K_z and K_x for E- and H-polarizations, respectively.

Once the excited current is found, the far zone scattered fields at the cylindrical point (ρ, ϕ) can be computed from the scattering integrals (8.25) and (8.28). In particular, using the large argument approximation of the Hankel function

$$H_0^{(2)}(k_0\rho) \sim \sqrt{\frac{2j}{\pi k_0 \rho}} e^{-jk_0\rho}, \quad k_0\rho \to \infty$$

and invoking the approximations

$$\left|\rho - \rho'\right| \simeq \rho - x'\cos\phi \tag{8.30}$$

$$\approx \rho \tag{8.31}$$

for the phase and amplitude, respectively, the far-zone scattered fields are found to be

$$E_z^s = -\frac{k_0 Z_0}{4}\sqrt{\frac{2j}{\pi k_0 \rho}} e^{-jk_0\rho} \int_{-w/2}^{w/2} K_z(x') e^{jk_0 x' \cos\phi} dx' \tag{8.32}$$

and

$$E_\phi^s = \frac{k_0 Z_0}{4} \sin\phi \sqrt{\frac{2j}{\pi k_0 \rho}} e^{-jk_o\rho} \int_{-w/2}^{w/2} K_x(x') e^{jk_0 x' \cos\phi} dx' \tag{8.33}$$

for E- and H-polarizations, respectively. The two-dimensional scattering echo widths are given by

$$\sigma_e = \frac{k_0}{4}\left| Z_0 \int_{-w/2}^{w/2} K_z(x') e^{jk_0 x' \cos\phi} dx' \right|^2 \tag{8.34}$$

and

$$\sigma_h = \frac{k_0}{4}\left| \sin\phi \int_{-w/2}^{w/2} K_x(x') e^{jk_0 x' \cos\phi} dx' \right|^2 \tag{8.35}$$

Typically, a solution of (8.26) and (8.29) can be accomplished numerically. However, approximate analytical solutions exist if the strip is electrically very wide or very narrow. These solutions are based on the physical optics and quasi-static approximations of the pertinent integral equations, respectively. They may be used to find closed form expressions for the echo width of the strip.

8.3.1.2 Very Wide Strips

For electrically wide conducting strips, the local electric current may be assumed to be that corresponding to an infinitely wide strip (i.e., $w \to \infty$). This is known as the physical optics approximation and is expressed as (for $\eta_s = 0$)

$$\mathbf{K} = 2\hat{n} \times \mathbf{H}^i \tag{8.36}$$

or more explicitly,

$$K_z(x) = 2Y_0 \sin\phi_0 e^{jk_0 x \cos\phi_0} \tag{8.37}$$

for TM incidence and

$$K_x(x) = 2e^{jk_0 x \cos\phi_0} \tag{8.38}$$

for TE incidence.

The above physical optics approximation is based on the fact that the only scattered fields from the infinitely wide strip are the reflected and transmitted ones. Consequently, the total magnetic field intensity is

$$\mathbf{H} = \mathbf{H}^+ = \mathbf{H}^i + \mathbf{H}^r \quad (y > 0), \quad \mathbf{H} = \mathbf{H}^- = \mathbf{H}^t \quad (y < 0) \tag{8.39}$$

with the dual also true for the electric field intensity. For the perfectly conducting strip $\mathbf{H}^t = 0$, and based on the usual boundary conditions, $\mathbf{K} = \widehat{n} \times (\mathbf{H}^+ - \mathbf{H}^-) = \widehat{n} \times (\mathbf{H}^i + \mathbf{H}^r) = 2\widehat{n} \times \mathbf{H}^i$.

However, the physical optics approximations may also be derived directly from the governing integral equations (8.26) and (8.29) when the strip is assumed to be infinitely wide. Under this assumption, we have

$$Y_0 e^{jk_0 x \cos\phi_0} = \eta_s K_z(x) + jk_0 \lim_{k_0 w \to \infty} \int_{-w/2}^{w/2} K_z(x') G_s(x; x') dx' \tag{8.40}$$

for TM incidence and

$$\begin{aligned} \sin\phi_0 e^{jk_0 x \cos\phi_0} &= \eta_s K_x(x) \\ &+ \frac{j}{k_0} \lim_{k_0 w \to \infty} \int_{-w/2}^{w/2} K_x(x') \left[\left(k_0^2 + \frac{\partial^2}{\partial x^2} \right) \right] G_s(x; x') dx' \end{aligned} \tag{8.41}$$

for TE incidence. The integrals on the right hand sides of (8.40) and (8.41) are identified as convolutions in the infinite domain since $G_s(x; x') = \frac{1}{4j} H_0^{(2)} (k_0 |x - x'|)$. Thus, upon invoking the transform transforms of both sides of (8.40) and (8.41), invoking the convolution theorem and using the transform pair

$$\mathcal{F}\{g(x)\} = \tilde{g}(k_z) = \int_{-\infty}^{\infty} g(x) e^{-jk_x x} dx \tag{8.42}$$

$$g(x) = \mathcal{F}^- 1\{\tilde{g}(k_x)\} = \frac{1}{2\pi} \int_{-\infty}^{\infty} \tilde{g}(k_x) e^{jk_x x} dk_x \tag{8.43}$$

we have

$$\widetilde{K}_z(k_x) = \frac{2\pi Y_o \delta(k_x - k_0 \cos\phi_o)}{\eta_s + jk_o \widetilde{G}_s(k_x)} \tag{8.44}$$

and

$$\widetilde{K}_x(k_x) = \frac{2\pi \sin\phi_o \delta(k_x - k_0 \cos\phi_o)}{\eta_s + \frac{j}{k_o}(k_o^2 - k_x^2)\widetilde{G}_s(k_x)} \tag{8.45}$$

where δ is the Dirac delta function.

By Fourier transforming the integral equations (8.40) and (8.41), we obtained algebraic equations to solve for the currents in the spectral domain. To obtain the current expressions in the spatial domain, we simply inverse transform (8.44) and (8.45) to get

$$K_z(x) = \frac{Y_0 e^{jk_0 x \cos\phi_0}}{\eta_s + jk_0 \widetilde{G}_s(k_0 \cos\phi_0)} \tag{8.46}$$

$$K_x(x) = \frac{\sin\phi_0 e^{jk_0 x \cos\phi_0}}{\eta_s + jk_0 \sin^2\phi_0 \widetilde{G}_s(k_0 \cos\phi_0)} \tag{8.47}$$

where use was made of the properties of the δ function. The Fourier transform of the Green's function $\widetilde{G}_s$ is given by

$$\widetilde{G}_s(k_x) = \frac{1}{2j\sqrt{k_o^2 - k_x^2}} \tag{8.48}$$

and when this is substituted in (8.46) and (8.47), we finally get

$$K_z(x) = \frac{2Y_0 \sin\phi_0 e^{jk_0 x \cos\phi_0}}{1 + 2\eta_s \sin\phi_0} \tag{8.49}$$

$$K_x(x) = \frac{2e^{jk_0 x \cos\phi_0}}{1 + 2\eta_s / \sin\phi_0} \tag{8.50}$$

which are explicit forms for the current density.

It is easy to see that the above results can, in general, be written as

$$\mathbf{K} = \frac{2\widehat{n} \times \mathbf{H}^i}{1 + 2\eta_s \sin\phi_0} \tag{8.51}$$

for E-polarization, and

$$\mathbf{K} = \frac{2\widehat{n} \times \mathbf{H}^i}{1 + 2\eta_s / \sin\phi_0} \tag{8.52}$$

for H-polarization. For perfectly conducting strips, $\eta_s = 0$, and we recover (8.37) and (8.38).

Calculation of the (physical optics) echo widths follows directly from (8.34) and (8.35). We find that

$$\sigma_{TM} = k_0 w^2 \left(\frac{\sin\phi_0}{1 + 2\eta_s \sin\phi_0} \right)^2 \text{sinc}^2 \left[(k_0 \frac{w}{2} (\cos\phi_0 + \cos\phi) \right] \tag{8.53}$$

and

$$\sigma_{TE} = k_0 w^2 \left(\frac{\sin\phi}{1 + 2\eta_s / \sin\phi_0} \right)^2 \text{sinc}^2 \left[k_0 \frac{w}{2} (\cos\phi_0 + \cos\phi) \right] \tag{8.54}$$

where $\text{sinc}(x) = \sin(x)/x$.

8.3.1.3 Very Narrow Strips

A general analysis of narrow conducting strips can be carried out analytically by employing a low frequency approximation to the integral equations developed in Sect. 3.1.

For a perfectly conducting strip $\eta_s = 0$ and the integral equations for the surface current densities are given by

$$Y_0 e^{jk_0 x \cos\phi_0} = \frac{k_0}{4} \int_{-w/2}^{w/2} K_z(x') H_0^{(2)}(k_0 |x - x'|) dx' \tag{8.55}$$

for TM incidence and

$$\sin\phi_0 e^{jk_0 x \cos\phi_0} = \frac{k_o}{4} \left(1 + \frac{1}{k_o^2} \frac{\partial^2}{\partial x^2} \right) \int_{-w/2}^{w/2} K_x(x') H_o^{(2)}(k_o |x - x'|) dx' \tag{8.56}$$

for TE incidence. These can be solved analytically for kw very small. Specifically, when $k_o w \ll 1$, we may introduce the small argument expansion for the Hankel function,

$$H_o^{(2)}(z) \simeq 1 - j \frac{2}{\pi} \ln \left(\frac{\gamma z}{2} \right) + \mathcal{O}(z^2, z^2 \ln z) \tag{8.57}$$

where $\gamma = 1.78108\ldots$ is the Euler's constant. Then, on retaining only terms to $\mathcal{O}(k_o w)$ in the Hankel function as well as the incident fields, we have

$$\int_{-w/2}^{w/2} K_z(x') \ln|x - x'|dx' = \frac{2\pi j}{k_0 Z_0} - \left[\ln\left(\frac{k_0\gamma}{2}\right) + j\frac{\pi}{2}\right] \int_{-w/2}^{w/2} K_z(x')dx' \quad (8.58)$$

for TM incidence and

$$\frac{\partial^2}{\partial x^2} \int_{-w/2}^{w/2} K_x(x') \ln|x - x'|dx' = 2\pi j k_0 \sin\phi_0 \quad (8.59)$$

for TE incidence. Further, by introducing the change of variables

$$\xi = \frac{x}{w/2}, \quad \xi' = \frac{x'}{w/2} \quad (8.60)$$

Equations (8.58) and (8.59), respectively, become

$$\int_{-1}^{1} K_z(\xi') \ln|\xi - \xi'|d\xi' = \frac{4j\pi}{k_0 w Z_0} - \left[\ln\left(\frac{k_0 w\gamma}{4}\right) + \frac{j\pi}{2}\right] \int_{-1}^{1} K_z(\xi')d\xi' \quad (8.61)$$

and

$$\frac{d^2}{d\xi^2} \int_{-1}^{1} K_x(\xi') \ln|\xi - \xi'|d\xi' = j\pi k_0 w \sin\phi_0 \quad (8.62)$$

To solve (8.61) and (8.62) we recall the following identities from Hilbert transform theory

$$\int_{-1}^{1} \frac{\ln|\xi - \xi'|}{\sqrt{1 - \xi'^2}} d\xi' = -\pi \ln 2, \quad \xi \in [-1, 1] \quad (8.63)$$

and

$$\frac{d^2}{d\xi^2} \int_{-1}^{1} \sqrt{1 - \xi'^2} \ln|\xi - \xi'|d\xi' = \pi, \quad \xi \in [-1, 1] \quad (8.64)$$

Since the right hand sides of (8.61) of (8.62) are independent of ξ, by comparison with (8.63) and (8.64), we deduce that

$$K_z(x) = \frac{\chi_e}{Z_0\sqrt{1-\left(\frac{x}{w/2}\right)^2}} \tag{8.65}$$

and

$$K_x(x) = \chi_h\sqrt{1-\left(\frac{x}{w/2}\right)^2} \tag{8.66}$$

where χ_e and χ_h are constants to be determined. By substituting (8.65)–(8.66) into (8.61)–(8.62) we obtain

$$\chi_e = \frac{8}{\pi k_0 w\left[1 - j\frac{2}{\pi}\ln\left(\frac{k_0 w\gamma}{8}\right)\right]} \tag{8.67}$$

and

$$\chi_h = jk_0 w\sin\phi_0 \tag{8.68}$$

As expected (8.65) and (8.66) display the familiar parabolic edge behaviors at the terminations of the conducting as shown in Fig. 8.3.

The scattering echo widths are computed from (8.34) to (8.35). However, in this case, we may use the approximation (8.31) for both amplitude and phase due to the small width of the strip. Thus,

$$\sigma_e = \frac{k_0}{4}\left|Z_0\int_{-w/2}^{w/2} K_z(x')dx'\right|^2 \tag{8.69}$$

$$\sigma_h = \frac{k_0}{4}\left|\sin\phi\int_{-w/2}^{w/2} K_x(x')dx'\right|^2 \tag{8.70}$$

and upon substituting for the current densities, we obtain the simple expressions

$$\sigma_e = k_0\left|\frac{\pi w}{4}\chi_e\right|^2 \tag{8.71}$$

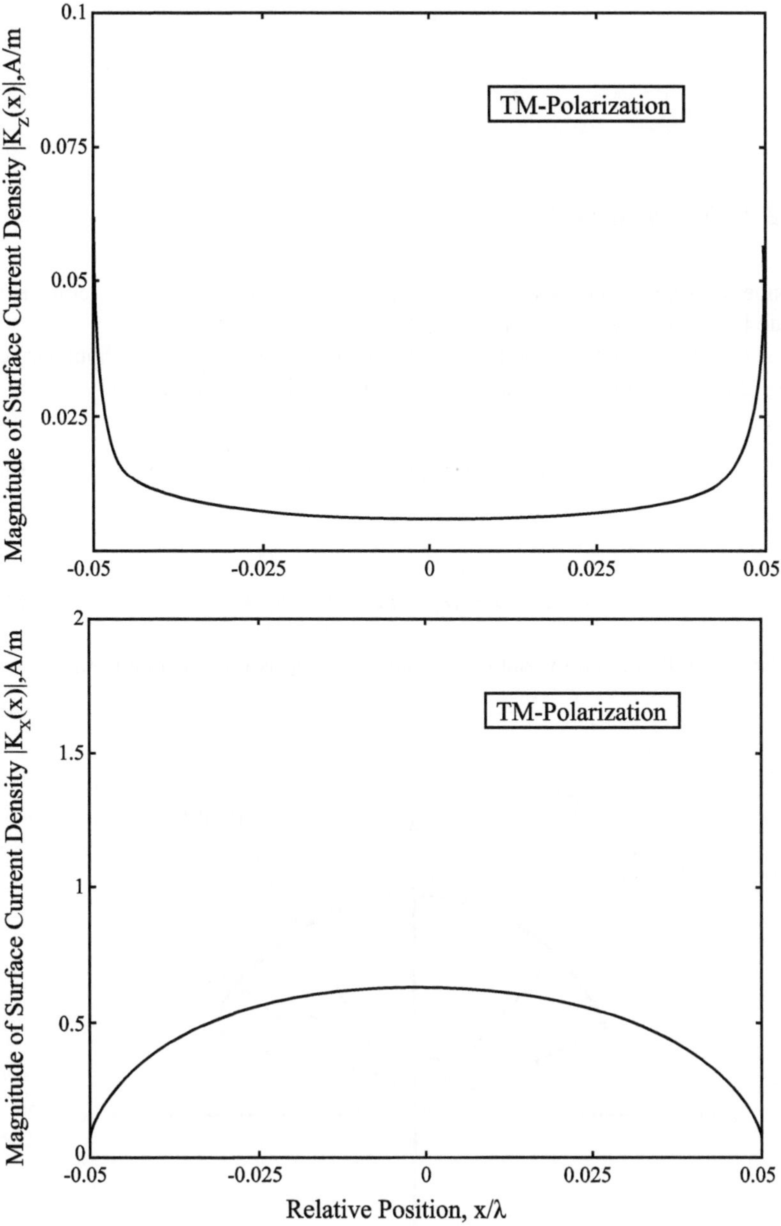

Fig. 8.3 Surface current densities on a 0.1λ thin conducting strip illuminated by normally incident plane waves for TM and TE polarizations

and

$$\sigma_h = k_0 \left| \frac{\pi w}{8} \chi_h \sin\phi \right|^2. \tag{8.72}$$

8.3.2 Cylindrical Strips

Consider a thin circular cylindrical shell of surface resistivity Z_s and radius a illuminated by a plane wave as depicted in Fig. 8.4.

The total tangential electric field on the strip satisfies the resistive boundary condition (3.60). Defining the phase reference at the origin, we may write

$$dl = a d\phi, \quad |\boldsymbol{\rho} - \boldsymbol{\rho}'| = 2a \sin(|\frac{\phi - \phi'}{2}|) \tag{8.73}$$

and

$$\widehat{k_i} \cdot \boldsymbol{\rho} = \boldsymbol{\rho}_0 - a\cos(\phi - \phi_0) \tag{8.74}$$

Also, since there is no variation in ρ and the strip is infinitesimally thin, we find that

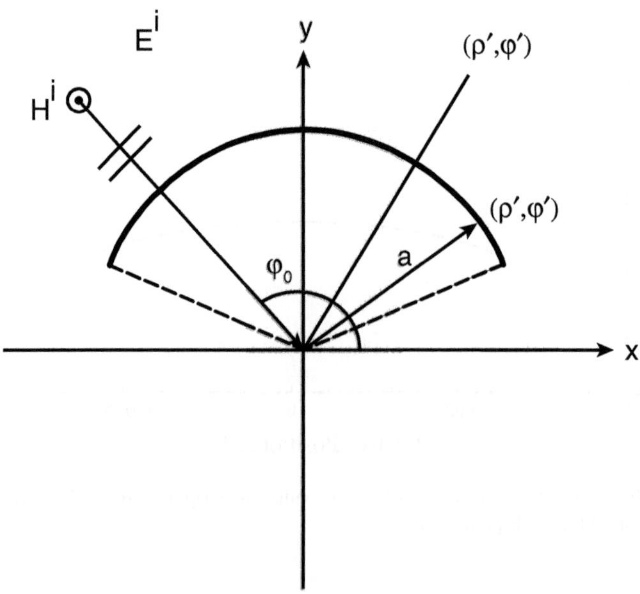

Fig. 8.4 A resistive cylindrical strip of constant curvature

$$E_z^i(\phi) = Z_s(\phi)K_z(\phi) + \frac{k_0 a Z_0}{4}\int_c K_z(\phi')H_0^{(2)}\left[2k_0 a \sin(\frac{|\phi-\phi'|}{2})\right]d\phi' \quad (8.75)$$

for E-polarization and

$$E_\phi^i(\phi) = Z_s(\phi)K_\phi(\phi) + \frac{k_0 a Z_0}{4}\left[1+\frac{1}{(k_0 a)^2}\frac{\partial^2}{\partial\phi^2}\right] \cdot \int_c K_\phi(\phi')H_0^{(2)}\left[2k_0 a \sin(\frac{|\phi-\phi'|}{2})\right]d\phi' \quad (8.76)$$

for H-polarization.

It is noted that if the radius of the strip is large compared to its width, we may modify the approximations (8.73) and (8.74) as

$$\begin{aligned}\lim_{a\to\infty}|\rho-\rho'| &= a|\phi-\phi'| \simeq |x-x'| \\ \lim_{a\to\infty} a\cos(\phi-\phi_0) &\simeq a\sin\phi_0 + a(\pi/2-\phi)\cos\phi_0 \\ &= a\sin\phi_0 + x\cos\phi_0\end{aligned} \quad (8.77)$$

and the formulation reduces to that of scattering from a flat strip (see (8.28) and (8.29)).

8.3.3 Cylindrical Reflector Antennas

Consider the circular cylindrical reflector shown in Fig. 8.5 illuminated by the line source

$$E_z = -I_e\frac{k_0 Z_0}{4}H_0^{(2)}(k_0\rho) \quad (8.78)$$

The total electric field $\mathbf{E}^T$ is evaluated in the far zone ($k_o\rho \gg 1$) as

$$E_z^T = -\frac{k_0 Z_0}{4}\sqrt{\frac{2j}{\pi k_0}} \cdot \left[I_e + a\int_c K_z(\phi')e^{jk_0 a\cos(\phi-\phi')}d\phi'\right]\frac{e^{jk_0\rho}}{\sqrt{\rho}} \quad (8.79)$$

with the normalized radiation pattern of the reflector antenna given by

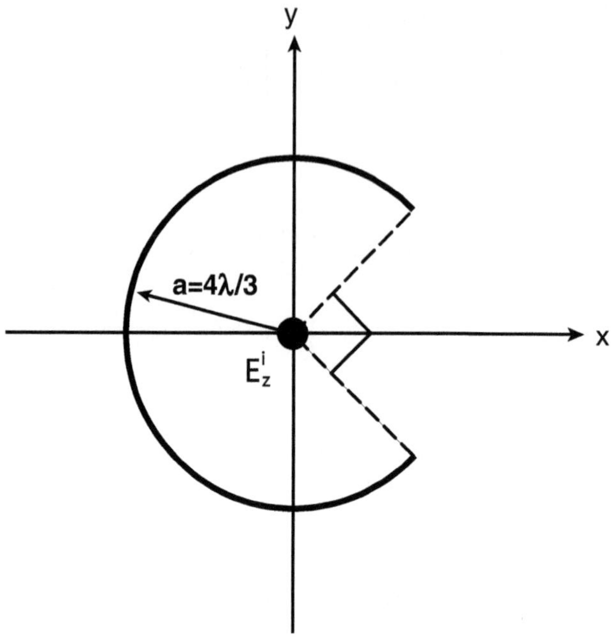

Fig. 8.5 A cylindrical reflector antenna excited by an infinite line source

$$F(\phi) = \left| 1 + \frac{a}{I_e} \int_C K_z(\phi') e^{jk_o a \cos(\phi - \phi')} d\phi' \right|^2 \tag{8.80}$$

For two-dimensional problems, the MFIE is most popular for TE polarizations.

8.4 The Linear Wire Antenna

Consider a cylindrical dipole antenna of arbitrary cross section in free space excited by an impressed field. Applying the physical surface equivalence principle, the current distribution over the antenna is responsible for the radiated fields. Denoting this by $\mathbf{E}^s$ (scattered field), we have

$$\mathbf{E}^s = \nabla(\nabla \cdot \pi^s) + k_0^2 \pi^s \tag{8.81}$$

where π^s is the electric Hertz potential

$$\pi^s(\mathbf{r}) = -j\frac{Z_0}{k_0} \int_V \mathbf{J}(\mathbf{r}') \frac{e^{-jk_0 R}}{R} dv' \tag{8.82}$$

and $R = |\mathbf{r} - \mathbf{r}'| = \sqrt{(x-x')^2 + (y-y')^2 + (z-z')^2}$. Substituting for π^s, in (8.81), we have

$$\mathbf{E}^s(\mathbf{r}) = -j\frac{Z_0}{k_0}(\nabla\nabla\cdot + k_0^2)\int_V \mathbf{J}(\mathbf{r}')\frac{e^{-jk_0R}}{4\pi R}dv' \tag{8.83}$$

We will now make the following assumptions in order to make the analysis simpler. If the wire antenna is of length ℓ and its largest cross sectional dimension is u, the following assumptions are in fact compatible with actual situations:

1. Thin dipole assumption: $k_0 u \ll 1$
2. Slender dipole assumption: $u \ll \ell$
3. Perfectly conducting or tubular arms.

In summary, the above assumptions lead to the consideration of a *lineal* z-directed *surface current* flowing on the antenna. Thus,

$$\begin{aligned}\mathbf{E}^s(\mathbf{r}) &= -j\frac{Z_0}{k_0}(\nabla\nabla\cdot + k_0^2)\int_S K_z(\mathbf{r}')\frac{e^{-jk_0R}}{4\pi R}dv' \\ &= -j\frac{Z_0}{k_0}(\nabla\nabla\cdot + k_0^2)\int_{-l/2}^{l/2}\oint_C K_z(s', z')\frac{e^{-jk_0R}}{4\pi R}ds'dz'\end{aligned} \tag{8.84}$$

where we assumed that the ϕ-directed component of the surface current is negligible in comparison with the z-directed component due to the slender dipole assumption. If we further assume that the cross sectional shape of the antenna is circular (Fig. 8.6a), then

$$K_z(s', z') = K_z(z') \tag{8.85}$$

and

$$\mathbf{E}^s(\mathbf{r}) = -j\frac{Z_0}{k_0}(\nabla\nabla\cdot + k_0^2)\int_{-l/2}^{l/2}\int_0^{2\pi} K_z(z')\frac{e^{-jk_0R}}{4\pi R}ad\phi'dz' \tag{8.86}$$

where a is the radius of the circular cross section and R is the distance between the source point (a, ϕ', z') and the observation point (ρ, ϕ, z). In order to construct an integral equation, we impose the boundary condition on the tangential component of the total field over the surface of the antenna

$$E_z^i + E_z^s = 0, \quad \rho = a \tag{8.87}$$

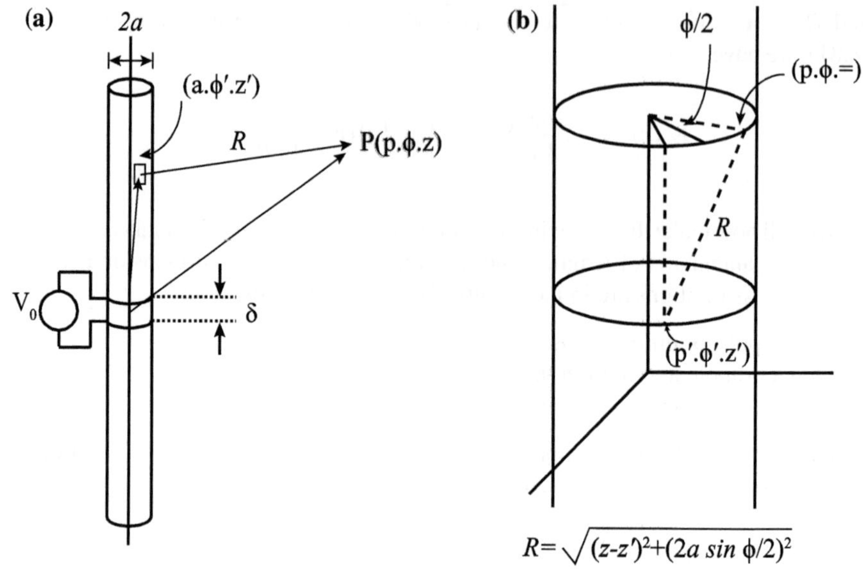

Fig. 8.6 **a** A linear dipole antenna of length ℓ and radius a with $a \ll \ell$. **b** Calculation of the distance R between the two points on the surface of the antenna for the extended kernel

where $\mathbf{E}^i$ is the impressed field assumed to be azimuthally symmetric. Also, introducing the *total* current flowing in the z-direction as

$$I(z) = 2\pi a K_z(z) \tag{8.88}$$

we have

$$E_z^i(a, \phi, z) = j\frac{Z_0}{k_0}(\nabla\nabla \cdot +k_0^2) \int_{-\ell/2}^{\ell/2} I(z')\frac{1}{2\pi} \int_0^{2\pi} \frac{e^{-jk_0R}}{4\pi R}\phi' dz' \tag{8.89}$$

where

$$R = \sqrt{(z - z')^2 + \left[2a \sin\frac{(\phi - \phi')}{2}\right]^2} \tag{8.90}$$

as indicated in Fig. 8.6b.

Since the integration over ϕ' is independent from the reference point for ϕ, we set $\phi = 0$ and obtain

$$E_z^i(z) = j\frac{Z_0}{k_0}(k_0^2 + \frac{\partial^2}{\partial z^2}) \int_{-\ell/2}^{\ell/2} I(z')G_e(z - z')dz' \tag{8.91}$$

where

$$G_e(z - z') = \frac{1}{2\pi}\int_0^{2\pi} \frac{e^{-jk_0R}}{4\pi R} d\phi' \tag{8.92}$$

is the *extended kernel* for the wire antenna. Also note that the second order operator $\nabla\nabla\cdot$ has reduced to a differentiation with respect to z, because the integrand is a function of z only.

A further simplification results if we impose the thin dipole assumption $k_0 a \ll 1$. Thus,

$$R \simeq \sqrt{(z - z')^2 + a^2} \tag{8.93}$$

and the integral equation becomes

$$E_z^i(z) = j\frac{Z_0}{k_0}(k_0^2 + \frac{\partial^2}{\partial z^2}) \int_{-\ell/2}^{\ell/2} I(z')G_r(z - z')dz' \tag{8.94}$$

where

$$G_r(z - z') = \frac{e^{-jk_0R}}{4\pi R} \tag{8.95}$$

is sometimes referred to as the *reduced* kernel. The reduced kernel has the advantage that it is nonsingular. The integram equation (8.94) is known as the Pocklington's integral equation for the wire antennas.

8.4.1 Source Modeling

There are two different ways in modeling source of excitation for linear wire antennas. If the antenna is driven as shown in Fig. 8.6a, the incident electric field is simply given by

$$\mathbf{E}^i = \begin{cases} V_0/\delta & |z| \le \delta/2 \\ 0 & \text{else} \end{cases} \tag{8.96}$$

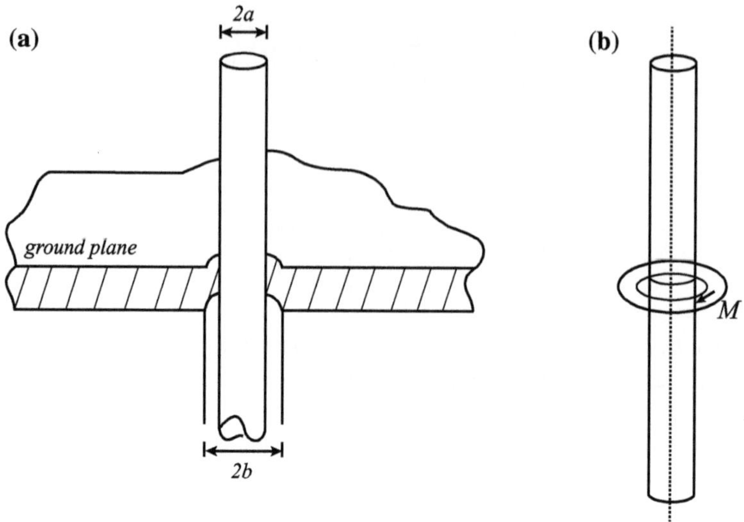

Fig. 8.7 **a** Excitation of a wire by a coaxial cable of inner and outer radii a and b operating in its dominant TEM mode. **b** The magnetic field generator $\mathbf{M} = 2\mathbf{E} \times \widehat{n}$ with the ground plane removed

This is known as the δ-gap model and is valid if the wire antenna is electrically narrow ($k_0 a \ll 1$) and $\delta/\lambda \ll 1$.

However, another way of exciting a linear antenna is to extend the inner conductor of a coaxial cable of inner and outer radii a and b as shown in Fig. 8.7. In this case, the monopole is radiating in the presence of a ground plane connected to the outer cable. Assuming that the dominant TEM wave is present in at the feed point, the electric field is given by

$$\mathbf{E} = \widehat{\rho}\frac{1}{2\rho \ln(b/a)} \tag{8.97}$$

This is a tangential electric field. Using the equivalent magnetic current

$$\mathbf{M} = \mathbf{E} \times \widehat{n} = -\widehat{\phi} M_\phi \tag{8.98}$$

we may assume that this current is responsible for the excitation of the monopole. Using the image theory, we may now remove the ground plane and assume that the magnetic current

$$\mathbf{M} = 2\mathbf{E} \times \widehat{n} = -\frac{1}{2\rho \ln(b/a)}\widehat{\phi} \tag{8.99}$$

is radiating in free space. This is known as *the magnetic frill model* for the source excitation.

Although the general expression for the radiated electromagnetic fields by this magnetic current is fairly complicated, the z-component of the electric field on the axis of the inner cylinder is given by

$$E_z^i = \frac{1}{2\ln(b/a)}\left(\frac{e^{-jk_0R_1}}{R_1} - \frac{e^{-jk_0R_2}}{R_2}\right) \tag{8.100}$$

where $R_1 = \sqrt{z^2+a^2}$ and $R_2 = \sqrt{z^2+b^2}$. This expression is useful for narrow antennas where the reduced kernel can be used.

8.4.2 Input Impedance

For the δ-gap model, the input impedance is simply given by

$$Z_{in} = \frac{V_0}{I(0)} \tag{8.101}$$

This expression is highly sensitive to the accuracy of the calculated input current $I(0)$.

For the magnetic frill model, we consider the total power radiated by the linear antenna radiated from a surface S exactly covering the antenna. The power should be equal to the input power if the wire antenna is assumed lossless. Thus

$$\oint_S \frac{1}{2}(\mathbf{E}^s \times \mathbf{H}^{s*})\cdot \mathbf{ds} = \frac{1}{2}Z_{in}|I(0)|^2 \tag{8.102}$$

$$Z_{in} = \frac{1}{|I(0)|^2}\oint_S (\mathbf{E}^s \times \mathbf{H}^{s*})\cdot \mathbf{ds} \tag{8.103}$$

The source integral on the right hand side can be written as

$$\begin{aligned}\oint_S (\mathbf{E}^s \times \mathbf{H}^{s*})\cdot \mathbf{ds} &= \oint_S (E_z^s\widehat{z} \times \widehat{\phi}H_\phi^{s*})\cdot \widehat{\rho}ds \\ &= -\int_{-\ell/2}^{\ell/2} E_z^i(a,z)\left[\int_0^{2\pi} K_z a d\phi'\right]dz \\ &= -\int_{-\ell/2}^{\ell/2} E_z^i(a,z)I^*(z)dz\end{aligned} \tag{8.104}$$

where we used the fact that $E_z^s = -E_z^i$ over the surface of the antenna. Thus, the input impedance is given by

$$Z_{in} = -\frac{1}{|I(0)|^2} \int_{-\ell/2}^{\ell/2} E_z^i(a, z) I^*(z) dz \tag{8.105}$$

This is a more accurate expression for the antenna input impedance as compared with (8.101).

8.5 Dielectric Scatterers

In this section, we discuss integral equation formulation of scattering problems involving dielectric bodies. An integral equation can be developed based on the volume equivalence principle. Assuming a nonmagnetic material body, the equivalent electric current inside the scatterer is defined as

$$\mathbf{J}_{eq} = j\omega\epsilon_0(\epsilon_r - 1)\mathbf{E} \tag{8.106}$$

where $\mathbf{E}$ is the total electric field inside the dielectric. Defining the electric susceptibility

$$\chi_e = \epsilon_r - 1 \tag{8.107}$$

the equivalent current can be expressed as

$$\mathbf{J}_{eq} = j\frac{k_0}{Z_0}\chi_e \mathbf{E} \tag{8.108}$$

We may now remove the dielectric body and assume that the equivalent current radiates in free space. The scattered field is given in terms of the Hertz potential as

$$\mathbf{E}^s = \nabla\nabla \cdot \pi^s + k_0^2 \pi^s \tag{8.109}$$

where

$$\pi^s = -j\frac{Z_0}{k_0} \int_V \mathbf{J}_{eq}(\mathbf{r}') G_0(\mathbf{r}; \mathbf{r}') dv' \tag{8.110}$$

and G_0 is the free space Green's function. In order to construct an integral equation, we write

$$\mathbf{E} = \mathbf{E}^i + \mathbf{E}^s \tag{8.111}$$

and substitute for $\mathbf{E}^s$. Thus, we have

$$\mathbf{E}^i(\mathbf{r}) = -j\frac{Z_0}{k_0\chi_e}\mathbf{J}_{eq}(\mathbf{r}) + j\frac{Z_0}{k_0}(k_0^2 + \nabla\nabla\cdot)\int_V \mathbf{J}_{eq}(\mathbf{r}')G_0(\mathbf{r};\mathbf{r}')dv' \quad \mathbf{r} \in V \quad (8.112)$$

This is the volumetric EFIE for scattering from a dielectric target.

Once the above integro-differential equation is solved for $\mathbf{J}_{eq}$, the diffracted field can be obtained from (8.109) to (8.111).

Exercises

8.1: Find the electric field integral equations (EFIE's) for the induced surface current on a perfectly conducting half plane illuminated by a plane wave for both polarizations $\mathrm{TM_Z}$ and $\mathrm{TE_Z}$.

8.2: Consider a thin conducting strip of width w, illuminated by an E-polarized plane wave ($\mathrm{TM_Z}$ case).

(a) Write the TE electric field integral equation formulation for scattering from the thin conducting strip.
(b) Using Fourier transform methods, show that the current distribution resuces to the physical optics approximation for wide strips.
(c) Find a closed form expression for the current distribution over a narrow strip ($w \ll \lambda$).
(d) Find the corresponding echo width for the narrow strip.

8.3: Repeat the above problem for the $\mathrm{TE_Z}$ case.

8.4: The complement of problem 2 above is a slot of width w in an infinite conducting plate. Using Babinet's principle, find the integral equations governing the equivalent surface magnetic current over the aperture of the slot.

8.5: A dielctric strip of relative permittivity ϵ_r, width w and thichness t is located on the $y = 0$ plane. Find the integral equations governing the volumetric equivalent currents due to a plane wave illumination. Angle of incidence is ϕ_0. Consider both principal polarizations.

8.6: A narrow slot of width Δ in a thin conducting plate of width w is fed by a parallel plate waveguide. Assuming that $\Delta \ll w$, the aperture field may be assumed to be the same as the dominant waveguide TEM mode with $\mathbf{H}^i = \hat{z}e^{-jky}$, $(y = 0)$.

(a) Using the surface equivalence principle, find the equivalent *magnetic* current density on the aperture.
(b) Calculate the electrostatic aperture voltage.
(c) Find the electric field integral equation (EFIE) for the induced *electric* current $\mathbf{K}$ induced over the conducting strip of width w.
(d) What is the edge behavior of the electric current $\mathbf{K}$?

Chapter 9
Method of Moments

The method of moments is a projective method in wich a functional equation in an infinite dimensional function space is approximated by a matrix equation in a finite dimensional subspace.

9.1 Formulation

Consider the linear ooperator equation

$$\mathcal{A}[f] = g \tag{9.1}$$

where $\mathcal{A}$ is the linear operator, g is a known function, and f is an unknown function to be determined. In the method of moments the unknown function f is represented approximately by a linear combination of a finite set of functions f_n in the domain of $\mathcal{A}$

$$f \simeq \sum_{n=1}^{N} c_n f_n \quad f_n \in \mathcal{D}_{\mathcal{A}} \tag{9.2}$$

where c_n are scalars to be determined. This approximation is the basic step in transforming the integral equation to a linear system of equations for the solution of $\{c_n\}$. The functions f_n are known as *expansion functions*. They have to be linearly independent. Substituting (9.2) into (9.1) gives

$$\sum_{n=1}^{N} c_n \mathcal{A}[f_n] \simeq g \tag{9.3}$$

© Springer International Publishing Switzerland 2015

K. Barkeshli, *Advanced Electromagnetics and Scattering Theory*,
DOI 10.1007/978-3-319-11547-4_9

where the linearity of the operator has been employed. Defining the residual error R

$$R = \sum_{n=1}^{N} c_n \mathcal{A}[f_n] - g \tag{9.4}$$

the coefficients c_n are computed so that the weighted residual error is set to zero with respect to an inner product

$$\langle w_m, R \rangle = 0 \quad m = 1, \ldots, N \quad w_m \in \mathcal{R}_{\mathcal{A}} \tag{9.5}$$

where the weighting functions w_n are defined in the range of $\mathcal{A}$. The above inner products are called the weighted residuals. This represents a system of linear equations

$$\sum_{n=1}^{N} c_n \langle w_m, \mathcal{A}[f_n] \rangle = \langle w_m, g \rangle \quad m = 1, \ldots, N \tag{9.6}$$

which can also be put in the matrix form

$$[A][c] = [g] \tag{9.7}$$

where $[A]$ is an $n \times N$ matrix of coefficients whose elements are given by

$$A_{mn} = \langle w_m, \mathcal{A}[f_n] \rangle \tag{9.8}$$

and $[g]$ is the column vector of size N with elements

$$g_m = \langle w_m, g \rangle \tag{9.9}$$

If $[A]$ is nonsingular, its inverse exists, and $[c]$ is given by

$$[c] = [A]^{-1}[g] \tag{9.10}$$

The solution f is then obtained from (9.2) as

$$f = [f]^t [A]^{-1} [g] \tag{9.11}$$

where $[f]^t$ is the row vector of basis functions.

The expansion functions $\{f_n\}$ should be linearly independent. If in addition, they are orthogonal and form a complete set for the function space, then the expansion (9.2) is actually a generalized Fourier series. In this case, N is usually infinity. But, orthogonality is not a requirement for the method of moments to give a valid solution. The functions f_n could span the whole domain of the solution, in which case, they are referred to as *entire-domain basis functions*. They may also be defined over a portion of the solution domain. In this case, they are called *sub-domain basis functions*. The choice of the basis functions and their domain of definition depends on the type of the problem at hand and the degree of accuracy needed for the solution.

The weighting functions $\{w_n\}$ should also be linearly independent. Three classical approaches have found utility in choosing the weighting functions w_n.

If the weighting functions are formally chosen to be Dirac delta functions

$$w_m = \delta(\mathbf{r} - \mathbf{r}_m), \quad m = 1, \ldots, N \tag{9.12}$$

then the elements of the coefficient matrix are given by

$$A_{mn} = \langle \delta(\mathbf{r} - \mathbf{r}_m), \mathcal{A}[f_n] \rangle = \mathcal{A}[f_n](\mathbf{r}_i) \tag{9.13}$$

The major advantage of this choice is that the integrations by inner products (9.8) and (9.9) become trivial since they are evaluated at discrete points. This is equivalent to satisfying (9.3) at discrete points in the region of interest. This is the simplest specialization of the moment method and is referred to as *point-matching* or *collocation*.

The second method may be considered as the specialization of moment method to the case of self-adjoint operators. The adjoint operator $\mathcal{A}^a$ is defined with respect to the inner product as

$$\langle w, \mathcal{A}[f] \rangle = \langle \mathcal{A}^a[w], f \rangle \quad f \in \mathcal{D}_{\mathcal{A}} \quad w \in \mathcal{D}^a_{\mathcal{A}} \tag{9.14}$$

and if the domains of $\mathcal{A}$ and $\mathcal{A}^a$ are the same, we can choose the weighting functions the same as the expansion functions

$$w_m = f_m, \quad m = 1, \ldots, N \tag{9.15}$$

For self-adjoint operators ($\mathcal{A} = \mathcal{A}^a$), this choice of weighting functions yields a symmetric matrix $[A]$ which may be desirable from a numerical standpoint. This technique is called *Galerkin's method*.

Another approach is to choose the weighting functions so that the root mean square norm of the residual error (9.4) is minimized. In order to identify these functions, we proceed as follows. The L^2 norm of the residual error is given by

$$||R||_2 = \langle R, R \rangle^{1/2} \tag{9.16}$$

We now choose $\{c_n\}$ in order to minimize (9.16). We thus write

$$\frac{\partial}{\partial c_m} \langle R, R \rangle = 0, \quad m = 1, \ldots, M \tag{9.17}$$

Carrying out the differentiation, we have

$$2\langle \frac{\partial R}{\partial c_m}, R \rangle = 0 \tag{9.18}$$

Comparing this result with (9.5), we immediately identify the weighting functions as

$$w_m = \frac{\partial R}{\partial c_m} = \frac{\partial}{\partial c_m}\left(\sum_{n=1}^{N} c_n \mathcal{A}[f_n] - g\right) \tag{9.19}$$

where use has been made of (9.4). Noting that g is independent of $\{c_m\}$, we find

$$w_m = \mathcal{A}[f_m], \quad m = 1, \ldots, N \tag{9.20}$$

Clearly, $\{w_m\}$ are now chosen from the range of $\mathcal{A}$. This choice for the weighting functions is more accurate, gives better convergence, but is more complicated to implement.

We will next discuss the solution of scattering integral equations by the method of moments. We consider two examples, namely, scattering from thin flat strips and radiation from a thin wire antenna.

9.2 Resistive Strips

In this section, we give a numerical solution for the electromagnetic scattering by resistive strips of constant curvature and arbitrary size by the method of moments. We begin with flat strips and then discuss the solution for circular cylindrical strips.

9.2.1 Flat Strips

The integral equation for the induced electric current over a resistive strip of width w illuminated by an E-polarized plane wave is given by (8.26)

$$Y_0 E_0 e^{jk_0 \cos\phi_0} = \eta_s K_z(x) + \frac{k_0}{4} \int_{-w/2}^{w/2} K_z(x') H_0^{(2)}(k_0|x - x'|) dx' \tag{9.21}$$

where η_s is the normalized surface resistivity of the strip.

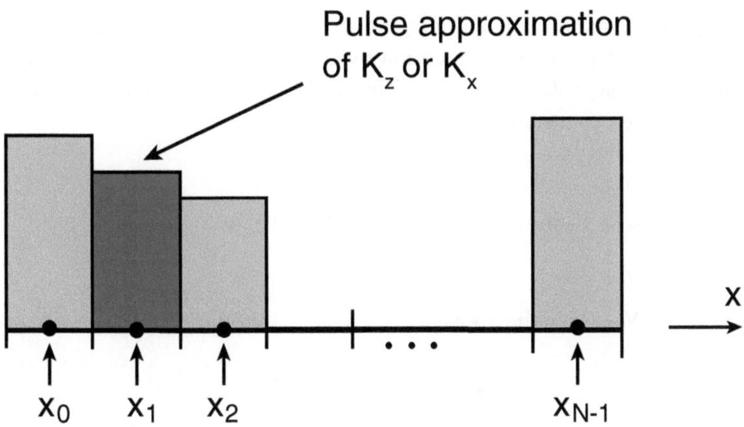

Fig. 9.1 Discretization of the surface current density using pulse basis expansion functions

To solve this integral equations, we divide the strip into N segments of equal width, with the segment centers denoted by

$$x_m = \left(m + \frac{1}{2}\right)\Delta, \quad m = 0, \ldots, N-1 \tag{9.22}$$

where $\Delta = w/N$ is the segment width. Next, we expand the surface current density K_z using sub-domain pulse basis expansion functions (Fig. 9.1)

$$K_z(x) = \sum_{n=0}^{N-1} \alpha_n P_n(x), \quad x_n = (n + \frac{1}{2})\Delta \tag{9.23}$$

where

$$P_n(x) = \begin{cases} 1, & |x - x_n| \leq \Delta x/2 \\ 0, & \text{else} \end{cases}$$

and $\{\alpha_n\}$ are the unknown expansion coefficients. Substituting (9.23) into the integral equation, we obtain

$$\begin{aligned} Y_0 E_0 e^{jk_0 x \cos\phi_0} &= \eta_s(x) \sum_n \alpha_n P_n(x) \\ &+ \frac{k_0}{4} \sum_n \alpha_n \int_{-w/2}^{w/2} P_n(x') H_0^{(2)}(k_0|x_m - x'|) dx' \end{aligned} \tag{9.24}$$

where the order of summation and integration was interchanged. In order to find the unknown coefficients, we use collocation by enforcing the above equation at points $x_m = (m + \frac{1}{2})\Delta, m = 0, \ldots, N - 1$. This yields a linear system of equations for the solution of the expansion coefficients. In particular, we have

$$V_m = \eta_{sm}\delta_{mn}\alpha_m + \frac{k_0}{4}\int\limits_{x_n-\Delta/2}^{x_n+\Delta/2} H_0^{(2)}(k_0|x_m - x'|)dx' \tag{9.25}$$

for $m = 0, \ldots, N - 1$. In the above,

$$V_m = Y_0 E_0 e^{jk_0 x_m \cos\phi_0} \tag{9.26}$$

and η_{sm} is the average normalized surface resistivity over the m-th segment and δ_{mn} is the Kronecker delta function.

Equation (9.25) represents a matrix equation of the form (9.7). The elements of the coefficient matrix $[A]$ are given by

$$A_{mn} = \eta_{sm}\delta_{mn} + \frac{k_0}{4}\int\limits_{x_n-\Delta/2}^{x_n+\Delta/2} H_0^{(2)}(k_0|x_m - x'|)dx' \tag{9.27}$$

Each element A_{mn} can be interpreted as the impedance seen by the current segment P_n whose radiation is evaluated at the center of the mth segment. Consequently, they are referred to as the *mutual impedance elements*, and $[A]$ is simply referred to as the *impedance matrix*. It is noted that A_{mn} is singular when $m = n$. These elements are called *self-cell interactions* and they form the diagonal elements of the impedance matrix. The elements A_{mn} can be evaluated as

$$A_{mn} \simeq \begin{cases} \eta_{sn} + \dfrac{k_0\Delta x}{4}\left[1 - \dfrac{j2}{\pi}\left(\ln(\dfrac{k_0\gamma\Delta}{4}) - 1\right)\right], & n = m \\ \dfrac{k_0\Delta x}{4} H_0^{(2)}(k_0|x_m - x_n|), & n \neq m \end{cases}$$

where we used (8.57) for the self cell interaction term ($n = m$). It is noted that the impedance matrix $[A]$ for this problem, is toeplitz in form. In other words, the elements of the matrix are related as follows

$$A_{mn} = A_{1,|m-n|+1}, \quad m \geq 2, n \geq 1 \tag{9.28}$$

This is due to the dependence of the integral in (9.28) to $|x_m - x_n|$. The property (9.28) implies that the element A_{mn} is a function of $|m - n|$ and thus $A_{12} = A_{21} = A_{23} = A_{32} = A_{56}$, $A_{10,12} = A_{2,4} = A_{20,22}$ etc. Consequently, the matrix is defined by only a single row of the matrix, each other row being obtained from the previous one by a cyclic shift of the elements to the right. The inversion of the matrix can therefore be carried out very efficiently using algorithms taking advantage of this property. We may formally write the solution for the expansion coefficients as

$$[\alpha_n] = [A]^{-1}[V] \tag{9.29}$$

and the surface current density may be expressed as

$$K_z(x) = [P]^T[A]^{-1}[V] \tag{9.30}$$

A similar analysis may be used to solve (8.29) by the method of moments. The calculated currents can be then used in (8.34) and (8.35) to obtain the echo widths based on the numerical solution of the integral equations.

Figures 9.2a, b and 9.3a, b show the surface current densities over a 6λ-wide perfectly conducting flat strip illuminated by TE and TM polarizations at normal incidences, respectively, as computed by the moment method. The physical optics solutions (Fig. 9.7) are also shown for comparison. It is seen that the numerical solution for the surface current density oscillates around the physical optics approximation in the region away from the strips terminations. However, the physical optics solution does not show the oscillatory behavior because the underlying assumption for this solution is that the surface is locally planar and infinite in extent (i.e., no diffraction from the edges is included).

When the strip is not perfectly conducting, the surface current density does not go to infinity near the strip edges for the TM polarization. Therefore, diffraction from the edges is not as strong and consequently the exact current density computed via the moment method is in close agreement with the physical optics approximation as shown in Fig. 9.4 for a 6λ strip having a resistivity $Z_s = Z_0$.

The backscattering echo widths for a 4λ-wide strip are shown in Fig. 9.5 for the two principal polarizations (for backscatter calculations, the angles of incidence and observation are the same). The echo widths are shown for the perfectly conducting strip as well as the parabolically tapered strip where the normalized resistivity is given by $\eta_s = (\frac{x}{w/2})^2$. For the latter case, it is seen that the echo width is reduced considerably because the current density has lower value due to losses from the finite surface resistivity. Resistive loading is, in fact, an effective method to reduce the scattering cross section of objects with sharp edges.

9.2.2 Circular Cylindrical Strips

We will now consider the moment method solution of (8.75) using the Galerkin's formulation. The current density is expanded in terms of a sub-domain expansion function as

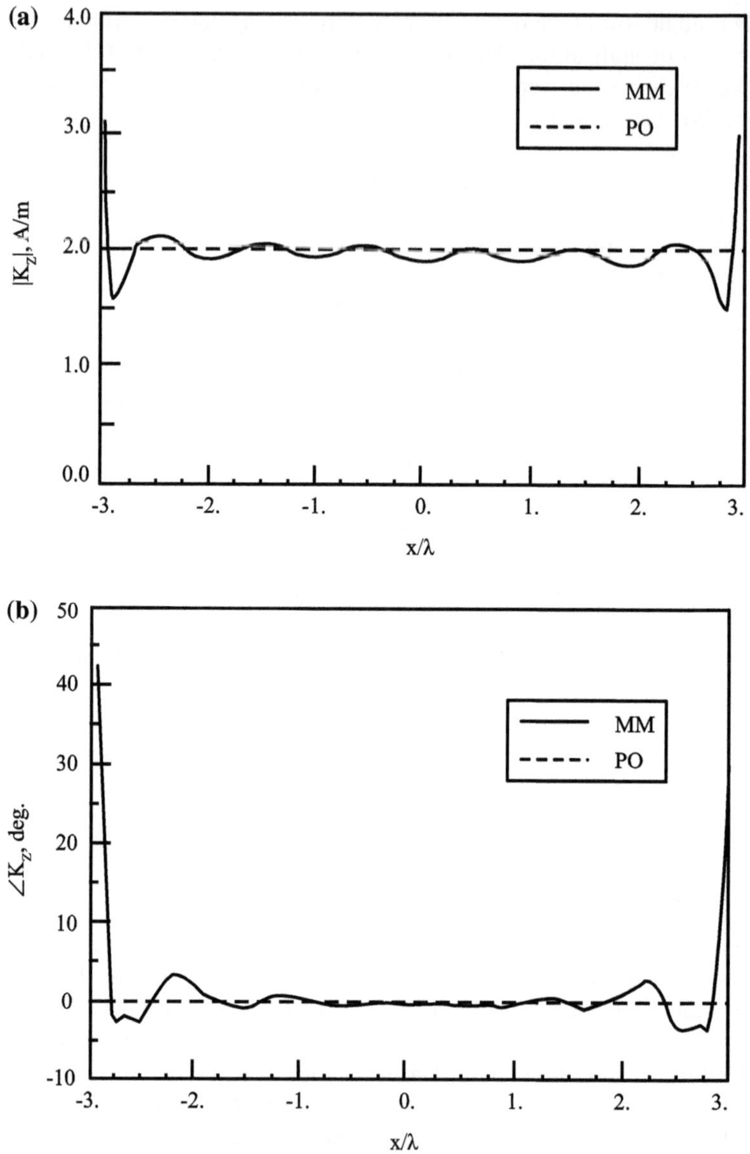

Fig. 9.2 Surface current density over a 6λ-wide perfectly conducting flat strip illuminated by a TM-polarized plane wave at normal incidence. The moment method and the physical optics approximation. **a** Magnitude, and **b** phase of the current density

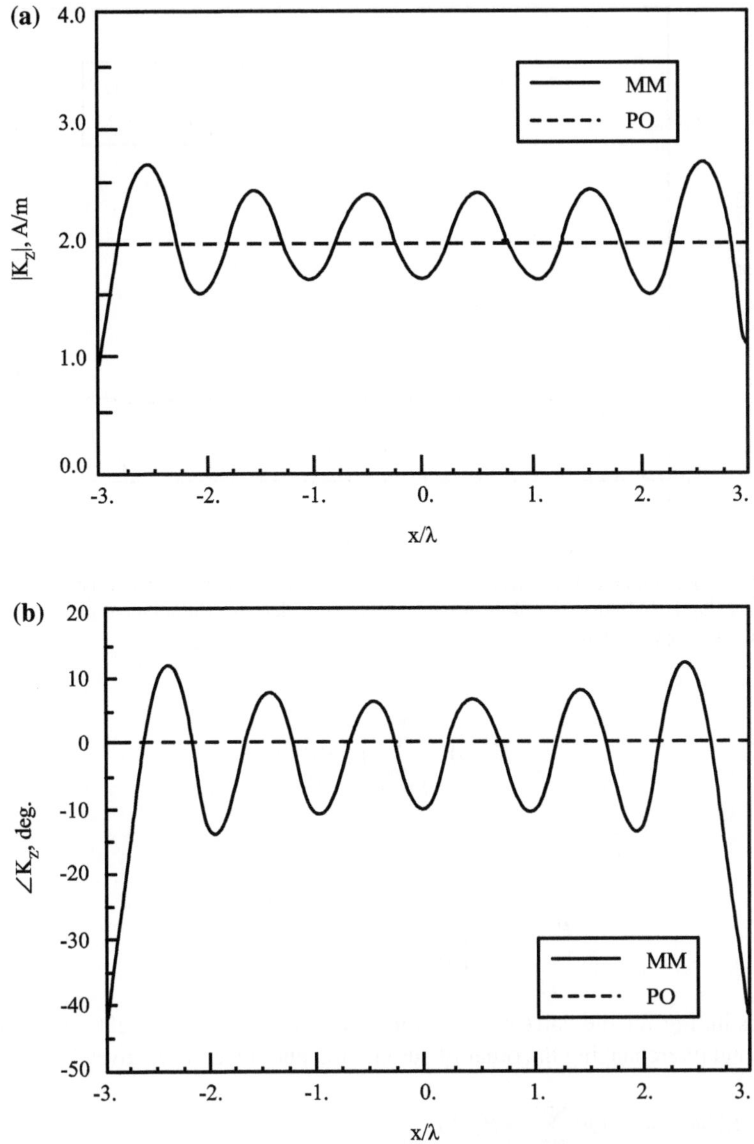

Fig. 9.3 Surface current density over a 6λ-wide perfectly conducting flat strip illuminated by a TE-polarized plane wave at normal incidence. The moment method and the physical optics approximation. **a** Magnitude, and **b** phase of the current density

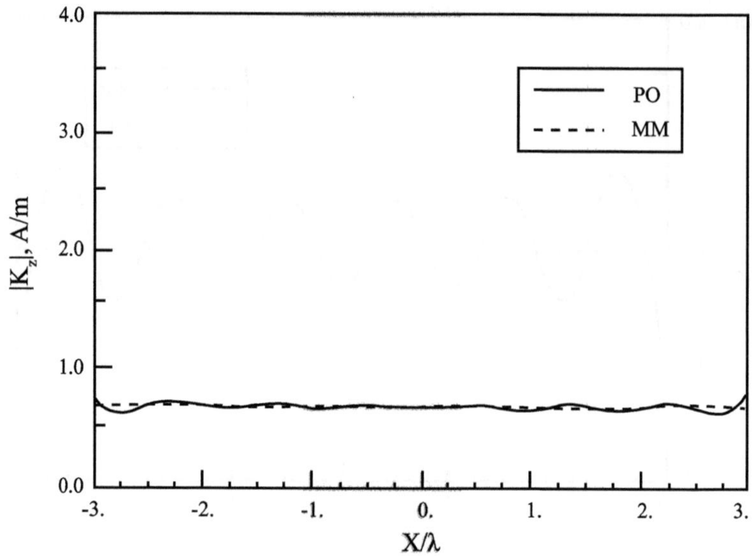

Fig. 9.4 The magnitude of the surface current density over a 6λ-wide flat strip of resistivity $Z_s = Z_0$ illuminated by a TM-polarized plane wave at normal incidences. The moment method and the physical optics approximation

$$K_z(\phi) = \sum_{n=0}^{N-1} \alpha_n P_n(\phi) \tag{9.31}$$

where

$$P_n(\phi) = \begin{cases} 1 & |\phi - \phi_n| \le \Delta\phi/2 \\ 0 & \text{else} \end{cases}$$

Substituting for the current expansion in the integral on the right hand side of (8.75) and interchanging the order of summation and integration, gives

$$\begin{aligned} E_z^i(\phi) &= Z_s(\phi) \sum_n \alpha_n P_n(\phi) \\ &+ \frac{k_o a Z_o}{4} \sum_n \alpha_n \int_c P_n(\phi') H_0^{(2)} \left[2k_o a \sin(\frac{|\phi - \phi'|}{2}) \right] d\phi' \end{aligned} \tag{9.32}$$

Applying Galerkin's method yields the system of equations

$$V_m = \Delta\phi Z_{sm} \delta_{mn} \alpha_m + \frac{k_o a Z_o}{4} \sum_n \xi_{mn} \alpha_n \tag{9.33}$$

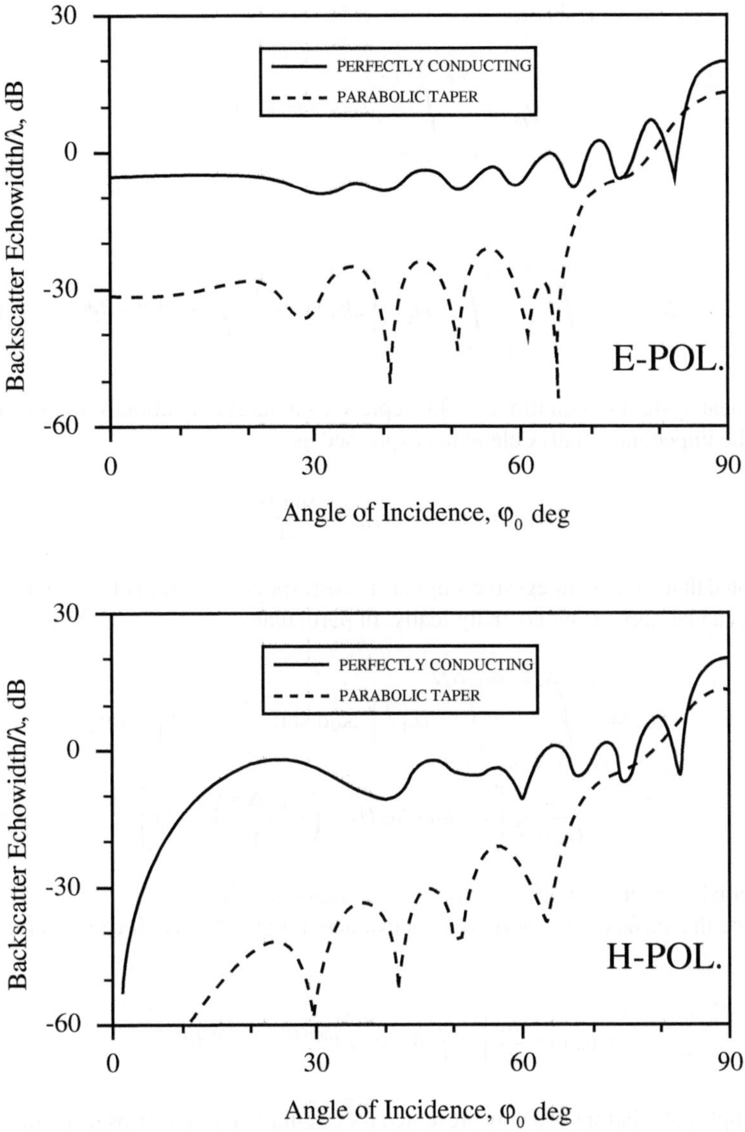

Fig. 9.5 Backscattering echo widths of a 4λ perfectly conducting and parabolically tapered resistive strips, $\eta_s = (\frac{x}{w/2})^2$ for TM and TE polarizations

for $m = 0, \ldots N-1$, where

$$V_m = \int_{\phi_m - \Delta\phi/2}^{\phi_m + \Delta\phi/2} e^{jk_o \cos(\phi - \phi_o)} d\phi \tag{9.34}$$

and

$$\xi_{mn} = \int_{\phi_m - \Delta/2}^{\phi_m + \Delta/2} \int_{\phi_n - \Delta/2}^{\phi_n + \Delta/2} H_0^{(2)} \left[2k_o a \sin(\frac{|\phi - \phi'|}{2}) \right] d\phi' d\phi \tag{9.35}$$

The linear system of equations (9.33) represents a matrix equation of the form (9.7) with the impedance matrix elements expressed as

$$A_{mn} = \Delta\phi Z_{sm} \delta_{mn} + \frac{k_0 a Z_0}{4} \xi_{mn} \tag{9.36}$$

It is noted that ξ has an integrable singularity corresponding to the self-cell interaction which can be approximated analytically. In particular,

$$\begin{aligned} \xi_{nn} &= \int_{\phi_n - \Delta/2}^{\phi_n + \Delta/2} \int_{\phi_n - \Delta/2}^{\phi_n + \Delta/2} H_0^{(2)} \left[2k_0 a \sin(\frac{|\phi - \phi'|}{2}) \right] d\phi' d\phi \\ &\simeq \frac{2}{(k_0 a)^2} \left[\sqrt{\pi} k_0 a \Delta\phi H_1^{(2)} \left(\frac{k_0 a \Delta\phi}{\sqrt{\pi}} \right) - 2j \right] \end{aligned} \tag{9.37}$$

The remaining terms ($n \neq m$) may be evaluated numerically.

Once the surface current density is evaluated, the echo width can be computed using

$$\sigma_e(\phi) = \frac{k_0}{4} \left| a \int_c K_z(\phi') e^{jk_0 a \cos(\phi - \phi_0)} d\phi' \right|^2 \tag{9.38}$$

Sample calculations are now presented for circularly curved strips using the above formulation[1]. Figure 9.6 shows the bistatic scattering patterns for a 2λ flat strip as it is uniformly bent to form a closed circular cylinder keeping its width (perimeter) constant. The strip is positioned symmetrically around the y-axis and illuminated by an E-polarized plane wave incident at 90°. It is noted that as the curvature $\kappa = 1/a$ increases from zero (flat strip), the main (specular) lobe drops and eventually disappears in the limit when the complete cylinder is achieved. The numerical result

[1] For more elaboration please refer to Fundamentals of the Physical Theory of Diffraction. By PyotrYa. Ufimtsev (John Wiley & Sons, 2007), Chapter 14.

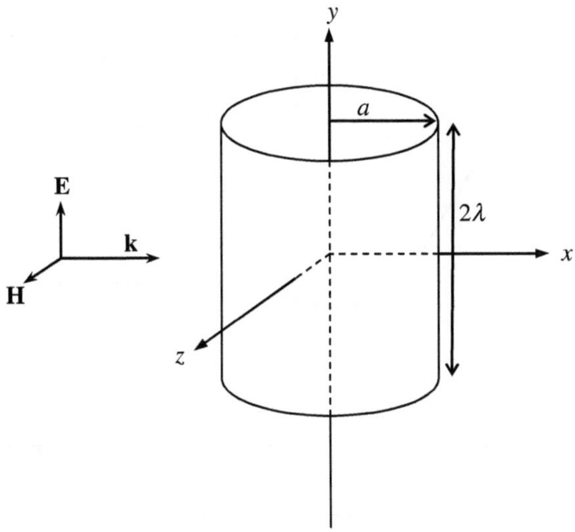

Fig. 9.6 A 2λ wide conducting strip as it is uniformly bent to form a circular cylinder

for the closed cylinder is in agreement with the classical eigen-function solution (Fig. 9.7).

Figure 9.8 shows the radiation pattern of an infinite electric line source in the presence of a $2\pi\lambda$ cylindrical resistive strip ($a = 4\lambda/3$). The line source is positioned at the center of the strip and radiates through a right angle slit. As expected, the nonzero resistivity reduces the directivity of the reflector.

9.3 The Linear Wire Antenna

Tha Pocklington integral equation for a linear wire antenna was derived in Sect. 8.4 as

$$E_z^i(z) = j\frac{Z_0}{k_0}(k_0^2 + \frac{\partial^2}{\partial z^2}) \int_{-\ell/2}^{\ell/2} I(z')\frac{e^{-jk_0R}}{4\pi R}dz' \tag{9.39}$$

where $R \simeq \sqrt{(z-z')^2 + a^2}$. We numerically solve this integral equation by the method of moments. We use the following sub-domain expansion functions to expand the linear current $I(z)$

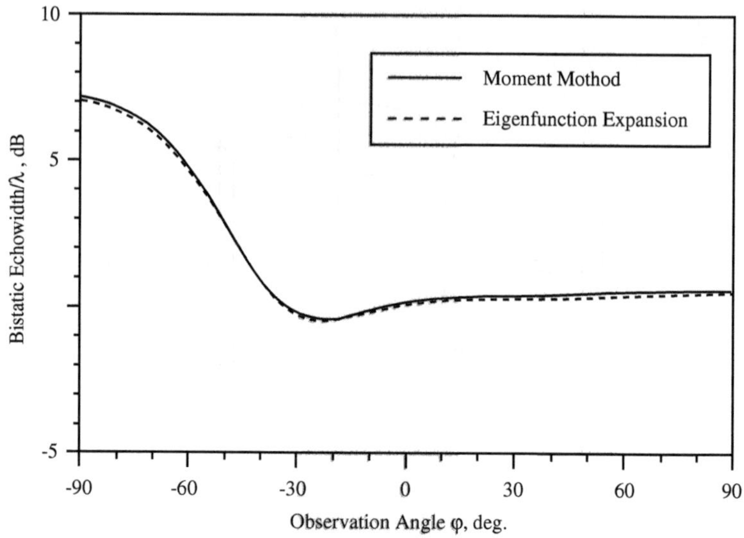

Fig. 9.7 Comparison of the bistatic echo widths for the circular cylinder calculated by the method of moments and the eigenfunctions expansion method

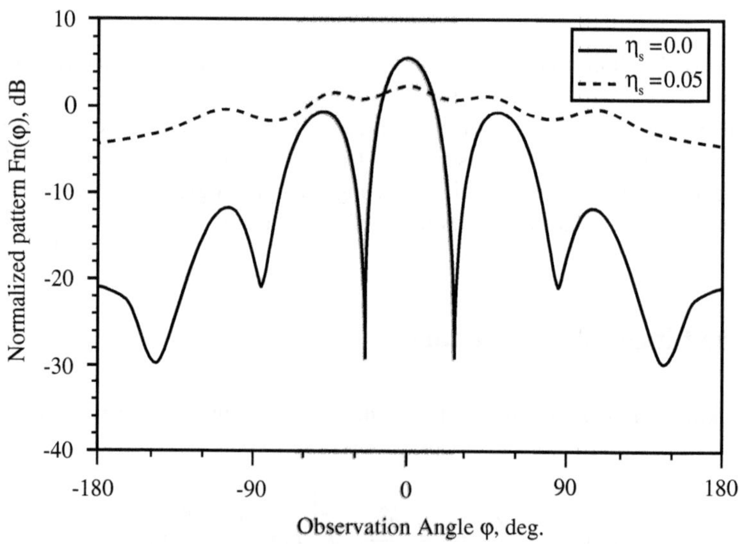

Fig. 9.8 Far field radiation pattern of the reflector antenna shown in Fig. 8.5 with normalized surface resistivities $\eta_s = 0$ and $\eta_s = 0.05$

$$f_n(z) = \begin{cases} \frac{\sin[k_0(\Delta z - |z - z_n|)]}{\sin k_0 \Delta z} & |z - z_n| \le \Delta z \\ 0 & \text{else} \end{cases}$$

where $z_n, n = 0, \ldots, N-1$ are the center points of the N equal segments over the antenna. These functions are called piecewise sinusoidal expansion functions. They have the advantage that they accurately model the zero current at the terminations of a thin wire antenna, and that they give analytical expressions for the elements of the impedance matrix. The current I is thus expressed as

$$I(z) = \sum_{n=0}^{N-1} I_n \frac{\sin[k_0(\Delta z - |z - z_n|)]}{\sin k_0 \sin \Delta z} P_{(2\Delta z)}(z - z_n) \tag{9.40}$$

where $\{I_n\}$ are the unknown expansion coefficients and $P_{(2\Delta z)}(z - z_n)$ are pulse basis functions defined as

$$P_{2\Delta z}(z - z_n) = \begin{cases} 1, & |z - z_n| \leq \Delta z \\ 0, & \text{else} \end{cases}$$

Substituting in (9.39), we have

$$E_z^i(z) = jZ_0 k_0 \sum_n I_n \int_{z_n - \Delta z}^{z_n + \Delta z} \frac{\sin[k_0(\Delta z - |z - z_n|)]}{\sin k_0 \Delta z} \left(1 + \frac{1}{k_0^2} \frac{\partial^2}{\partial z^2}\right) \frac{e^{-jk_0 R}}{4\pi R} dz' \tag{9.41}$$

It can be shown by integration by parts that

$$\int_{z_n - \Delta z}^{z_n + \Delta z} \sin[k_0(\Delta z - |z - z_n|)] \left(1 + \frac{1}{k_0^2} \frac{\partial^2}{\partial z^2}\right) \frac{e^{-jk_0 R}}{4\pi R} dz'$$
$$= \frac{1}{k_0} \left(\frac{e^{-jk_0 R_{1n}}}{R_{1n}} + \frac{e^{-jk_0 R_{2n}}}{R_{2n}} - 2\cos(k_0 \Delta z) \frac{e^{-jk_0 R_0}}{R_0} \right) \tag{9.42}$$

where

$$\begin{aligned} R_{1n} &= \sqrt{(z - z_n - \Delta z)^2 + a^2} \\ R_{2n} &= \sqrt{(z - z_n + \Delta z)^2 + a^2} \\ R_0 &= \sqrt{(z - z_n)^2 + a^2} \end{aligned} \tag{9.43}$$

Thus, we have

$$-j4\pi Y_0 E_z^i(z) = \frac{1}{\sin k_0 \Delta z} \left(\frac{e^{-jk_0 R_{1n}}}{R_{1n}} + \frac{e^{-jk_0 R_{2n}}}{R_{2n}} - 2\cos(k_0 \Delta z) \frac{e^{-jk_0 R_0}}{R_0} \right) \tag{9.44}$$

In order to form the system of linear equations, we use point-matching at $z_m, m = 0, \ldots, N-1$ to find

$$\sum_n Z_{mn} I_n = V_m, \quad m = 0, \ldots, N-1 \tag{9.45}$$

where

$$Z_{mn} = \frac{1}{\sin k_0 \Delta z} \left(\frac{e^{-jk_0 R_{1mn}}}{R_{1mn}} + \frac{e^{-jk_0 R_{2mn}}}{R_{2mn}} - 2\cos(k_0 \Delta z) \frac{e^{-jk_0 R_{0mn}}}{R_{0mn}} \right) \tag{9.46}$$

with

$$\begin{aligned} R_{1mn} &= \sqrt{(z_m - z_n - \Delta z)^2 + a^2} \\ R_{2mn} &= \sqrt{(z_m - z_n + \Delta z)^2 + a^2} \\ R_{0mn} &= \sqrt{(z_m - z_n)^2 + a^2} \end{aligned} \tag{9.47}$$

and

$$V_m = -j4\pi Y_0 E_z^i(z_m), \quad m = 0, \ldots, N-1 \tag{9.48}$$

The coefficients $\{I_n\}$ are, thus, given by

$$[I_n] = [Z]^{-1}[V] \tag{9.49}$$

where the elements of $[Z]$ and $[V]$ are given by (9.46) and (9.48). It can be easily verified that the impedance matrix $[Z]$ is also Toeplitz in this case.

For the source excitation, we may use (8.96) or (8.100), depending on the way the antenna is actually excited. The input impedance can be calculated from (8.101) or (8.105) depending on the source model used. In the latter case, the input impedance can be calculated as

$$Z_{in} \simeq \frac{1}{I_{(N+1)/2}} \sum_{n=0}^{N-1} E_z^i(n\Delta z) \cdot I_n^* \tag{9.50}$$

where $I_{(N+1)/2}$ is the coefficient of the current at the driving position.

9.4 The Dielectric Cylinder

An infinitely long inhomogeneous dielectric cylinder parallel to the z-axis is illuminated by a plane wave. The geometry and permittivity of the cylinder is independent of z.

Consider the transverse magnetic case where the incident plane wave is given by

$$\mathbf{E}^i(x, y) = \widehat{z} E_0 e^{jk_0(x \cos\phi_0 + y \sin\phi_0)} \tag{9.51}$$

Due to the translational symmetry of the problem along the z-axis, the equivalent polarization current is also z-directed and is given by

$$\mathbf{J}_{eq} = \widehat{z} J_{eq} = \widehat{z} j\omega\epsilon_0[\epsilon_r(x, y) - 1]E_z = j\frac{k_0}{Z_0}\chi_e(x, y)\mathbf{E} \tag{9.52}$$

where ϵ_r and χ_r are the relative permitivitty and the electric susceptibility of the cylinder, respectively. The electric field integral equation (EFIE) for the equivalent current is expressed as

$$\begin{aligned} E_z^i(x, y) &= \frac{J_{eq}}{j\omega\epsilon_0\chi_e} \\ &+ \frac{k_0 Z_0}{4}\int_S J_{eq}(x', y')H_0^{(2)}(k_0\sqrt{(x - x')^2 + (y - y')^2})ds' \end{aligned} \tag{9.53}$$

which is a Fredholm integral equation of the second type. The integral on the right hand side should be interpreted as the Cauchy principle value.

As a simple solution, we divide the cross-sectional area S of the cylinder into N cells and the unknown current distribution is expanded using pulse basis functions as

$$\mathbf{J}_{eq} \simeq \widehat{z}\sum_{n=1}^{N-1} c_N P_n(x, y) \tag{9.54}$$

where

$$P_n(x, y) = \begin{cases} 1 & (x, y) \in S_n \\ 0 & (x, y) \notin S_n \end{cases} \tag{9.55}$$

Substituting into the integral equation, we get

$$\begin{aligned} E_z^i(x, y) \simeq \sum_n c_n \Bigg[& \frac{P_n(x, y)}{j\omega\epsilon_0\chi_e} \\ & + \frac{k_0 Z_0}{4}\int_{S_n} H_0^{(2)}(k_0\sqrt{(x - x')^2 + (y - y')^2})dx'dy' \Bigg] \end{aligned} \tag{9.56}$$

Using point-matching, the system of algebraic equations takes the form

$$[Z][c] = [V] \tag{9.57}$$

where

$$V_m = E_0 e^{jk_0(x_m \cos\phi_0 + y_m \sin\phi_0)}, \quad m = 0, \ldots, N-1 \tag{9.58}$$

and

$$Z_{mn} = \begin{cases} \frac{k_0 Z_0}{4} \int_{S_n} H_0^{(2)}(k_0\sqrt{(x'-x_m)^2 + (y'-y_m)^2})dx'dy' & m \neq n \\ -j\frac{Z_0}{k_0 \chi_m} + \frac{k_0 Z_0}{4} ⨍_{S_m} H_0^{(2)}(k_0\sqrt{(x'-x_m)^2 + (y'-y_m)^2})dx'dy' & m = n \end{cases} \tag{9.59}$$

In the above, χ_m is the average value of χ_e in cell m. If we approximate the cell area S_n with a circular cell of the same area, the above integrals can be evaluated in closed form. Thus

$$Z_{mn} \simeq \begin{cases} \frac{Z_0 \pi a_n}{2} J_1(k_0 a_n) H_0^{(2)}(k_0 R_{mn}) & m \neq n \\ -j\frac{Z_0}{k_0 \chi_m} + \frac{Z_0 \pi a_m}{2} H_1^{(2)}(k_0 a_m) & m = n \end{cases} \tag{9.60}$$

where a_n is the radius of the equivalent circle of the same area as S_n (that is $a_n = \sqrt{S_n/\pi}$).

This approach gives good results for TM scattering from dielectric cylinders. The accuracy can be improved by applying Galerkin's method. However, the electric field integral equation formulation is not suitable for the TE case, especially when the permeability is high. In this case, the magnetic field integral equation (MFIE) should be used.

Exercises

9.1: Consider the integral equation

$$\int_{-1}^{1} f(x') \ln|x - x'| dx' = -1, \quad |x| \leq 1$$

(a) Solve the above equation by the method of moments. Use pulse basis functions and collocation method.
(b) Solve the above equation using Galerkin's method and collocation.

(c) Compare the above results with the exact analytical solution.

9.2: A thin perfectly conducting rod of radius a and length ℓ is maintained at the potential V_0. Using the method of moments, find the surface static charge density over the rod.

9.3: Find the capacitance of a parallel plate capacitor. The electrodes are $a \times a$ conducting plates separated by a distance d. Plot the capacitance versus the ratio d/a. Estimate the error incurred when neglecting the fringing fields and thus using the approximate formula $C_0 = \epsilon a^2/d$.

9.4: Consider a thin conducting strip of width $w = 1\lambda$, illuminated by a plane wave. Using the EFIE formulation and the method of moments

(a) find the surface current distribution for the E-polarized plane wave at normal incidence.
(b) calculate the bistatic and monostatic echo widths as a function of observation angle.
(c) find the surface current distribution for the H-polarized plane wave at normal incidence.
(d) calculate the bistatic and monostatic echo widths as a function of observation angle.
(e) compare the results obtained in parts (a) and (c) and discuss the current behavior at the edges. Also compare the behavior of the results obtained in parts (b) and (d) at low grazing angles.

9.5: A narrow slot of width Δ in a thin conducting plate of width w is fed by a parallel plate waveguide. Assuming that $\Delta \ll w$, the aperture field may be assumed to be the same as the dominant waveguide TEM mode with $\mathbf{H}^i = \hat{z}e^{-jky}$, $(y = 0)$.

(a) Using the surface equivalence principle, find the equivalent *magnetic* current density on the aperture.
(b) Find the electric field integral equation (EFIE) for the induced *electric* current $\mathbf{K}$ induced over the conducting strip of width w.
(c) Solve the above integral equation by the moment method. Take $w = 1\lambda$ and $\Delta = 0.1\lambda$.

9.6: A dielctric strip of relative permittivity ϵ_r, width w and thichness t is located on the $y = 0$ plane.

(a) Find the integral equations governing the volumetric equivalent currents due to a TM_Z plane wave illumination. Angle of incidence is ϕ_0.
Consider a dielectric strip with

$$\epsilon_r = 4.2 \quad w = 1\lambda_0 \quad t = 0.05\lambda_0$$

(b) Use method of Moments with pulse basis functions and point-matching to set up the system of algebraic equations for the expansion coefficients. Since the strip is electrically thin, you may subdivide the strip along its width using $0.05\lambda_0 \times 0.05\lambda_0$ cells.

(c) Find the induced equivalent current and the bistatic echo width for the cases: $\phi_0 = 0°$ and $\phi_0 = 60°$. Plot the bistatic echo widths for the observation range $0 \leq \phi_s \leq 2\pi$.

(d) Find the monostatic echo width ($\phi_0 = \phi_s$) in the in the range $0 \leq \phi_0 \leq \pi$.

9.7: A linear wire antenna of length $\ell = 0.48\lambda$ and radius $a = 0.005\lambda$ is excited by a 50Ω coaxial cable. The inner radius of the cable is a. Using the magnetic frill model for the source excitation, apply the method of moments to find

(a) the current distrubution,
(b) the input impedance, and
(c) use different values of N to examine the convergence of the above solutions.

9.8: Consider the two-dimensional problem shown below. If the object has a square cross section of $a \times a$, use four cells as shown and obtain the echo width.

Chapter 10
Periodic Structures

The reflection and transmission of plane wave incident on periodic structures are of great importance in many areas of engineering and physics. Examples include microwave mesh reflectors, optical gratings, and crystal structures.

Scattered waves from periodic structures are characterized by grating modes resulting from periodic interference of waves in different directions.

The starting point in solving the problems of periodic structures in Floquet's theorem.

10.1 Floquet's Theorem

Floquet's theorem states that the wave in periodic structures consists of an infinite number of *space harmonics*.

Consider a wave propagating in periodic structures as characterized by periodic boundary conditions or periodically varied dielectric constant. It is noted that the fields at a point in an infinite periodic structure differ from the fields one period L away by a complex constant. Because in an infinite periodic structure, the fields at z and at $z + L$ should differ only by a constant attenuation and phase shift

$$u(z + L) = Cu(z), \quad C = e^{-j\beta L}, \beta \in \mathcal{C}$$

Similarly,

$$u(z + 2L) = Cu(z + L)$$

$$u(z + mL) = C^m u(z)$$

Consider now a function

$$R(z) = e^{j\beta z} u(z) \tag{10.1}$$

© Springer International Publishing Switzerland 2015
K. Barkeshli, *Advanced Electromagnetics and Scattering Theory*,
DOI 10.1007/978-3-319-11547-4_10

Then

$$R(z+L) = e^{j\beta(z+L)}u(z+L) = e^{j\beta z}e^{j\beta L}e^{-j\beta L}u(z) = R(z) \qquad (10.2)$$

So, $R(z)$ is a periodic function of z with period L. Its Fourier series representation is

$$R(z) = \sum_{n=-\infty}^{\infty} A_n e^{-j(\frac{2n\pi}{L})z} \qquad (10.3)$$

Thus

$$\begin{aligned} u(z) &= e^{-j\beta z}R(z) \\ &= \sum_{n=-\infty}^{\infty} A_n e^{-j\beta_n z} \end{aligned} \qquad (10.4)$$

where β_n are the Floquet's mode numbers and are given by

$$\beta_n = \beta + 2n\pi/L. \qquad (10.5)$$

10.2 Scattering From Strip Gratings

Periodic diffraction gratings constitute an important class of frequency selective structures. They find numerous applications at microwave, millimeter wave, and optical frequencies. These include Bragg cells, optical switches and polarizers, surface waveguides and accelerators. The scattering characteristics of the grating is controlled by its geometrical dimensions.

Consider a periodic grating of period L consisting of thin conducting strips of width w. The grating is illuminated by an H-polarized plane wave

$$\mathbf{H}^i = \hat{z}e^{j\mathbf{k}\cdot\mathbf{r}} \qquad (10.6)$$

incident from the free space at an angle ϕ_0. The problem is to find the scattered field in the external region.

According to the Floquet's theorem, the periodicity of the structures results in field quantities which are periodic in x. Thus, the scattered magnetic field in the external region is written as

$$\mathbf{H}^{s>} = \hat{z} \sum_{n=-\infty}^{\infty} A_n e^{j(k_{xn}x - k_{yn}y)} \qquad (10.7)$$

where the superscript > denotes the external field and

$$k_{xn} = k_x + 2n\pi/L \tag{10.8}$$

are the wave number of Floquet's modes. Also, k_{xn} and k_{yn} satisfy the compatibility condition

$$k_{xn}^2 + k_{yn}^2 = k_0^2 \tag{10.9}$$

in accordance with the Helmholtz wave equation. To express the amplitudes of the Floquet's modes, A_n, in terms of physical field quantities, we invoke the equivalence principle. Thus, introducing the equivalent magnetic current over the aperture

$$\mathbf{M}^> = \mathbf{E}^> \times \widehat{n}|_{y=0^+} \tag{10.10}$$

where $\widehat{n}$ is the outward unit normal to the plane of grating, we have

$$\mathbf{M}^> = -\widehat{z}\frac{Z_0}{k_0}\sum_{n=-\infty}^{\infty} k_{yn} A_n e^{jk_{xn}x} \tag{10.11}$$

The above summation is a Fourier expansion whose coefficients may be calculated by using orthogonality. We thus find

$$A_n = -\frac{k_0 Y_0}{k_{yn} L}\int_{-w/2}^{w/2} M_z^>(x) e^{-jk_{xn}x} dx \tag{10.12}$$

so that substituting in (10.7), we have

$$H_z^>(x) = -jk_0 Y_0 \int_{-w/2}^{w/2} M_z^>(x') G^>(x, y; x', 0) dx' \tag{10.13}$$

where Y_0 is the free space intrinsic admittance and

$$G^>(x, y; x', 0) = \sum_{n=-\infty}^{\infty} \frac{e^{j[k_{xn}(x-x')-k_{yn}y]}}{jk_{yn}L} \tag{10.14}$$

is the *free space periodic Green's function*. Similarly, the scattered magnetic field in the $y < 0$ region can be written in terms of the equivalent magnetic current

$$\mathbf{M}^< = \mathbf{E}^< \times n'|_{y=0^-} \tag{10.15}$$

Imposing the continuity of the tangential electric current in the aperture, we find that

$$\mathbf{M}^{<} = \mathbf{E}^{<} \times \widehat{n}' = -\mathbf{E}^{<} \times \widehat{n} = -\mathbf{E}^{>} \times \widehat{n} = -\mathbf{M}^{>}, \quad y = 0 \tag{10.16}$$

Thus, the scattered magnetic field in $y < 0$ is given by

$$H_z^{<} = jk_0 Y_0 \int_{-w/2}^{w/2} M_z^{>}(x') G^{<}(x, y; x', 0) dx' \tag{10.17}$$

where

$$G^{<}(x, y; x', 0) = \sum_{p=0}^{\infty} \frac{e^{j[k_{xn}(x-x')+k_{yn}y]}}{jk_{yn}L} \tag{10.18}$$

To obtain an integral equation for the equivalent magnetic current, we enforce the continuity of the tangential magnetic field over the aperture

$$\widehat{n} \times [\mathbf{H}^i + \mathbf{H}^r + \mathbf{H}^{>}], \quad y = 0 \tag{10.19}$$

where $\mathbf{H}^i$ and $\mathbf{H}^r$ are incident and reflected magnetic fields, respectively. Thus

$$2H_0 e^{jk_x x} - jk_0 Y_0 \int_{-w/2}^{w/2} M_z(x\prime) G(x; x') dx' = jk_0 Y_0 \int_{-w/2}^{w/2} M_z^{>}(x') G(x; x') \tag{10.20}$$

or, equivalently

$$H_0 e^{jk_x x} = jk_0 Y_0 M_z(x') G(x; x') dx' \tag{10.21}$$

where $G(x; x') = G^{>}(x, 0; x', 0) = G^{<}(x, 0; x', 0)$. This is the desired integral equation for the unknown magnetic current M_z.

An analytical solution is available for (10.21) when $k_0 w \ll 1$. In this case we choose

$$M_z(x) = \frac{\chi_p^h e^{jk_x x}}{\sqrt{(w/2)^2 - x^2}} \tag{10.22}$$

in accordance with the edge condition. Substituting in (10.21), multiplying both sides by M_z^* and integrating over the aperture, we find that

$$\chi_p^h = 4H_0 \left[\frac{k_0 w}{L} \sum_{n=-\infty}^{\infty} \frac{J_0^2(n\pi w/L)}{k_{yn}} \right]^{-1} \tag{10.23}$$

For wide apertures, the integral equation (10.21) can be solved numerically. The numerical treatment of this problem is based on the efficient computation of $G(x; x')$. The convergence of the series (10.14) is slow when the observation point lies on the plane of grating, $y = 0$

$$G(x; x') = \sum_{n=-\infty}^{\infty} \frac{e^{jk_{xn}(x-x')}}{jk_{yn}L} \tag{10.24}$$

This is due to the K_{yn} term in the denominator. To accelerate the convergence of the series, we employ Kummer's transformation.

Kummer's transformation is an asymptotic technique to accelerate the convergence of weakly convergent series. It is based on the analogy in the theory of Fourier series that a function which is wide-band in one domain is narrow-band in the other domain and its Fourier series representation can be summed up faster.

Suppose the single sum

$$S = \sum_{n=-\infty}^{\infty} f(n) \tag{10.25}$$

is a slowly converging series. Let $f(n)$ be asymptotic to a function $f_1(n)$ for large n. We write (10.25) as

$$S = \sum_n [f(n) - f_1(n)] + \sum_n f_1(n) \tag{10.26}$$

and use Poisson transformation of the last series in the above equation to arrive at

$$S = \sum_n [f(n) - f_1(n)] + \sum_n F_1(2n\pi) \tag{10.27}$$

where F_1 is the Fourier transform of f_1. With the appropriate choice of f_1, the slowly converging series (10.25) is transformed into the sum of two highly convergent series.

In the present situation, we introduce the asymptotic function

$$f_1 = \sum_n \frac{e^{jk_{xn}(x-x')}}{\sqrt{u^2 + k_{xn}^2}} \tag{10.28}$$

and write (10.24) as

$$G(x; x') = \frac{1}{jL} \sum_n e^{jk_{xn}(x-x')} \left(\frac{1}{k_{yn}} + \frac{1}{j\sqrt{u^2 + k_{xn}^2}} \right) + \frac{1}{L} \sum_n \frac{e^{jk_{xn}(x-x')}}{\sqrt{u^2 + K_{xn}^2}} \tag{10.29}$$

The parameter u is called the *smoothing parameter*. It can be shown that the first term converges as the inverse cubic power of k_{xn}. The second term is nonsingular and can be summed by employing the Poisson transformation

$$\frac{1}{L}\sum_n \frac{e^{jk_{xn}(x-x')}}{\sqrt{u^2+K_{xn}^2}} = \frac{1}{\pi}\sum_n e^{jk_x nL} K_0(u|x-x'-nL|) \tag{10.30}$$

where K_0 is the modified Bessel function which has a rapidly decaying behavior. Thus, the Green's function can be written in the form

$$\begin{aligned} G(x;x') &= \frac{1}{jL}\sum_n e^{jk_{xn}(x-x')}\left(\frac{1}{k_{yn}}+\frac{1}{j\sqrt{u^2+K_{xn}^2}}\right) \\ &+\frac{1}{\pi}\sum_n e^{jk_x nL} K_0(u|x-x'-nL|) \end{aligned} \tag{10.31}$$

A reasonable choice for the smoothing parameter which seems to ensure good convergence for both the spatial and the spectral sums is

$$u \simeq \pi/L. \tag{10.32}$$

Exercises

10.1: Consider the function

$$F(x) = \sum_{n=-\infty}^{\infty} f(x+2n\pi)$$

(a) Show that $F(x)$ is a 2π-periodic function.
(b) Show that its Fourier coefficients are given by

$$c_k = \frac{1}{2\pi}\int_{-\pi}^{\pi} F(x)e^{-jkx}dx = \frac{\hat{f}(k)}{2\pi}$$

(c) From the Fourier representation of $F(x) = \sum c_k e^{jkx}$ at $x = 0$, find Poisson's summation formula

$$\sum_{n=-\infty}^{\infty} f(2n\pi) = \sum_{-\infty}^{\infty} \frac{\hat{f}}{2\pi}$$

10.2: A plane wave is incident on a sinusoidally periodic conducting plate. The plate height profile is given by the function

$$y = \xi(x) = -h \cos \frac{2\pi x}{L}$$

where h is the height amplitude and L is the period of variations. If $h/L \ll 1$, all scattered fields may be assumed to be outward travelling. This is known as the Rayleigh hypothesis.

(a) Applying the Floquet's theorem, find the general form of the scattered electric field in the region above the surface. Write the mode propagation constants in terms of the mode number and period L.
(b) Enforcing appropriate boundary conditions on the conducting surface, find an equation for the expansion coefficients.
(c) Using a Fourier series analysis, convert the above equation to a system of algebraic equations, solution of which gives the Floquet's coefficients.

Chapter 11
Inverse Scattering

In the direct scattering problem, it is predicting the scattered field from a target illuminated by an incident field, which is of interest.

The inverse scattering problem is that of determining the nature of the target from the knowledge of the scattered field. Figure 11.1 shows a target illuminated by a number of transmitters. The scattered field is measured by the receivers both in bistatic and monostatic modes. The task is to determine the shape and type of the target under test.

The methods used for the invese problem depend on the electrical size of the inhomogeneity. If D is the characteristic dimension of the scatterer and k is the wavenumber, the quantity kD gives a measure of electrical length of the target. When $ka \ll 1$, scattering is weak, and we may apply low frequency methods such as Rayleigh and Born approximations. On the other hand, when $kD \gg 1$, we may use high frequency asymptotic techniques such as geometrical or physical optics methods.

In many cases, inverse scattering of an object of resonant size with kD of the order of unity is of interest. A common approach to solve such problems has been through the method of integral equations in combination with field equivalence principles. Inversion of such integral equations is, however, an ill-posed problem and much work has been devoted to prove uniqueness for the solution of such problems. In the presence of limited data, it is inevitable that any inversion method reduces to solving some optimization problem.

11.1 Dielectric Bodies

11.1.1 Born Approximation

Consider a non-magnetic dielectric cylindrical target of permitivity ϵ and maximum cross sectional dimension D illuminated by a TM_z polarized plane wave

© Springer International Publishing Switzerland 2015

K. Barkeshli, *Advanced Electromagnetics and Scattering Theory*,
DOI 10.1007/978-3-319-11547-4_11

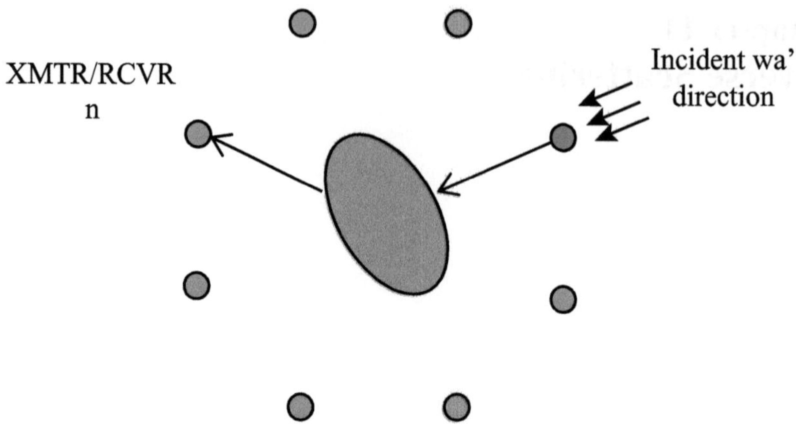

Fig. 11.1 A target illuminated by a number of transmitters with the scattered fields measured by the receivers

$$\mathbf{E}^i = \hat{z}E_0 e^{jk_0\hat{\rho_i}\cdot\boldsymbol{\rho}} \tag{11.1}$$

where $\hat{\rho_i}$ is the unit vector in the incident direction. The equivalent volumetric current is given by

$$\mathbf{J}_{eq} = j\omega\epsilon_0(\epsilon_r - 1)\mathbf{E} \tag{11.2}$$

where $\mathbf{E} = \mathbf{E}^i + \mathbf{E}^s$ is the total internal electric field. Assuming that the cylinder's contrast is low, the equivalent current is given by

$$\mathbf{J}_{eq} = j\omega\epsilon_0\chi_e\mathbf{E}^i \tag{11.3}$$

where χ_e is the electric susceptibility of the medium. This approximation is valid if

$$k_0 D\sqrt{\chi_e} < \pi \tag{11.4}$$

The first order Born approximation is given by

$$\mathbf{E}^s = -\frac{k_0 Z_0}{4}\hat{z}\int\limits_S j\omega\epsilon_0\chi_e(\rho')E_z^i(\rho')H_0^{(2)}(k_0|\boldsymbol{\rho} - \boldsymbol{\rho}'|)ds' \tag{11.5}$$

where $\boldsymbol{\rho}$ and $\boldsymbol{\rho}'$ are the observation and source point position vectors in planar coordinates, respectively. Using the asymptotic expansion for the Hanlkel function, the scattered field far from the cylinder is expressed as

$$\mathbf{E}^s = -j\frac{k_0^2}{4}\sqrt{\frac{2j}{\pi k_0 \rho}}e^{-jk_0\rho}\widehat{z}\int_S \chi_e(\rho')E_0 e^{-jk_0(\widehat{\rho_i}\cdot\rho' - \widehat{\rho_s}\cdot\rho')}ds' \tag{11.6}$$

where $\widehat{\rho_s}$ is the unit vector in the scattered field direction. We now define the normalized scattered field

$$E_n^s = \frac{E_z^s}{E_0\sqrt{\dfrac{2j}{\pi k_0 \rho}}e^{-jk_0\rho}}$$
$$= -j\frac{k_0^2}{4}\int_S \chi_e(\boldsymbol{\rho}')e^{jk_0(\widehat{\rho_s}-\widehat{\rho_i})\cdot\boldsymbol{\rho}'}ds' \tag{11.7}$$

which is independent of ρ and E_0. Since

$$\rho' = x'\widehat{x} + y'\widehat{y}, \quad \widehat{\rho_i} = \widehat{x}\cos\phi_i + \widehat{y}\sin\phi_i, \quad \widehat{\rho_s} = \widehat{x}\cos\phi_s + \widehat{y}\sin\phi_s \tag{11.8}$$

we may right

$$\begin{aligned} k_0(\widehat{\rho_s} - \widehat{\rho_i}) &= k_0(\cos\phi_s - \cos\phi_i)\widehat{x} + k_0(\sin\phi_s - \sin\phi_i)\widehat{y} \\ &= K_x\widehat{x} + K_y\widehat{y} \end{aligned} \tag{11.9}$$

where

$$K_x = k_0(\cos\phi_s - \cos\phi_i), \quad K_y = k_0(\sin\phi_s - \sin\phi_i) \tag{11.10}$$

Thus, (11.7) can be written as

$$E_n^s = -j\frac{k_0^2}{4}\int_S \chi_e(x', y')e^{j(K_x x' + K_y y')}dx'dy' \tag{11.11}$$

Clearly, the normalized scattered field is the inverse Fourier transform of the electric susceptibility χ_e, and

$$\chi_e(x, y) = \frac{4j}{k_0^2}\frac{1}{4\pi^2}\int E_n^s(K_x, K_y)e^{-j(K_x x' + K_y y')}dK_x dK_y \tag{11.12}$$

The scattered field is a function of the incidence and observation angles. In order to calculate the above Fourier transform, the spectral variables K_x and K_y must vary from $-\infty$ to ∞. This requires a variation of k_0 from 0 to ∞. But, in practice, we would like to fix the frequency and change the angles of incidence and observation ϕ_i and ϕ_s. Since the vector $\mathbf{K}$ in the $K_x K_y$-plane is given by

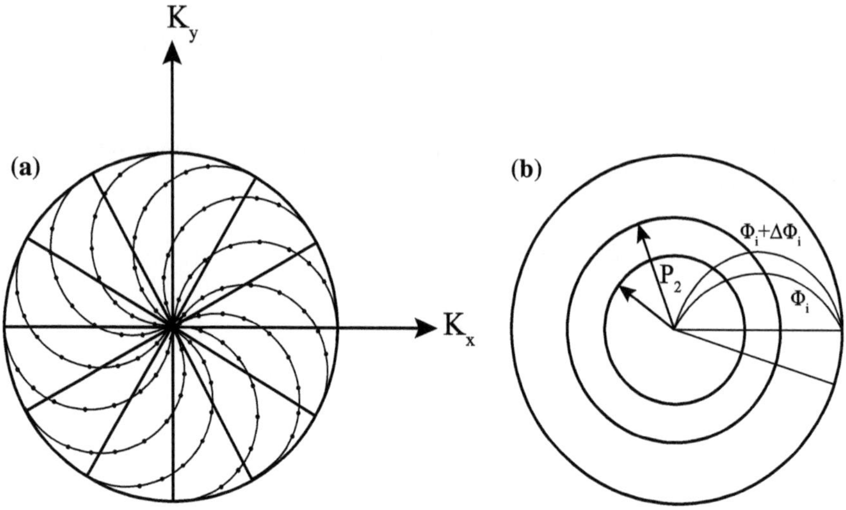

Fig. 11.2 **a** Sampling of points in the circular region of radius $2k_0$ in $K_x - K_y$-plane. **b** Calculation of the differential surface element $dK_x dK_y$

$$\begin{aligned}\mathbf{K} &= K_x\widehat{x} + K_y\widehat{y} = k_0(\widehat{x}\cos\phi_s + \widehat{y}\sin\phi_s) - k_0(\widehat{x}\cos\phi_i + \widehat{y}\sin\phi_i)\\ &= k_0\widehat{\rho_s} - k_0\widehat{\rho_i}\end{aligned} \tag{11.13}$$

changing the incident angle in the range $-180° < \phi_i < 180°$ and the observation angle in the range $\phi_i < \phi_s < \phi_i + 180°$ would cover a circular area of radius $2k_0$ in the $K_x K_y$-plane. This is shown in Fig. 11.2a, where increments $\Delta\phi_i$ and $\Delta\phi_s$ are chosen so that the circular region is spanned.

The transform (11.12) can therefore be calculated in a discrete manner as

$$\begin{aligned}\chi_e(x, y) &= \frac{4j}{k_0^2}\frac{1}{4\pi^2}\sum_{\phi_i=-\pi}^{\pi}\sum_{\phi_s=\phi_i}^{\phi_i+\pi} E_n^s(\phi_i, \phi_s)e^{-jk_0x(\cos\phi_s-\cos\phi_i)}\\ &\quad\cdot e^{-jk_0x(\sin\phi_s-\sin\phi_i)}\Delta K_x\Delta K_y\end{aligned} \tag{11.14}$$

The element of differential area $dK_x dK_y$ is approximately given by (Fig. 11.2b)

$$\Delta K_x\Delta K_y \simeq k_0^2\Delta\phi_i\pi(\rho_2^2 - \rho_1^2) \tag{11.15}$$

where the angles are in radians and

$$\rho_2 = \sqrt{2[1 - \cos(\phi_s - \phi_i + \Delta\phi_s)]},\quad \rho_1 = \sqrt{2[1 - \cos(\phi_s - \phi_i)]} \tag{11.16}$$

Thus, we finally have

$$\chi_e(x, y) = \frac{j}{180\pi} \sum_{\phi_i=-\pi}^{\pi} \sum_{\phi_s=\phi_i}^{\phi_i+\pi} E_n^s(\phi_i, \phi_s) e^{-jk_0 x(\cos\phi_s - \cos\phi_i)} \cdot e^{-jk_0 x(\sin\phi_s - \sin\phi_i)} \sin(\phi_s - \phi_i) \Delta\phi_s \Delta\phi_i \tag{11.17}$$

As an example, consider a dielectric cylinder of relative permitivity $\epsilon_r =$ and radius $a = 0.5\lambda_0$ illuminated by TM_z plane waves. The scattered electric field is given by (11.7). The far-field scattered field is given by

$$\begin{aligned}\mathbf{E}^s &= \hat{z}\sqrt{\frac{2j}{\pi k_0 \rho}} e^{-jk_0\rho} \sum_{n=-\infty}^{\infty} a_n e^{jn\phi_s} \\ &= \hat{z}\sqrt{\frac{2j}{\pi k_0 \rho}} e^{-jk_0\rho} \left(a_0 + 2\sum_{n=0}^{\infty} a_n \cos n\phi_s \right)\end{aligned} \tag{11.18}$$

where $\{a_n\}$ are given by (Sect. 7.2). The normalized scattered field is given by

$$E_n^s = a_0 + 2\sum_{n=0}^{\infty} a_n \cos n\phi_s \tag{11.19}$$

The above expression is for the case when the angle of incidence is π. If the incident angle is ϕ_i, then the normalized scattered field is expressed as

$$E_n^s = a_0 + 2\sum_{n=0}^{\infty} a_n \cos n(\phi_s - \phi_i) \tag{11.20}$$

We use the above expression as the measured normalized scattered field. In order to carry out the simulations, we consider a square area of side L and discretize it by differential elements of area $\Delta x \times \Delta y$. We then calculate E_n^s for various ϕ_i and ϕ_s from (11.20). Finally, the susceptibility χ_e is calculated from the expression (11.17). This will reconstruct the shape and the electric permitivity of the cylinder. For the simluations, we choose

$$L = 3\lambda_0, \quad \Delta x = \Delta y = 0.1\lambda_0, \quad \Delta\phi_i = \Delta\phi_s = 30^\circ \tag{11.21}$$

Figure 11.3 shows the reconstructed circular cylinder. There is a 9 % error in the estimation of ϵ_r.

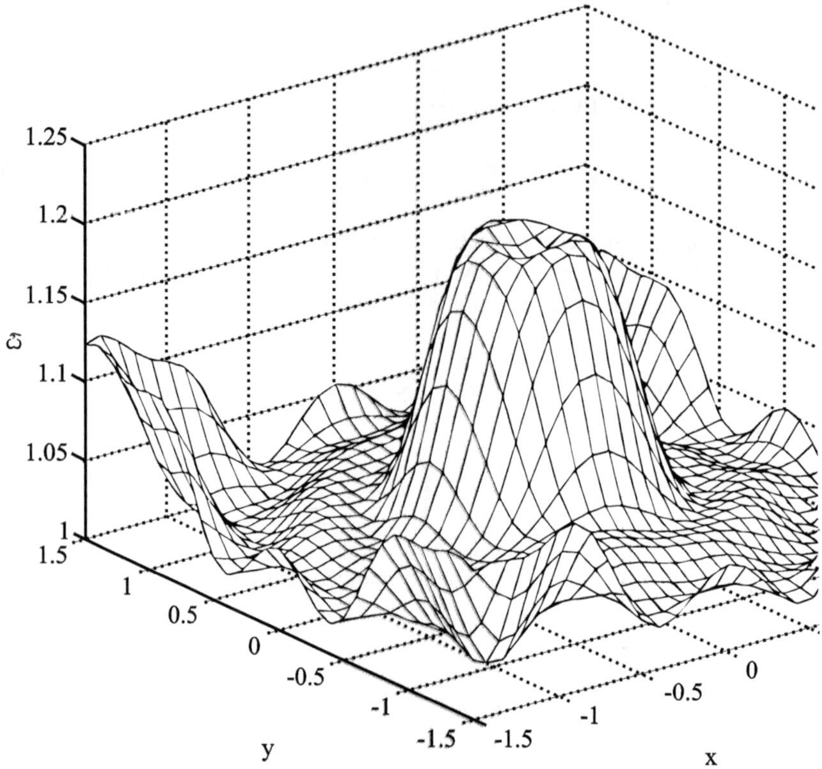

Fig. 11.3 The reconstructed relative permitivity ϵ_r of the low-contrast dielectric cylinder using the expression (11.20)

11.2 Perfectly Conducting Bodies

In this section, we discuss methods for the reconstruction of the shape of a conducting object. We begin with the physical optics inverse scattering method which is applicable in high frequencies. We then discuss the inversion of scattered data for conducting objects of resonant size. For these cases, the inversion algorithm is based on the integral equation formulation and an optimization algorithm.

11.2.1 Physical Optics Inverse Scattering

Consider an electrically large perfectly conducting body illuminated by a plane wave

$$\mathbf{E}_i = E_0\widehat{e}_i e^{-jk_0\widehat{i}\cdot\mathbf{r}} \tag{11.22}$$

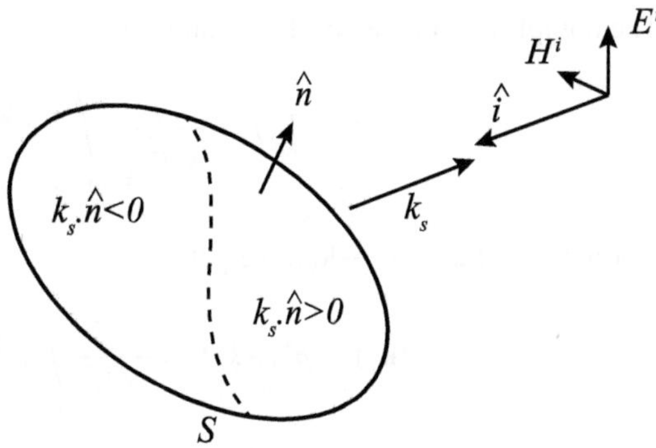

Fig. 11.4 An electrically large perfectly conducting target illuminated by a plane wave

in free space as shown in Fig. 11.4. The physical optics current is defined

$$\mathbf{K} \simeq \mathbf{K}_{po} = \begin{cases} 2\widehat{n} \times \mathbf{H}^i & \text{in the lit region} \\ 0 & \text{in the shadow region} \end{cases} \tag{11.23}$$

and the monostatic radar cross section under the physical optics approximation is given by (Sect. 7.2)

$$\sigma^b = \frac{k_0^2}{\pi} \left| \int\limits_{S_{\text{lit}}} 2(\widehat{i} \cdot \widehat{n}) e^{j2k_0\widehat{s}\cdot\mathbf{r}'} ds' \right| \tag{11.24}$$

Consider the vector $k_s = 2k_0\widehat{s} = -2k_0\widehat{i}$ and define the *the normalized complex scattering amplitude*

$$\rho(\mathbf{k}_s) = \frac{-j}{\sqrt{4\pi}} \int\limits_{\mathbf{k}_s\cdot\mathbf{n}>0} e^{k_s\cdot r'} k_s \cdot ds' \tag{11.25}$$

where $\mathbf{n}$ is the unit vector normal to the surface of the target. Since $\mathbf{k}_s$ is opposite to $\widehat{i}$, $\mathbf{k}_s \cdot \mathbf{n} > 0$ denotes the lit part of the conducting surface. It is noted that the

$$\rho(\mathbf{k}_s)\rho^*(\mathbf{k}_s) = \sigma^b \tag{11.26}$$

If the target is illuminated from the opposite direction, then the normalized complex scattering amplitude is given by

$$\rho(-\mathbf{k}_s) = \frac{j}{\sqrt{4\pi}} \int\limits_{\mathbf{k}_s\cdot\mathbf{n}<0} e^{-j\mathbf{k}_s\cdot\mathbf{r}'} \mathbf{k}_s \cdot \mathbf{ds}' \tag{11.27}$$

The complex conjugate of this quantity is

$$\rho^*(-\mathbf{k}_s) = \frac{-j}{\sqrt{4\pi}} \int\limits_{\mathbf{k}_s\cdot\mathbf{n}<0} e^{j\mathbf{k}_s\cdot\mathbf{r}'}\mathbf{k}_s \cdot \mathbf{ds}' \tag{11.28}$$

Adding $\rho(\mathbf{k}_s)$ and $\rho^*(-\mathbf{k}_s)$, we get

$$\begin{aligned} \rho(\mathbf{k}_s) + \rho^*(-k_s) &= \frac{-j}{\sqrt{4\pi}} \oint\limits_S e^{j\mathbf{k}_s\cdot\mathbf{r}'}\mathbf{k}_s \cdot \mathbf{ds}' \\ &= \frac{-j}{\sqrt{4\pi}} \int\limits_V \nabla \cdot \left(e^{j\mathbf{k}_s\cdot\mathbf{r}'}\mathbf{k}_s\right) dv' \\ &= \frac{2k_0^2}{\sqrt{\pi}} \int\limits_V e^{j\mathbf{k}_s\cdot\mathbf{r}'} dv' \end{aligned} \tag{11.29}$$

where we used divegence theorem and V is the volume of the scattering body.

We now define the *complex scattering amplitude*

$$\begin{aligned} \Gamma(\mathbf{k}_s) &= \frac{\sqrt{\pi}}{2k_0^2}[\rho(\mathbf{k}_s) + \rho^*(-\mathbf{k}_s)] \\ &= \int\limits_V e^{j\mathbf{k}_s\cdot\mathbf{r}'} dv' \end{aligned} \tag{11.30}$$

and the characteristic function

$$\gamma(\mathbf{r}) = \begin{cases} 1 & \mathbf{r} \in V \\ 0 & \mathbf{r} \notin V \end{cases} \tag{11.31}$$

Clearly, Γ is the inverse Fourier transform of γ, and thus

$$\gamma(\mathbf{r}) = \frac{1}{(2\pi)^3} \int \Gamma(\mathbf{k}_s) e^{-j\mathbf{k}_s\cdot\mathbf{r}} d\mathbf{k}_s \tag{11.32}$$

The above is known as *Bojarski's identity*. This identity gives the position of a perfectly conducting object from a knowledge of the normalized complex scattering amplitude $\rho(\mathbf{k}_s)$. It can be used for electrically large objects and is applicable if ρ is measured for all values of $\mathbf{k}_s$ implying all frequencies and all incident angles.

Appendix A
Vector Analysis

A.1 Orthogonal Coordinate Systems

A.1.1 Cartesian (Rectangular) Coordinate System

The unit vectors are denoted by $\widehat{x}, \widehat{y}, \widehat{z}$ in the Cartesian system. By convention, $(\widehat{x}, \widehat{y}, \widehat{z})$ triplet form a right-handed system and obey the following cyclic relations

$$\widehat{x} \times \widehat{y} = \widehat{z}, \quad \widehat{z} \times \widehat{x} = \widehat{y}, \quad \widehat{y} \times \widehat{z} = \widehat{x} \tag{A.1}$$

Differential length is defined as

$$\mathbf{d}\ell = dx\,\widehat{x} + dy\,\widehat{y} + dz\,\widehat{z} \tag{A.2}$$

Differential surface areas with normal vectors $\widehat{x}$, $\widehat{y}$, and $\widehat{z}$ are respectively defined as

$$\begin{aligned} d\mathbf{s}_x &= dy\,dz\,\widehat{x} \\ d\mathbf{s}_y &= dx\,dz\,\widehat{y} \\ d\mathbf{s}_z &= dx\,dy\,\widehat{z} \end{aligned} \tag{A.3}$$

The differential volume element is

$$dv = dx\,dy\,dz \tag{A.4}$$

A.1.2 Cylindrical Coordinate System

The unit vectors are denoted by $\widehat{\rho}, \widehat{\phi}, \widehat{z}$ in the cylindrical system. By convention, $(\widehat{\rho}, \widehat{\phi}, \widehat{z})$ triplet form a right-handed system and obey the following cyclic relations

© Springer International Publishing Switzerland 2015
K. Barkeshli, *Advanced Electromagnetics and Scattering Theory*,
DOI 10.1007/978-3-319-11547-4

$$\widehat{\rho} \times \widehat{\phi} = \widehat{z}, \quad \widehat{z} \times \widehat{\rho} = \widehat{\phi}, \quad \widehat{\phi} \times \widehat{z} = \widehat{\rho} \tag{A.5}$$

Differential length is defined as

$$\mathbf{d}\ell = d\rho\,\widehat{\rho} + \rho\,d\phi\,\widehat{\phi} + dz\,\widehat{z} \tag{A.6}$$

Differential surface areas with normal vectors $\widehat{\rho}$, $\widehat{\phi}$, and $\widehat{z}$ are respectively defined as

$$\begin{aligned} d\mathbf{s}_\rho &= \rho\,d\phi\,dz\,\widehat{\rho} \\ d\mathbf{s}_\phi &= d\rho\,dz\,\widehat{\phi} \\ d\mathbf{s}_z &= \rho\,d\rho\,d\phi\,\widehat{z} \end{aligned} \tag{A.7}$$

The differential volume element is

$$dv = \rho\,d\rho\,d\phi\,dz \tag{A.8}$$

A.1.3 Spherical Coordinate System

The unit vectors are denoted by $\widehat{r}, \widehat{\theta}, \widehat{\phi}$ in the spherical system. By convention, $(\widehat{r}, \widehat{\theta}, \widehat{\phi})$ triplet form a right-handed system and obey the following cyclic relations

$$\widehat{r} \times \widehat{\theta} = \widehat{\phi}, \quad \widehat{\phi} \times \widehat{r} = \widehat{\theta}, \quad \widehat{\theta} \times \widehat{\phi} = \widehat{r} \tag{A.9}$$

Differential length is defined as

$$\mathbf{d}\ell = dr\,\widehat{r} + r\,d\theta\,\widehat{\theta} + r\sin\theta\,d\phi\,\widehat{\phi} \tag{A.10}$$

Differential surface areas with normal vectors $\widehat{r}, \widehat{\theta}$, and $\widehat{\phi}$ are respectively defined as

$$\begin{aligned} d\mathbf{s}_r &= r^2 \sin\theta\,d\theta\,d\phi\,\widehat{r} \\ d\mathbf{s}_\theta &= r\sin\theta\,dr\,d\phi\,\widehat{\theta} \\ d\mathbf{s}_\phi &= r\,dr\,d\theta\,\widehat{\phi} \end{aligned} \tag{A.11}$$

The differential volume element is

$$dv = r^2 \sin\theta\,dr\,d\theta\,d\phi \tag{A.12}$$

A.2 Coordinate Transformations

Consider a point P whose coordinates in Cartesian, cylindrical, and spherical coordinate systems, respectively are (x, y, z), (ρ, ϕ, z), and (r, θ, ϕ). The Cartesian and cylindrical coordinates are related by

$$\begin{aligned} x &= \rho \cos\phi \\ y &= \rho \sin\phi \end{aligned} \tag{A.13}$$

The Cartesian and spherical coordinates are related by

$$\begin{aligned} x &= r \sin\theta \cos\phi \\ y &= r \sin\theta \sin\phi \\ z &= r \cos\theta \end{aligned} \tag{A.14}$$

The spherical and cylindrical coordinates are related by

$$\begin{aligned} \rho &= r \sin\theta \\ z &= r \cos\theta \end{aligned} \tag{A.15}$$

A.3 Vector Transformations

Consider a three-dimensional vector **F**. We can express **F** in terms of the unit vectors in Cartesian, cylindrical, and spherical systems respectively as:

$$\begin{aligned} \mathbf{F} &= F_x\widehat{x} + F_y\widehat{y} + F_z\widehat{z} \\ \mathbf{F} &= F_\rho\widehat{\rho} + F_\phi\widehat{\phi} + F_z\widehat{z} \\ \mathbf{F} &= F_r\widehat{r} + F_\theta\widehat{\theta} + F_\phi\widehat{\phi} \end{aligned} \tag{A.16}$$

A.3.1 Cartesian-Cylindrical Vector Transformations

Cartesian to cylindrical components are related by

$$\begin{aligned} F_\rho &= F_x \cos\phi + F_y \sin\phi \\ F_\phi &= -F_x \sin\phi + F_y \cos\phi \\ F_z &= F_z \end{aligned} \tag{A.17}$$

Cylindrical to Cartesian components are related by

$$\begin{aligned} F_x &= F_\rho \cos\phi - F_\phi \sin\phi \\ F_y &= F_\rho \sin\phi + F_\phi \cos\phi \\ F_z &= F_z \end{aligned} \tag{A.18}$$

A.3.2 Cartesian-Spherical Vector Transformations

Cartesian to spherical components are related by

$$\begin{aligned} F_r &= F_x \sin\theta \cos\phi + F_y \sin\theta \sin\phi + F_z \cos\theta \\ F_\theta &= F_x \cos\theta \cos\phi + F_y \cos\theta \sin\phi - F_z \sin\theta \\ F_\phi &= -F_x \sin\phi + F_y \cos\phi \end{aligned} \tag{A.19}$$

Spherical to Cartesian components are related by

$$\begin{aligned} F_x &= F_r \sin\theta \cos\phi + F_\theta \cos\theta \cos\phi - F_\phi \sin\phi \\ F_y &= F_r \sin\theta \sin\phi + F_\theta \cos\theta \sin\phi - F_\phi \cos\phi \\ F_z &= F_r \cos\theta - F_\theta \sin\theta \end{aligned} \tag{A.20}$$

A.3.3 Cylindrical-Spherical Vector Transformations

Cylindrical to spherical components are related by

$$\begin{aligned} F_r &= F_\rho \sin\theta + F_z \cos\theta \\ F_\theta &= F_\rho \cos\theta - F_z \sin\theta \\ F_\phi &= F_\phi \end{aligned} \tag{A.21}$$

Spherical to cylindrical components are related by

$$\begin{aligned} F_\rho &= F_r \sin\theta + F_\theta \cos\theta \\ F_\phi &= F_\phi \\ F_z &= F_r \cos\theta - F_\theta \sin\theta \end{aligned} \tag{A.22}$$

Appendix B
Vector Calculus

In the following, scalar functions are denoted by lower-case letters and vector functions are denoted by upper-case bold letters. The notations for vector components are same as shown in (A.17).

B.1 Differential Operators

B.1.1 Cartesian System

$$\nabla f = \frac{\partial f}{\partial x}\widehat{x} + \frac{\partial f}{\partial y}\widehat{y} + \frac{\partial f}{\partial z}\widehat{z} \tag{B.1}$$

$$\nabla \cdot \mathbf{F} = \frac{\partial F_x}{\partial x} + \frac{\partial F_y}{\partial y} + \frac{\partial F_z}{\partial z} \tag{B.2}$$

$$\nabla \times \mathbf{F} = \left(\frac{\partial F_z}{\partial y} - \frac{\partial F_y}{\partial z}\right)\widehat{x} + \left(\frac{\partial F_x}{\partial z} - \frac{\partial F_z}{\partial x}\right)\widehat{y} + \left(\frac{\partial F_y}{\partial x} - \frac{\partial F_x}{\partial y}\right)\widehat{z} \tag{B.3}$$

$$\nabla^2 f = \frac{\partial^2 f}{\partial x^2} + \frac{\partial^2 f}{\partial y^2} + \frac{\partial^2 f}{\partial z^2} \tag{B.4}$$

© Springer International Publishing Switzerland 2015

K. Barkeshli, *Advanced Electromagnetics and Scattering Theory*,
DOI 10.1007/978-3-319-11547-4

B.1.2 Cylindrical System

$$\nabla f = \frac{\partial f}{\partial \rho}\widehat{\rho} + \frac{1}{\rho}\frac{\partial f}{\partial \phi}\widehat{\phi} + \frac{\partial f}{\partial z}\widehat{z} \tag{B.5}$$

$$\nabla \cdot \mathbf{F} = \frac{1}{\rho}\frac{\partial}{\partial \rho}\left(\rho F_\rho\right) + \frac{1}{\rho}\frac{\partial F_\phi}{\partial \phi} + \frac{\partial F_z}{\partial z} \tag{B.6}$$

$$\begin{aligned}\nabla \times \mathbf{F} = &\left(\frac{1}{\rho}\frac{\partial F_z}{\partial \phi} - \frac{\partial F_\phi}{\partial z}\right)\widehat{\rho} + \left(\frac{\partial F_\rho}{\partial z} - \frac{\partial F_z}{\partial \rho}\right)\widehat{\phi} \\ &+ \frac{1}{\rho}\left(\frac{\partial}{\partial \rho}\left(\rho F_\phi\right) - \frac{\partial F_\rho}{\partial \phi}\right)\widehat{z}\end{aligned} \tag{B.7}$$

$$\nabla^2 f = \frac{1}{\rho}\frac{\partial}{\partial \rho}\left(\rho\frac{\partial f}{\partial \rho}\right) + \frac{1}{\rho^2}\frac{\partial^2 f}{\partial \phi^2} + \frac{\partial^2 f}{\partial z^2} \tag{B.8}$$

B.1.3 Spherical System

$$\nabla f = \frac{\partial f}{\partial r}\widehat{r} + \frac{1}{r}\frac{\partial f}{\partial \theta}\widehat{\theta} + \frac{1}{r\sin\theta}\frac{\partial f}{\partial \phi}\widehat{\phi} \tag{B.9}$$

$$\nabla \cdot \mathbf{F} = \frac{1}{r^2}\frac{\partial}{\partial r}\left(r^2 F_r\right) + \frac{1}{r\sin\theta}\frac{\partial}{\partial \theta}\left(F_\theta \sin\theta\right) + \frac{1}{r\sin\theta}\frac{\partial F_\phi}{\partial \phi} \tag{B.10}$$

$$\begin{aligned}\nabla \times \mathbf{F} = &\frac{1}{r\sin\theta}\left[\frac{\partial}{\partial \theta}\left(F_\phi \sin\theta\right) - \frac{\partial F_\theta}{\partial \phi}\right]\widehat{r} \\ &+ \left[\frac{1}{r\sin\theta}\frac{\partial F_r}{\partial \phi} - \frac{1}{r}\frac{\partial}{\partial r}\left(r F_\phi\right)\right]\widehat{\theta} + \frac{1}{r}\left[\frac{\partial}{\partial r}\left(r F_\theta\right) - \frac{\partial F_r}{\partial \theta}\right]\widehat{\phi}\end{aligned} \tag{B.11}$$

$$\nabla^2 f = \frac{1}{r^2}\frac{\partial}{\partial r}\left(r^2\frac{\partial f}{\partial r}\right) + \frac{1}{r^2\sin\theta}\frac{\partial}{\partial\theta}\left(\sin\theta\frac{\partial f}{\partial\theta}\right) + \frac{1}{r^2\sin^2\theta}\frac{\partial^2 f}{\partial\phi^2} \tag{B.12}$$

B.2 Differentiation

$$\nabla\cdot(f\mathbf{F}) = f\nabla\cdot\mathbf{F} + \mathbf{F}\cdot\nabla f \tag{B.13}$$

$$\nabla\cdot(\mathbf{E}\times\mathbf{F}) = \mathbf{F}\cdot\nabla\times\mathbf{E} - \mathbf{E}\cdot\nabla\times\mathbf{F} \tag{B.14}$$

$$\nabla\times(f\mathbf{F}) = \nabla f\times\mathbf{F} + f\nabla\times\mathbf{F} \tag{B.15}$$

$$\nabla\times(\mathbf{E}\times\mathbf{F}) = \mathbf{E}\nabla\cdot\mathbf{F} - \mathbf{F}\nabla\cdot\mathbf{E} + (\mathbf{F}\cdot\nabla)\,\mathbf{E} - (\mathbf{E}\cdot\nabla)\,\mathbf{F} \tag{B.16}$$

$$\nabla\,(\mathbf{E}\cdot\mathbf{F}) = (\mathbf{E}\cdot\nabla)\,\mathbf{F} + (\mathbf{F}\cdot\nabla)\,\mathbf{E} + \mathbf{E}\times(\nabla\times\mathbf{F}) + \mathbf{F}\times(\nabla\times\mathbf{E}) \tag{B.17}$$

$$\nabla\times\nabla\times\mathbf{F} = \nabla\,(\nabla\cdot\mathbf{F}) - \nabla^2\mathbf{F} \tag{B.18}$$

$$\nabla\cdot(\nabla\times\mathbf{F}) = 0 \tag{B.19}$$

$$\nabla\times\nabla f = 0 \tag{B.20}$$

B.3 Integration Theorems

In the following V is a three-dimensional volume with differential volume element dv, S is a closed two-dimensional surface enclosing V with differential surface element ds and outward unit normal vector $\widehat{n}$. An open surface is denoted by Ω and its bounding contour by C.

$$\int_V \nabla \cdot \mathbf{F}\, dv = \oint_S \mathbf{F} \cdot \widehat{n}\, ds \qquad \text{(Divergence Theorem)} \tag{B.21}$$

$$\int_V \nabla f\, dv = \oint_S f \widehat{n}\, ds \tag{B.22}$$

$$\int_V \nabla \times \mathbf{F}\, dv = \oint_S \widehat{n} \times \mathbf{F}\, ds \tag{B.23}$$

$$\int_V \left(f\nabla^2 g - g\nabla^2 f \right) dv = \oint_S (f\nabla g - g\nabla f) \cdot \widehat{n}\, ds \tag{B.24}$$

$$\int_\Omega (\nabla \times \mathbf{F}) \cdot \widehat{n}\, ds = \oint_C \mathbf{F} \cdot \mathbf{d}\ell \qquad \text{(Stoke's Theorem)} \tag{B.25}$$

$$\int_\Omega \widehat{n} \times \nabla f\, ds = \oint_C f\, \mathbf{d}\ell \tag{B.26}$$

Appendix C
Bessel Functions

C.1 Gamma Function

In the 18th century, Swiss mathematician Leonhard Euler concerned himself with the problem of interpolating the factorial function ($n! = n(n-1)(n-2)\ldots 3.2.1$) between non-integer values. This problem eventually led him to gamma function defined as

$$\Gamma(z) = \int_0^\infty e^{-t} t^{z-1} \, dt \tag{C.1}$$

The integral in (C.1) is convergent for all complex numbers except negative integers and zero. It can be shown that for positive integers

$$\Gamma(n) = (n-1)! \tag{C.2}$$

Two useful properties of the gamma function are

$$\Gamma(z)\Gamma(1-z) = \frac{\pi}{\sin(\pi z)} \quad \text{(Reflection Formula)} \tag{C.3}$$

$$2^{2z-1}\Gamma(z)\Gamma\left(z+\frac{1}{2}\right) = \sqrt{\pi}\,\Gamma(2z) \quad \text{(Duplication Formula)} \tag{C.4}$$

It is important to note that the reflection formula in (C.3) is only valid for non-integer values of z.

© Springer International Publishing Switzerland 2015

K. Barkeshli, *Advanced Electromagnetics and Scattering Theory*,
DOI 10.1007/978-3-319-11547-4

C.2 Bessel Functions

Bessel functions are the solutions of Bessel's equation

$$\frac{d^2y}{dz^2} + \frac{1}{z}\frac{dy}{dz} + \left(1 - \frac{\nu^2}{z^2}\right) y = 0 \tag{C.5}$$

where ν which may be complex is called the order of the equation and the solution.

C.2.1 Bessel Functions of the First Kind $J_\nu(z)$

The series representation for the Bessel function of the first kind of order ν is

$$J_\nu(z) = \sum_{k=0}^{\infty} \frac{(-1)^k}{k!\Gamma(k+\nu+1)} \left(\frac{z}{2}\right)^{2k+\nu} \tag{C.6}$$

It can be easily verified that $J_\nu(z)$ satisfies the differential equation in (C.5). Bessel functions of integer order are of particular interest in electromagnetic problems. For integer values of ν the expression in (C.6) is simplified to

$$J_n(z) = \sum_{k=0}^{\infty} \frac{(-1)^k}{k!(k+n)!} \left(\frac{z}{2}\right)^{2k+n}, \quad n = 0, 1, 2, \ldots \tag{C.7}$$

Furthermore it can be shown that for $n = 0, 1, 2, \ldots$

$$J_{-n}(z) = (-1)^n J_n(z) \tag{C.8}$$

Figure C.1 shows the plot for Bessel functions of the first kind of orders 0, 1, 2, and 3. As it can be seen from (C.6) the analytical expression for Bessel functions is rather complicated. In many electromagnetic boundary value problems, depending on the application, only Bessel functions of small arguments (for near-field analysis) and large arguments (for far-field analysis) are of interest. In these cases it is much more convenient to use concise asymptotic expressions which have been derived for Bessel functions. It can be shown that

$$\lim_{z\to 0^+} J_n(z) \simeq \begin{cases} 1 & \text{if } n = 0 \\ \frac{(\frac{z}{2})^2}{n!} & \text{if } n = 1, 2, 3, \ldots \end{cases} \tag{C.9}$$

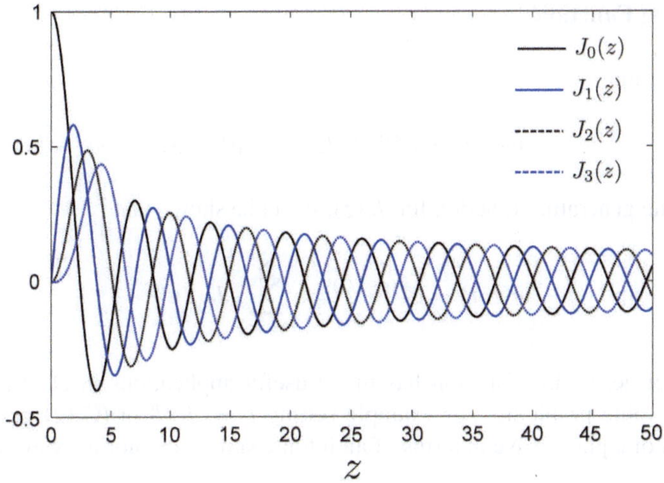

Fig. C.1 Bessel function of the first kind

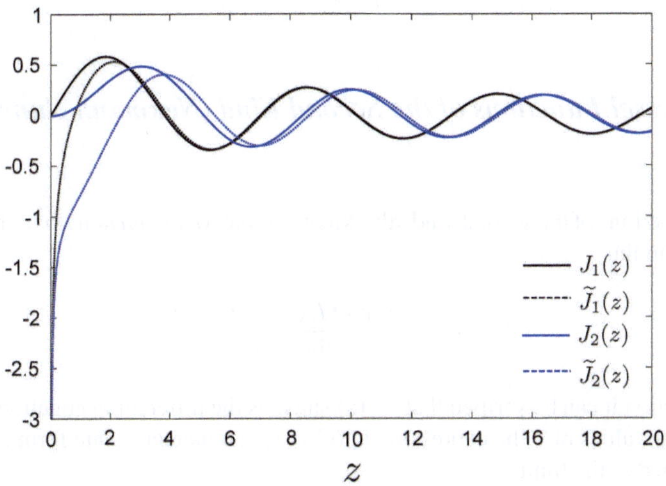

Fig. C.2 Large argument asymptotic expressions for $J_1(z)$ and $J_2(z)$

and

$$\lim_{z\to\infty} J_n(z) \simeq \sqrt{\frac{2}{\pi z}} \cos\left(z - \frac{\pi}{4} - \frac{n\pi}{2}\right) \tag{C.10}$$

Figure C.2 compares the values for $J_1(z)$ and $J_2(z)$ with their large argument asymptotic expression from (C.10) which have been denoted by $\tilde{J}_1(z)$ and $\tilde{J}_2(z)$ respectively.

Generating Function

The expression

$$w(z,t) = e^{\frac{z}{2}(t-1/t)}, \quad 0 < |t| < \infty \tag{C.11}$$

is called the generating function for $J_n(z)$. It can bc shown that

$$w(z,t) = e^{\frac{z}{2}(t-1/t)} = \sum_{n=-\infty}^{\infty} J_n(z)t^n \tag{C.12}$$

The Bessel generating function has many useful applications in electromagnetic boundary value problems. For example setting $t = je^{j\phi}$ in (C.12) gives us the expansion of a plane wave in terms of an infinite sum of cylindrical waves

$$e^{jz\cos\phi} = \sum_{n=-\infty}^{\infty} j^n J_n(z)e^{jn\phi} \tag{C.13}$$

C.2.2 Bessel Functions of the Second Kind (Neumann Functions) $N_\nu(z)$

Bessel functions of the second kind, also known as the *Neumann functions* are defined by the formula

$$N_\nu(z) = \frac{\cos(\nu\pi)J_\nu(z) - J_{-\nu}(z)}{\sin(\nu\pi)} \tag{C.14}$$

By inspection it can be verified that $N_\nu(z)$ satisfies the differential equation in (C.5). For integer values of ν the expression in (C.14) is an indeterminate form and has to be evaluated at the limit

$$N_n(z) = \lim_{\nu\to n} N_\nu(z), \quad n = 1, 2, 3, \ldots \tag{C.15}$$

The limit in (C.15) can be evaluated using l'Hospital's rule, leading to

$$\begin{aligned} N_n(z) = \frac{2}{\pi}J_n(z)\ln\left(\frac{z}{2}\right) - \frac{1}{\pi}\sum_{k=0}^{n-1}\frac{(n-k-1)!\left(\frac{z}{2}\right)^{2k-n}}{k!} \\ -\frac{1}{\pi}\sum_{k=0}^{\infty}\frac{(-1)^k\left(\frac{z}{2}\right)^{2k+n}[\psi(k+n+1)+\psi(k+1)]}{k!(n+k)!} \end{aligned} \tag{C.16}$$

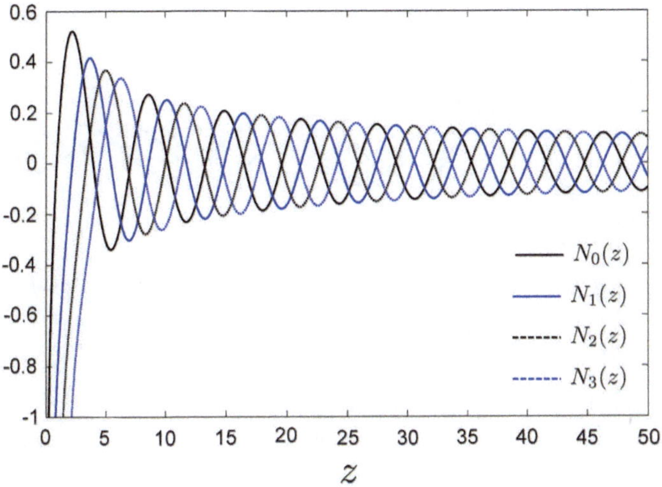

Fig. C.3 Neumann functions of orders 0, 1, 2, and 3

where $\psi(z)$ is the digamma function defined as

$$\psi(z) = \frac{d}{dz} \ln \Gamma(z) = \frac{\Gamma'(z)}{\Gamma(z)} \tag{C.17}$$

Similar to Bessel functions of the first kind, it can be shown that for $n = 0, 1, 2, \ldots$

$$N_{-n}(z) = (-1)^n N_n(z) \tag{C.18}$$

Figure C.3 shows the plot for Neumann functions of orders 0, 1, 2, and 3. As it can be seen from the plot, N_n contains a singularity at zero due to the logarithm term in (C.16). The asymptotic expressions for $N_n(z)$ are

$$\lim_{z\to 0^+} N_n(z) \simeq \begin{cases} \frac{2}{\pi} \ln\left(\frac{\gamma z}{2}\right) & \text{if } n = 0 \\ -\frac{(n-1)!}{\pi}\left(\frac{2}{z}\right)^n & \text{if } n = 1, 2, 3, \ldots \end{cases} \tag{C.19}$$

where $\gamma = 1.78107\ldots$ is Euler's constant and

$$\lim_{z\to\infty} N_n(z) \simeq \sqrt{\frac{2}{\pi z}} \sin\left(z - \frac{\pi}{4} - \frac{n\pi}{2}\right) \tag{C.20}$$

Figure C.4 compares the values for $N_1(z)$ and $N_2(z)$ with their large argument asymptotic expression from (C.20) which have been denoted by $\tilde{N}_1(z)$ and $\tilde{N}_2(z)$ respectively.

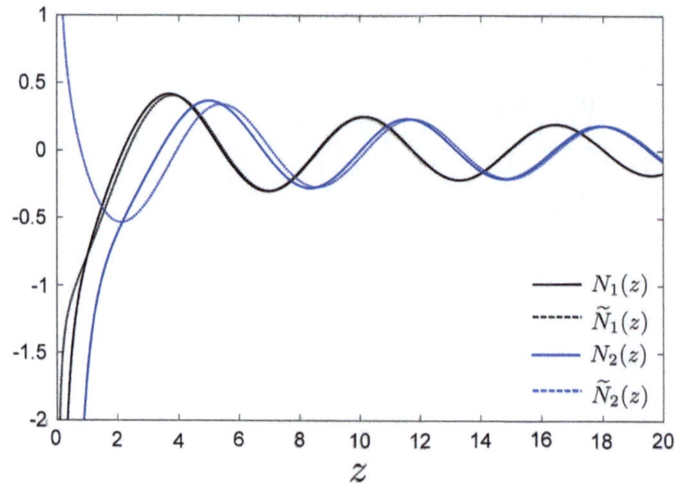

Fig. C.4 Large argument asymptotic expressions for $N_1(z)$ and $N_2(z)$

C.2.3 Bessel Functions of the Third Kind (Hankel Functions)

Bessel functions of the first kind and Neumann functions are the orthogonal eigenfunctions of the Bessel equation, hence any linear combination of them is also a solution to (C.5). Two particularly important such combinations are the Hankel function of the first kind of order n defined as

$$H_n^{(1)}(z) = J_n(z) + jN_n(z) \tag{C.21}$$

and the Hankel function of the second kind of order n defined as

$$H_n^{(2)}(z) = J_n(z) - jN_n(z) \tag{C.22}$$

Hankel functions are often encountered in electromagnetic radiation problems. For large arguments, the asymptotic expressions for Hankel functions are

$$\lim_{z\to\infty} H_n^{(1)}(z) \simeq \sqrt{\frac{2}{\pi z}} \exp\left[j\left(z - \frac{\pi}{4} - \frac{n\pi}{2}\right)\right] \tag{C.23}$$

$$\lim_{z\to\infty} H_n^{(2)}(z) \simeq \sqrt{\frac{2}{\pi z}} \exp\left[-j\left(z - \frac{\pi}{4} - \frac{n\pi}{2}\right)\right] \tag{C.24}$$

C.2.4 Formulas for Bessel Functions

In the following $Z_n(z)$ represents $J_n(z)$, $N_n(z)$, $H_n^{(1)}(z)$, $H_n^{(2)}(z)$ or any linear combinations of these functions.

Recurrence and Differentiation Formulas

$$2Z_n'(z) = Z_{n-1}(z) - Z_{n+1}(z) \tag{C.25}$$

$$\frac{2n}{z} Z_n(z) = Z_{n-1}(z) + Z_{n+1}(z) \tag{C.26}$$

$$J_n(z)N_n'(z) - N_n(z)J_n'(z) = \frac{2}{z\pi} \tag{C.27}$$

$$J_n(z)J_{-n}'(z) - J_{-n}(z)J_n'(z) = -\frac{2}{z\pi}\sin(n\pi) \tag{C.28}$$

Integrals

$$\int z^{n+1} Z_n(z)\, dz = z^{n+1} Z_{n+1}(z) \tag{C.29}$$

$$\int z^{1-n} Z_n(z)\, dz = -z^{1-n} Z_{n-1}(z) \tag{C.30}$$

$$\int z Z_n^2(kz)\, dz = \frac{z^2}{2}\left[Z_n^2(kz) - Z_{n-1}(kz)Z_{n+1}(kz)\right] \tag{C.31}$$

$$J_n(z) = \frac{1}{\pi}\int_0^{\pi} \cos(n\phi - z\sin\phi)\, d\phi \tag{C.32}$$

C.2.5 Formulas for Bessel Functions

...differentiation formulas

FSC
www.fsc.org
MIX
Papier aus verantwortungsvollen Quellen
Paper from responsible sources
FSC® C105338

If you have any concerns about our products,
you can contact us on
ProductSafety@springernature.com

In case Publisher is established outside the EU,
the EU authorized representative is:
Springer Nature Customer Service Center GmbH
Europaplatz 3, 69115 Heidelberg, Germany

Printed by Libri Plureos GmbH
in Hamburg, Germany